ADVANCES IN

CHROMATOGRAPHY

Volume 14

ADVANCES IN CHROMATOGRAPHY

Volume 14

Edited by

J. CALVIN GIDDINGS
EXECUTIVE EDITOR

DEPARTMENT OF CHEMISTRY
UNIVERSITY OF UTAH
SALT LAKE CITY, UTAH

ELI GRUSHKA
GAS CHROMATOGRAPHY

DEPARTMENT OF CHEMISTRY
STATE UNIVERSITY OF NEW YORK AT BUFFALO
BUFFALO, NEW YORK

JACK CAZES
MACROMOLECULAR CHROMATOGRAPHY

WATERS ASSOCIATES, INC.
MILFORD, MASSACHUSETTS

PHYLLIS R. BROWN
BIOCHEMICAL CHROMATOGRAPHY

DEPARTMENT OF CHEMISTRY
UNIVERSITY OF RHODE ISLAND
KINGSTON, RHODE ISLAND

MARCEL DEKKER, Inc., New York and Basel

MARCEL DEKKER, INC.

270 Madison Avenue, New York, New York 10016

LIBRARY OF CONGRESS CATALOG CARD NUMBER 65-27435

ISBN 0-8247-6436-6

Current printing (last digit):
10 9 8 7 6 5 4 3 2 1

PRINTED IN THE UNITED STATES OF AMERICA

CONTRIBUTORS TO VOLUME 14

ANDREW J. CLIFFORD, Department of Nutrition, University of California Davis, California

ANTONIO De CORCIA, Instituto di Chemica Analitica, Universita di Roma, Rome, Italy

C. L. de LIGNY, Laboratory for Analytical Chemistry, University of Utrecht, The Netherlands

J. K. HAKEN, Department of Polymer Science, The University of New South Wales, New South Wales, Australia

ISTVÁN HALÁSZ, Angewandte Physikalische Chemie, Universität Saarbrücken, West Germany

ARNALDO LIBERTI, Instituto di Chimica Analitica, Universita di Roma, Rome, Italy

DAVID C. LOCKE, Department of Chemistry, Queens College, CUNY, Flushing, New York

IMRICH SEBESTIAN, Angewandte Physikalische Chemie, Universität Saarbrücken, West Germany

BENGT STENLUND, Department for Development of Chemical Products, The Finnish Pulp and Paper Research Institute, Helsinki, Finland

W. J. A. VANDENHEUVEL, Merck Sharp & Dohme Research Laboratories, Rahway, New Jersey

A. G. ZACCHEI, Merck Sharp & Dohme Research Laboratories, West Point, Pennsylvania

CONTENTS

CONTENTS OF OTHER VOLUMES

Volume 3

Volume 4

Volume 5

Volume 6

Volume 7

Volume 8

Volume 13

ADVANCES IN

CHROMATOGRAPHY

Volume 14

Chapter 1

NUTRITION: AN INVITING FIELD TO HIGH-PRESSURE LIQUID CHROMATOGRAPHY

Andrew J. Clifford

Department of Nutrition
University of California
Davis, California

I. INTRODUCTION

The recent development of commercial high-pressure liquid chromatography systems has provided new instrumentation for the separation, characterization, identification, and quantification of minute amounts of essential dietary components. The application of high-pressure liquid chromatography to nutrition will facilitate studies of the relationship among nutrients, their precise biological roles, and the dietary requirements necessary for optimal growth and development, maintenance of health, and the prevention and treatment of disease. High-pressure liquid chromatography provides a rapid, accurate, and sensitive technique for the separation and analysis of subnanomole quantities of a wide range of complex, high-molecular-weight, nonvolatile, thermally labile, polar compounds which are vital for metabolic and nutritional studies.

Nutrition may be basically defined as the science of food, the nutrients and other substances therein, their action, interaction, and balance in relation to health and disease, and the processes by which the organism ingests, digests, absorbs, transports, utilizes, and excretes food substances. Since the science of nutrition relies on chemical and biochemical techniques, the organized study of nutrition has been confined to the twentieth century. However, interest in the relationship between man and his food was stimulated much earlier; the history of nutrition can be divided into three eras: the naturalistic era from 400 B.C. to A.D. 1750; the chemical-analytical era from 1750 to 1900; and the biological era from 1900 to the present.

Early man considered food essential for survival but made no discernment about the relative value of different foods. In 400 B.C. Hippocrates considered food one universal nutrient and believed that weight loss during starvation was caused by insensible perspiration. By the sixteenth century the principle of diet and longevity was well established and in 1747 the first controlled nutrition experiment was conducted by Lind who found that lemon or lime juice was effective in curing scurvy whereas other substances, such as seawater and vinegar, were ineffective.

During the chemical-analytical era in the study of nutrition, Lavoisier, Black, and Priestly conducted studies on respiration, oxidation, and calorimetry. Early in the nineteenth century, methods were developed for determining carbon, hydrogen, and nitrogen in organic compounds. Analyses of these elements in foods led Liebig to hypothesize that the nutritive value of food was dependent on its nitrogen content and that an adequate diet should contain protein, carbohydrate, and fat. This hypothesis was tested by Dumas and Lunin who constructed synthetic foods containing protein, carbohydrate, and fat in proportions believed to be present in cow's milk. Infants and mice fed the synthetic diet died while those who were fed milk thrived. These and other reports on the use of purified diets led to the conclusion that the addition of minute amounts of natural foods was necessary for growth and maintenance of health. Thus, it became apparent that foods contained more than carbohydrate, protein, fat, and minerals, but the nature of these other substances was unknown until McCollum and Eijkman, early in the twentieth century, showed that at least two vitamins, fat-soluble A and water-soluble B, were essential to life. During this period very rapid progress was made in identifying essential dietary components and it was discovered that diseases such as beriberi, scurvy, rickets, and pellagra, previously believed to be caused by toxic substances or to be infectious in nature, were in reality the result of an absence of nutrients required in very small amounts.

Discoveries of many dietary factors with vitaminlike properties were made in the early part of the biological era, and by 1940 four fat-soluble and eight water-soluble vitamins had been identified as essential components of the human diet, while several others were characterized as essential for various animal species. The chemical structure of these vitamins was established, many were synthesized, and knowledge of their biological roles accumulated rapidly. However, since 1940 only two essential vitamins, folic acid and vitamin B_{12}, have been identified. During this same era the noncombustible component, or mineral ash, of diets was investigated and it too proved to be a complex mixture of elements, 17 of which have been identified as dietary essentials for humans.

Since 1955, with the development of the electron microscope, ultracentrifuge, microchemical techniques, and the use of radioactive isotopes, emphasis has been directed to the study of nutritional needs and metabolism of individual cells and even the subcellular components of the cells. This information is leading to a more complete understanding of the nutritional needs and metabolism of individual cells, and the vital role that nutrients play in the growth, development, and maintenance of cells, tissue, organs, and ultimately the whole complex human body. Thus, a defect in nutrition at the cellular level or the absence of any one of the known dietary essentials, regardless of the amount needed, can adversely affect the health and functioning of the whole body.

The advent of high-pressure liquid chromatography, which makes possible the quantitative evaluation of substances in micro amounts, has in essence extended the analytical era of nutrition. The introduction of capillary columns filled with ion exchange resins has made possible the determination of micronutrients in the range of 1 to 10 nmoles. Although the application of high-pressure liquid chromatography to nutrition studies is still in its infancy, this chromatographic technique has been used frequently in nucleic acid and purine research. High-pressure liquid chromatography was developed initially for the separation of compounds such as nucleotides, nucleosides, and purine bases [1] which absorb strongly in the UV range of the light spectrum. A very recent application of high-pressure liquid chromatography in nutrition studies has been in the area of micronutrients such as vitamins. Thus, the discussion of the application of high-pressure liquid chromatography to nutrition studies in this chapter will be confined to purines and vitamins, and suggestions will be made for further applications of this technique to nutritional studies.

II. DIET PURINES AND HIGH-PRESSURE LIQUID CHROMATOGRAPHY

A. The Significance of Diet Purines

Hyperuricemia appears to be a necessary but not sufficient cause for the development of gouty arthritis and may be due to increased endogenous production or reduced excretion of uric acid [2-4]. It is exacerbated by diets high in fat [5], protein [6-13], or nucleic acids [2, 6, 7, 13-19]. As summarized by Wyngaarden [20], Seegmiller [21], and Smith [22], the purine intake has the greatest dietary influence on blood uric acid levels.

Recent studies of yeast and other new and unconventional sources of dietary protein for humans have shown that (a) ingestion of proteins from these sources results in a substantial elevation of blood uric acid levels [16, 19]; (b) the blood uric acid response of different individuals to the same amount of total nucleic acid and to individual purines varies considerably and may be related to inherent susceptibilities of individuals to develop clinical gout [23]; and (c) each individual purine characteristically affects purine salvage enzyme activity and the urine purine excretion pattern [24].

Several recent studies [13-19, 23-25] have shown the role of diet and nucleic acid in altering plasma uric acid levels, as well as uric acid excretion in urine of humans. These studies are summarized in Table 1 which shows that plasma uric acid increased 0.5 to 1.0 mg/100 ml for each gram of ribonucleic acid (RNA) consumed in the diet while the urine uric acid excretion increased 113 to 164 mg/24 hr for each gram of RNA consumed in the diet by normal healthy human adults (controls). Similar studies conducted on hyperuricemic patients showed an increase of 1.5 mg uric acid/100 ml plasma for each gram of RNA consumed in the diet and an increase of 120 mg of uric acid in the urine per day per gram of RNA consumed in the diet. Thus, it appears that hyperuricemic persons had a 50% greater increase in blood uric acid per unit of RNA consumed than did controls, whereas the urine uric acid output per gram of RNA consumed was similar for hyperuricemic and control humans. These studies were carried out with levels of RNA ranging from 0 to 8 g/day per person added to a purine-free basic formula diet adequate and constant in all known essential nutrients.

Ribonucleic acid is a polynucleotide occurring mainly in the cytoplasm of cells and only to a limited extent in the nucleus. The constituents of the individual nucleotides are purine (adenine, guanine) and pyrimidine (cytosine, uracil) bases, a sugar (D-ribose), and phosphoric acid. Since high-pressure liquid chromatography facilitates the reliable quantification of adenine, guanine, hypoxanthine, and xanthine in food items, several studies [23, 24] have been conducted to evaluate the hyperuricemic effect of these individual purines on normal, hyperuricemic, and gouty humans. Gouty

TABLE 1

Regression Equations Describing the Relationship Among Dietary RNA Intake and Blood Serum and Urine Uric Acid Levels in Human Adults

Type of subject	Regression equation on diet RNA (g/day) (X) on		Ref.
	Serum uric acid (mg/dl) (Y)	Urine uric acid (mg/day) (Y)	
Normal	Y = 4.84 + (0.65) (g RNA)[a]	Y = 367.8 + (147.2) (g RNA)[a]	Waslien et al. [16]
Normal	Y = 5.05 + (0.56) (g RNA)[b]	Y = 645.0 + (163.6) (g RNA)[a]	Edozien et al. [19]
Normal	Y = 3.25 + (0.90) (g RNA)[c]	Y = 377 + (113) (g RNA)[b]	Zöllner and Griebsch [25]
Hyperuricemic[d]	Y = 4.40 + (1.46) (g RNA)	Y = 286 + (120) (g RNA)	Zöllner and Griebsch [25]

[a]Represents grams of RNA consumed per person per day.
[b]Calculated from the data of Edozien et al. [19].
[c]Represents grams of RNA consumed per 70-kg person per day.
[d]Persons having a serum uric acid level in excess of 6.5 mg/100 ml.

patients were selected on the basis of having had at least one episode of clinical gout and having shown a positive response to pharmacologic treatment of gout. Hyperuricemic individuals were chosen as having a plasma uric acid value in excess of 8.0 mg/100 ml; controls were selected on the basis of an absence of a family history of gout and having a plasma uric acid level lower than 8.0 mg/100 ml. Adenine, guanine, hypoxanthine, and xanthine were given singly at the rate of 0.1 mmole/kg of body weight and plasma uric acid was monitored at several time intervals thereafter until the highest values were recorded.

Table 2 shows the increase in plasma uric acid due to the individual purines in each type of subject. It is clear that hypoxanthine caused the greatest increase in blood uric acid per unit weight of purine given, followed by adenine, xanthine, and finally guanine. This was true for all subjects. The data also show that hypoxanthine produced a greater increase in the uric acid level in gouty patients than it did in control or hyperuricemic patients, whereas adenine, guanine, and xanthine produced similar increases in all three types of patients. Hypoxanthine markedly increased the urine uric acid/creatinine ratio for all patients whereas adenine increased the ratio for the control and the hyperuricemic group but did not alter the values for the gouty group. Guanine and xanthine did not alter the urine uric acid/creatinine ratio in any patient. Urine uric acid/creatinine ratio was not different among the three groups of patients before or after loading with individual purines. The data suggest that each individual purine characteristically elevated the level of uric acid in blood and urine of humans, and that hypoxanthine alone had a much greater hyperuricemic effect on gouty patients compared with normals or hyperuricemics. The biochemical mechanism for this difference is currently unknown, as is the true clinical significance of these observations, although it has been suggested that the oral administration of hypoxanthine may be of value in predicting, preclinically, a genetic tendency to gout.

Consistent with the idea that each individual purine has its own characteristic hyperuricemic effect was the concept that nucleoproteins from different sources may produce different hyperuricemic effects when fed to normal human adults. Nucleoproteins were isolated from powdered yeast, bacterial cells, and frozen porcine liver which was homogenized in water and treated with papain to digest the greater proportion of the protein fraction. The digestion was terminated by heating the mixture at 90° C for 15 min to inactivate the papain, and after cooling to 12° C the mixture was clarified by centrifugation. The supernatant was further cooled to 4° C, acidified with HCl to a pH of 3.1, and the precipitated nucleoproteins were again separated by centrifugation and lyophilized. Single doses of the isolated nucleoproteins were fed to human adults as described in Table 3 and blood and urine uric acid were monitored. Nucleoproteins from the three sources produced different degrees of hyperuricemia per unit of

TABLE 2

Effect of Oral Purine Loading on Blood and Urine Uric Acid in Control, Hyperuricemic, and Gouty Humans[a]

	Control	Hyperuricemic	Gout
Number of subjects	6	11	8
Age (yr)	45 ± 5	45 ± 3	49 ± 5
Weight (kg)	87 ± 6	100 ± 2[d]	87 ± 3
Fasting SUA (mg %)	6.3 ± 0.5	8.5 ± 0.2[d]	8.3 ± 0.7[d]
Change[b] in SUA due to			
Adenine	1.8 ± 0.3[e]	1.6 ± 0.1[e]	2.0 ± 0.2[e]
Guanine	-0.2 ± 0.1[f]	0.1 ± 0.2[f]	0.2 ± 0.3[f]
Hypoxanthine	2.4 ± 0.2[e]	2.8 ± 0.2[e]	4.1 ± 0.4[e]
Xanthine	0.7 ± 0.1[f]	0.7 ± 0.1[f]	0.7 ± 0.1[f]
Urine uric acid/creatinine[c]			
Before loading	0.39 ± 0.03	0.40 ± 0.04	0.34 ± 0.04
After loading with			
Adenine	0.46[g] ± 0.02	0.53[g] ± 0.02	0.40 ± 0.02
Guanine	0.35 ± 0.02	0.50 ± 0.06	0.34 ± 0.03
Hypoxanthine	0.55[g] ± 0.07	0.58[g] ± 0.08	0.68[g] ± 0.09
Xanthine	0.40 ± 0.04	0.45 ± 0.03	0.38 ± 0.03

[a]All values in this table are means ±1 standard error.

[b]Change in serum uric acid (SUA) was determined by subtracting the fasting value from the highest subsequent value after purine intake and expressed as mg/100 ml. All purines except guanine increased serum uric acid above testing values ($P < .05$).

[c]Urine uric acid/creatinine ratios are based on four patients per treatment group. Complete 24-hr urine collections were made on the control and gouty subjects while spot urine samples were taken from the hyperuricemic subjects.

[d]Within rows, means different from controls ($P < .01$).

[e, f]Within columns, means not sharing a common superscript letter are different ($P < .05$).

[g]Greater than the ratio before loading ($P < 0.05$). Values are means ±1 standard error.

TABLE 3

Effect of Nucleoproteins from Liver, Yeast, and Bacterial Sources on Blood and Urine Uric Acid in Human Adults

	Liver	Yeast	Bacteria
Dry weight (%)	95	100	96
Total nitrogen (%)	9.94	11.95	13.46
Total nucleic acids (%)	31	60	62
Nucleoprotein intake (g)	20	10	10
Number of subjects	5	6	8
Age	30 ± 2	30 ± 2	30 ± 2
Weight (kg)	70 ± 2	70 ± 2	70 ± 2
Fasting blood uric acid (mg/dl)	5.65 ± 0.77	5.67 ± 0.43	5.65 ± 0.29
Increase in blood uric acid (mg/100 ml)	1.89 ± 0.09[a]	2.13 ± 0.08[b]	2.41 ± 0.08[c]
Urine uric acid before nucleoprotein (mg/day)	529 ± 74	508 ± 128	530 ± 146
Increase in urine uric acid due to nucleoprotein (mg/day)	317 ± 93	309 ± 71	402 ± 148

[a, b, c]Within rows, means with different letter superscripts are different ($P < .01$). Values are means ± 1 standard error.

nucleoprotein consumed. Bacterial nucleoproteins produced the greatest increase in blood uric acid (2.41 mg/dl), followed by yeast (2.13 mg/dl) and liver (1.89 mg/dl). Each nucleoprotein markedly increased urine uric acid concentrations and although there were no statistical differences between the nucleoproteins, the bacterial nucleoprotein appeared to produce a greater elevation than either yeast or liver nucleoproteins.

Recent studies [24] have shown that the addition of graded levels of adenine to purified amino acid diets of growing rats caused a marked reduction in growth rate and food consumption (Table 4). Similar studies have also shown that the addition of graded levels of adenine to the diet caused a marked increase in kidney weight and urine output volume (Table 5), whereas guanine, hypoxanthine, and xanthine did not affect the animals in this manner. Table 6 shows the effect of diet purines on the pattern of purines excreted in urine. The major purine excreted in urine of control rats was hypoxanthine followed by guanine, adenine, and xanthine. The addition of adenine to the diet caused a slight reduction in the amount of hypoxanthine excreted; it did not alter guanine or adenine excretion but caused a 30-fold increase in the amount of xanthine excreted in urine. Addition of guanine, hypoxanthine, or xanthine to the diets did not alter the purine excretion pattern in urine.

In addition to the effects of individual purines on growth and urine purine excretion patterns, the effects of these same purines on purine salvage enzyme activity (Table 7) were evaluted. The purine salvage enzymes whose activity was measured included adenosine kinase (E.C. 2.7.1.20), adenosine deaminase (E.C. 2.5.4.4), hypoxanthine phosphoriboxyltransferase (E.C. 2.4.2.8), adenine phosphoribosyltransferase (E.C. 2.4.2.7), and 5'-nucleotidase (E.C. 3.1.3.5). Rats that had been fed purified amino acid diets supplemented with adenine, guanine, hypoxanthine, and xanthine were sacrificed by decapitation and the liver was removed, washed with ice-cold saline, homogenized in 4 volumes of sodium phosphate buffer (0.15 M, pH 6.8, 0-3° C), and centrifuged at 100,000 × g for 60 min to obtain a cell-free supernatant on which the enzyme assays were performed as outlined by Shenoy and Clifford [26]. Addition of adenine to the diet did not affect the activity of adenine phosphoriboxyltransferase, but enhanced the activities of 5'-nucleotidase, adenosine kinase, hypoxanthine phosphoribosyltransferase, and adenosine deaminase (Table 7). The addition of guanine, hypoxanthine, or xanthine did not alter the activity of any salvage enzyme.

The addition of purines to purified amino acid diets markedly altered renal histology, as seen in Fig. 1 which shows a phase contrast micrograph of kidney from a rat fed adenine [24]. Dietary adenine caused an accumulation of crystalline material within the tubules, localized areas of inflammatory response, sloughing off of the lining of the luminal cells, and a loss of cellular structure. The mitochondria became rounded and accumulated a dark staining material which, by using high-pressure liquid chromatography, we found to be 2,8-dihydroxyadenine. The kidney slices were obtained from rats fed purified amino acid diets containing 0.5% adenine for a two-week period. Adenosine, guanine, hypoxanthine, or xanthine additions to the diet did not alter renal histology. Somewhat similar observations were made by Elion and co-workers [27] several years ago but the comparisons of the different individual purines under conditions of amino acid feeding were not reported previously.

TABLE 4

Effect of Diet Ribonucleic Acid (Experiment I) and Diet Purines (Experiment II) on Growth, Food Intake, and Urine Volume in the Growing Rat

Diet	Initial weight (g)	Food consumed (g/day)	Weight gain (g/day)	Urine volume (ml/day)
Experiment I[a]				
Control (C)	102 ± 3	9.9 ± 0.6	3.88 ± 0.55	7.23 ± 0.98
C + 10% RNA	103 ± 2	9.5 ± 0.3	2.88[b] ± 0.29	10.73[b] ± 0.68
C + 10% cellulose	109 ± 3	10.2 ± 1.0	3.75 ± 0.48	6.82 ± 0.44
Experiment II[c]				
Control (C)	87.75 ± 8.09	8.65 ± 0.76	3.45 ± 0.41	6.78 ± 1.08
C + adenine	98.00 ± 2.30	5.25[d] ± 0.35	-0.50[d] ± 0.52	19.12[d] ± 1.74
C + adenosine	93.25 ± 5.15	9.85 ± 0.66	4.30 ± 0.41	8.87 ± 0.43
C + guanine	101.25 ± 5.85	9.30 ± 0.63	3.58 ± 0.46	7.46 ± 0.78
C + hypoxanthine	90.00 ± 3.16	9.55 ± 0.78	3.88 ± 0.54	6.42 ± 0.76
C + xanthine	97.25 ± 3.01	9.60 ± 0.29	4.08 ± 0.28	6.91 ± 0.87

[a]The composition of the control diet was that described by Rogers and Harper, J. Nutr., 87, 267 (1965). C + 10% cellulose served as a negative control so that the reduced weight gain on feeding RNA was not caused by the lower level of carbohydrate in the diet.

[b]Values different from control ($P < 0.05$). Values are means ± 1 standard error (four rats per treatment, fed for 10 days), experiment I.

[c]Individual purines were added singly at the level of 0.75 g purine/100 g of the control diet described above. The purines were reagent-grade compounds purchased from commercial sources.

[d]Values different from control values ($P < 0.01$). Values are means ± 1 standard error (four rats per treatment, fed for 14 days), experiment II.

TABLE 5

The Effect of Graded Levels of Diet Adenine on Growth, Food Intake, Kidney Weight, Urine Volume, and Plasma Urea in the Growing Rat

Diet	Weight gain (g/day)	Food intake (g/100 g body wt./day)	Kidney weight (g/100 g body wt.)	Urine volume (ml/100 g body wt./day)	Plasma urea (mmoles/liter)
Control[a]	6.0 ± 0.1	12.1 ± 0.3	1.02 ± 0.07	5.1 ± 0.3	6.0 ± 0.7
Control + 0.1% adenine	5.7 ± 0.5	12.2 ± 0.2	0.95 ± 0.02	6.0 ± 0.7	5.9 ± 0.5
Control + 0.2% adenine	5.8 ± 0.5	13.0 ± 0.2	1.24 ± 0.05	14.8 ± 1.4[b]	5.6 ± 0.5
Control + 0.3% adenine	4.1 ± 0.3[b]	11.8 ± 0.2	1.70 ± 0.08[b]	31.0 ± 3.9[b]	11.4 ± 0.8[b]
Control + 0.4% adenine	1.0 ± 0.4[b]	11.3 ± 0.3	3.38 ± 0.14[b]	34.2 ± 1.5[b]	23.1 ± 0.9[b]
Control + 0.5% adenine	0.5 ± 0.1[b]	11.2 ± 0.4	3.48 ± 0.21[b]	42.1 ± 2.1[b]	19.2 ± 2.5[b]
Control + 0.75% adenine	-0.8 ± 0.1[b]	8.6 ± 0.3[b]	5.32 ± 0.34[b]	34.6 ± 1.1[b]	22.8 ± 0.8[b]

[a]Composition (%) of the control diet: 20 casein, 8 corn oil, 63.98 glucose, 6 mineral mix (g/kg of mix: 318.2 $CaCO_3$, 344.7 KH_2PO_4, 63.6 $CaHPO_4$, 178.2 NaCl, 26.5 $FeSO_4 \cdot 7\ H_2O$, 64.7 $MgSO_4$, 0.84 KI, 0.26 $ZnCO_3$, 0.31 $CuSO_4 \cdot 5\ H_2O$, 2.5 $MnSO_4 \cdot H_2O$), and 2.0 vitamin mix (g/kg of mix: 25 inositol, 5 ascorbic acid, 2.5 calcium pantothenate, 1.5 thiamine hydrochloride, 1.5 pyridoxine hydrochloride, 1.5 nicotinic acid, 1.25 menadione, 0.5 riboflavin, 0.5 p-aminobenzoic acid, 0.03 folic acid, 71.5 choline chloride (70% solution), 0.0125 biotin, 2.7×10^6 I.U. α-tocopherol, 0.75×10^6 I.U. vitamin A palmitate, 0.075×10^6 I.U. cholecalciferol, 0.0015 cyanocobalamin, 880 glucose); β-hydroxytoluene was used as an antioxidant at 0.02% of the complete diet.

[b]Values different from control ($P < 0.05$). Values are means ± 1 standard error (six rats per treatment, fed for 14 days).

TABLE 6

Diet Purines and Urine Excretion of Purines[a]

Diet	Urine purines (μmoles/10-day period)			
	Hypoxanthine	Xanthine	Guanine	Adenine
Control (C)	16.1 ± 1.6	1.9 ± 0.2	7.1 ± 1.2	2.5 ± 0.4
C + adenine	11.8 ± 1.2	56.8[b] ± 6.5	8.1 ± 1.5	1.5 ± 0.2
C + adenosine	22.1 ± 2.1	10.0[b] ± 1.0	8.2 ± 0.8	4.3 ± 0.4
C + guanine	22.7 ± 2.3	3.2 ± 0.4	6.8 ± 1.2	3.0 ± 0.4
C + hypoxanthine	16.5 ± 1.6	2.2 ± 0.3	3.4 ± 0.6	1.4 ± 0.2
C + xanthine	19.6 ± 2.0	2.3 ± 0.3	3.0 ± 0.5	1.8 ± 0.2

[a]Entries are means of four rats per treatment fed the diets described in Table 4 for 10 days.

[b]Values different from corresponding values for control animals ($P < 0.01$).

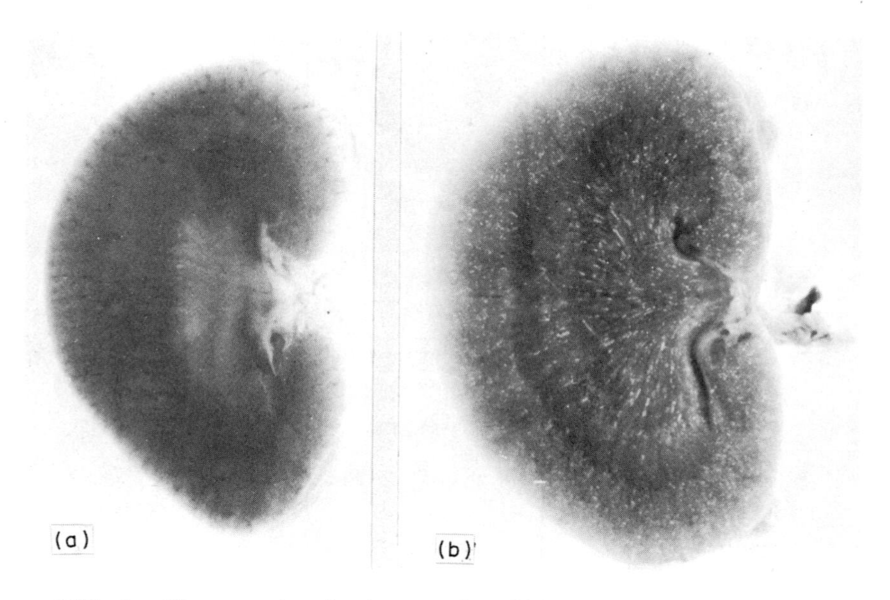

FIG. 1. Phase contrast micrographs of kidneys from rats fed a purified amino acid control diet (A) and the same diet supplemented with 0.4% adenine for 6 days (B). Magnification 3.5 ×. (From D. L. Story, M.S. thesis paper, University of California, Davis, California, 1974).

TABLE 7

Effect of Diet Purines on Purine Salvage Enzyme Activity in Rat Liver[a]

Diet	Hypoxanthine phosphoribosyl-transferase	Adenine phosphoribosyl-transferase	5'-Nucleotidase	Adenosine kinase	Adenosine deaminase
Control (C)	650 ± 57	120 ± 5	191 ± 12	2327 ± 71	401 ± 5
C + adenine	640 ± 12	185[b] ± 7	267[b] ± 14	2709 ± 262	619[b] ± 52
C + guanine	638 ± 35	106 ± 4	193 ± 11	2561 ± 42	364 ± 20
C + hypoxanthine	729 ± 19	119 ± 6	198 ± 13	2449 ± 42	379 ± 64
C + xanthine	768[b] ± 40	130 ± 4	190 ± 7	2790[b] ± 10	384 ± 17

[a]Enzyme activity is expressed as micromoles of product formed per gram of liver protein per hour. Values are means (three rats per treatment) ± 1 standard error.

[b]Within columns, values different from corresponding controls ($P < 0.01$).

B. Measurement of Diet Purines

Prior to the introduction of high-pressure liquid chromatography, food purines were measured by analytical procedures developed in 1905 by Kruger and Schmidt [28] in which the UV absorption of a perchloric acid extract of foods was measured. As a result of these analyses, foods were divided into two categories: foods whose purines ranged from 0 to 15 mg/100 g were classified "low purine" and foods whose purine content ranged from 50 to 150 mg/100 g were classified "high purine." Low-purine foods included vegetables, fruits, milk, cheese, eggs, and cereals. High-purine foods included meats, fish, seafoods, beans, peas, and lentils. Several years later foods were divided into three groups [29] as follows: group I, having 0 to 50 mg purine; group II, having 50 to 150 mg purine; and group III, having 150 to 800 mg of purine per 100 g of food. These data gave a measure of the total UV absorbing substances (purines) in the food but did not distinguish the individual purines making up the total purines. From the preceding discussion it is clear that a measure of total purines without considering the individual purines making up this total is of limited value.

The introduction of high-pressure liquid chromatography whereby it became possible to separate and identify adenine, guanine, hypoxanthine, and xanthine has made it possible to reevaluate the purine content of many new and unconventional food sources, in addition to many of the conventional and already used food items. Over the past several years many foods ordinarily consumed in the diet have been assayed for their content of adenine, guanine, hypoxanthine, xanthine, RNA, and protein in my laboratory; these data are presented in Table 8. Many of the food items were obtained in local supermarkets and some were obtained fresh and promptly frozen until ready for analyses. The food items were homogenized in 10 volumes of 1.0 N perchloric acid and the homogenate was placed in a boiling water bath for 30 min to convert all nucleotides and nucleosides to free bases. The samples were then filtered with Millipore filters (0.02 μm) and were separated on cation exchange columns using the LCS-1000 system (Varian Aerograph) and a nongradient elution with 0.5 M ammonium dihydrogen phosphate buffer, pH 2.0, at a flow rate of 12 ml/hr, column pressure of 1600 psi, and temperature of 40°C. The order of elution in this system [30] was hypoxanthine, xanthine, guanine, and adenine. The area of each peak was quantitated by triangulation and all samples were standardized against known standards of pure reagent-grade adenine, guanine, hypoxanthine, and xanthine. RNA determinations were made by the procedure of Drysdale and Munro [31] and protein by micro-Kjeldahl. The total purines in Table 8 represent the sum of adenine, guanine, hypoxanthine, and xanthine.

Table 8 shows that the proportions of the individual purines making up the total purine content vary considerably from one food item to another.

TABLE 8

Purine, RNA, and Protein Content of Selected Foods[a]

	Adenine (mg/100 g)	Guanine (mg/100 g)	Hypoxanthine (mg/100 g)	Xanthine (mg/100 g)	Total purines (mg/100 g)	RNA (mg/100 g)	Protein (%)
Organ meats							
Beef liver	62	74	61	0	197	268	20
Beef kidney	42	47	63	61	213	134	18
Beef heart	15	16	38	102	171	49	19
Beef brain	12	12	26	112	162	61	11
Pork liver	59	77	71	82	289	259	22
Chicken liver	72	78	71	22	243	402	20
Chicken heart	32	41	12	138	223	187	18
Lamb liver	30	43	54	18	147	88	22
Lamb heart	30	23	20	98	171	50	19
Fresh seafoods							
Anchovies	8	185	6	121	411	341	20
Clams	14	24	12	86	136	85	17
Mackerel	11	26	5	152	194	203	23
Salmon	26	80	11	133	250	289	23

Fresh seafoods (Continued)							
Sardines	6	118	6	215	345	343	23
Squid	18	15	24	78	135	100	15
Canned seafoods							
Anchovies	0	39	14	268	321	6	30
Clams	30	5	7	20	62	44	20
Herring	15	180	6	177	378	82	17
Mackerel	23	109	16	98	246	122	26
Oysters	39	22	30	16	107	239	9
Salmon	23	39	13	13	88	26	26
Sardines	19	95	30	255	399	590	24
Shrimp	16	12	15	191	234	10	22
Tuna	27	13	11	91	142	5	29
Dried legumes							
Garbanza bean	17	14	18	7	56	356	21
Cranberry bean	21	19	23	12	75	248	17
Split peas	88	74	11	22	195	173	21
Red bean	54	51	15	42	162	140	20
Lentils	104	82	20	16	222	484	28
Blackeye peas	77	80	32	41	230	306	22

TABLE 8 (Continued)

	Adenine (mg/100 g)	Guanine (mg/100 g)	Hypoxanthine (mg/100 g)	Xanthine (mg/100 g)	Total purines (mg/100 g)	RNA (mg/100 g)	Protein (%)
Dried legumes (Continued)							
Large lima bean	42	41	14	52	149	293	21
Baby lima bean	46	39	25	34	144	190	19
Pinto bean	57	54	16	44	171	485	20
Small white bean	59	74	25	44	202	305	18
Great northern bean	56	68	25	64	213	284	18

[a]From Clifford and Story [24].

For example, beef liver had virtually no detectable amounts of xanthine and considerable amounts of guanine, with adenine and hypoxanthine occurring in equal and intermediate amounts, respectively. Beef heart, beef brain, pork liver, chicken heart, and lamb heart are characterized by their high concentration of xanthine and much smaller amounts of guanine, adenine, and hypoxanthine. Beef heart and beef brain are characterized by small amounts of guanine while chicken liver had the highest concentrations of adenine and hypoxanthine of all the organ meats analyzed.

Analyses of the purine composition of fresh seafoods showed that anchovies and sardines were characterized by their high concentrations of guanine and xanthine and corresponding low concentrations of adenine and hypoxanthine. Mackerel and salmon also had very high concentrations of xanthine and relatively low concentrations of the remaining purines with the exception of salmon, which had relatively high concentrations of guanine. These data are of special interest in view of the fact that anchovies and sardines have been characterized as high-purine foods and are eliminated from therapeutic low-purine diets. It is clear from the data presented in Table 8 that although the total purines were high, some of the individual purine components which can produce deleterious effects on the organism, such as adenine and hypoxanthine, occurred in very low concentrations.

An evaluation of the purine content of canned seafoods showed essentially the same individual purine pattern as was obtained with fresh seafoods. Anchovies and sardines again had high concentrations of xanthine and relatively low concentrations of adenine and hypoxanthine. Herring and mackerel had high concentrations of guanine and xanthine and relatively low concentrations of adenine and hypoxanthine. The discrepancies between the RNA content and total purines probably reflect the fact that the RNA in canned seafoods was already hydrolyzed by the very active tissue RNAs. In fact, the concentrations of hypoxanthine has been used for many years as an index of the wholesomeness of fish and of the degree of deterioration prior to canning or freezing of these food items.

The purine composition of dried legumes differed markedly from one legume to another. For example, lentils had extremely high concentrations of adenine and low concentrations of xanthine (Table 8).

Several local hospital diets formulated as low (200 mg purines/100 g food) and high (500-1000 mg purines/100 g food) purine diets were analyzed and the data in Table 9 show that the low- and high-purine diets both contained very large quantities of adenine, and relatively small amounts of hypoxanthine, guanine, and xanthine. Also shown in this table is the fact that the total purine concentrations were high in the high-purine diet but the RNA concentrations of the low-purine diet was higher than that of the high-purine diet, and the protein content of both diets was similar. Sources of single-cell protein from liver, yeast, and bacteria were also analyzed for

TABLE 9

Purine Composition of Diets and Single-Cell Proteins

	Adenine (mg/100 g)	Guanine (mg/100 g)	Hypoxanthine (mg/100 g)	Xanthine (mg/100 g)	Total purines (mg/100 g)	RNA (mg/100 g)	Protein (%)
Low-purine diet	184	35	17	12	247	644	20
High-purine diet	223	46	47	33	348	440	22
Nucleoproteins from							
Liver	2850	3640	997	0	7,487	17,664	62
Yeast	7160	8830	2442	0	18,432	45,056	84
Bacteria	6320	8040	2123	0	16,483	37,683	75

purines; they were found to be extremely high in total purines and devoid of xanthine, but had high levels of guanine and lesser amounts of adenine and hypoxanthine.

In summary, individual diet purines characteristically affect blood and urine uric acid levels, growth, the activities of purine salvage enzymes, purine excretion patterns in urine, and renal histology. Adenine caused marked alterations in growth, in purine patterns in urine, in the rate of de novo purine synthesis, and in the relative specific activities of purine salvage enzymes, whereas hypoxanthine, guanine, and xanthine had little or no effect on these parameters. In human studies, both adenine and hypoxanthine markedly increased blood uric acid level while guanine and xanthine had little effect. Furthermore, hypoxanthine had a much greater effect on humans with gout than on hyperuricemic or normal humans which suggests a possible diagnostic value of a purine tolerance test for preclinical gout. The importance of these observations is further emphasized with the introduction into the general human diet of new and unconventional sources of protein, such as single-cell proteins from yeast or bacteria, leaf protein concentrate, fish protein concentrate, and a variety of conventional foods. The introduction of high-pressure liquid chromatography has now made it possible to better evaluate food purines and enhance our understanding of their metabolism so that the contribution of many new and unconventional sources of proteins toward meeting the protein requirements of man might no longer be compromised by their high content of nucleic acids.

III. DIET, VITAMINS, AND HIGH-PRESSURE LIQUID CHROMATOGRAPHY

A. The Significance of Folic Acid Cofactors

Folic acid is required for growth, reproduction, and the prevention and treatment of megaloblastic anemia. Nutritional folate deficiency occurs even in affluent populations, but particularly in stress situations such as premature infancy [32], pregnancy [33], and alcoholism [34]. Megaloblastic anemia is a metabolic disease caused by a reduction in the ability to double the DNA complement in order for the cell to divide, and in the vast majority of cases is caused by folate or vitamin B_{12} deficiency, or both [35]. The formation and development of every human cell are dependent on an adequate supply of folic acid. Folic acid thus governs the synthesis of the precursors of DNA, the chemical constituents responsible for genetic transfers and protein synthesis, and may therefore have far greater fundamental importance than is currently recognized.

Folic acid does not exist as a single chemical entity but occurs as a variety of structurally related compounds. Dietary folate has been shown to consist mainly of polyglutamic acid forms of folic acid [35, 36]. Since man is totally unable to synthesize his estimated daily requirement of about 50 μg [37, 38], he is dependent on efficient digestion and absorption of folic acid and folic acid peptides.

Recent developments in the area of high-pressure liquid chromatography have for the first time made it possible to separate and quantitate the various folic acid cofactors by chromatographic procedures which are extremely sensitive, rapid, and reproducible. Owing to the multiplicity of naturally occurring chemical forms of folate and the lack of technology to measure these forms individually, most of the current theories on folate metabolism and factors affecting metabolism and requirements have only indirect supporting evidence. Oral contraceptives, anticonvulsant drugs, chemotherapeutic drugs, drugs related to malignancy and allergy, vitamin B_{12} deficiency or pernicious anemia, riboflavin deficiency, pregnancy, and old age are among the conditions and environmental factors that are believed to influence adversely folate metabolism and requirements. By methods currently available only total folates can be measured with any degree of reliability. In order to evaluate the mechanisms of these defects and to develop reliable procedures with which to evaluate folic acid status, methodology was required to measure individual folate forms as they occur naturally.

As its tetrahydro derivative folic acid serves as a carrier of the hydroxymethyl and -formyl groups. The N^5-methyl derivative of tetrahydrofolic acid is an intermediate in the de novo synthesis of the methyl group of methionine. Although originally synthesized as a dihydro compound and functional as the tetrahydro compound, fully oxidized folic acid is obtained with some isolation procedures because of the spontaneous nonenzymatic oxidation of several of the derivatives. Mammalian cells have available at all times the system for reduction of oxidized folic acid or the dihydro compound to the tetrahydro derivative; NADH is utilized in the reduction which appears to require the presence of ascorbic acid. Because of the metabolic role of folic acid, its deficiency is expressed primarily as a failure to make the purines and thymine required for DNA synthesis. Therefore, it is indicated that folic acid is required for the synthesis of purines and pyrimidines which are key compounds in nucleic acid and protein synthesis.

An additional interesting, important, and still puzzling aspect of folate function concerns its relationship to vitamin B_{12}. Several lines of evidence bear out this relationship: the appearance of megaloblastic anemia in either folate or vitamin B_{12} deficiency; the reversal of megaloblastic anemia in vitamin B_{12} deficiency by large doses of folate; the amelioration of megaloblastic changes in folic acid deficiency by large doses of vitamin B_{12} [39]; the increased plasma concentrations of N^5-methyltetrahydrofolate in patients with vitamin B_{12} deficiency [40]; the excretion of excessive amounts of formiminoglutamic acid (FIGLU) after histidine loading in patients with either vitamin B_{12} or folate deficiency [41, 42]; and the reduced amounts of total vitamin B_{12} in the liver of patients with folate deficiency [43].

A plausible explanation for most of these overlapping effects has been summarized by Rosenberg [44]. The explanation depicts the conversion of tetrahydrofolate (THF) via $N^{5,10}$-methylene-THF to N^5-methyl-THF and the regeneration of THF as a result of the transfer of the methyl group to homocysteine to form methionine, a reaction catalyzed by a vitamin B_{12}-dependent methyltransferase. If methionine biosynthesis was the only quantitatively significant reaction using N^5-methyl-THF, vitamin B_{12} deficiency would interfere with the folate cycle and lead to the accumulation of N^5-methyl-THF and the depletion of other folate derivatives. This depletion could become severe enough to interfere with other reactions requiring THF, such as the synthesis of purines or pyrimidines and the conversion of formiminoglutamate. Under these circumstances THF deficiency could be relieved by administration of either folate or vitamin B_{12}, but only the latter would complete the folate cycle.

This scheme, if totally correct, would obviate the need for additional vitamin B_{12}-dependent mechanisms to explain the megaloblastic changes observed in vitamin B_{12} deficiency and would account for the specific disorders of folate metabolism observed in vitamin B_{12} deficiency. However, it does not explain the low vitamin B_{12} content of livers from folate-deficient subjects [43], nor the hematologic response of folate-deficient patients to vitamin B_{12} [39]. The latter observations suggest an alternate feedback regulatory system between folic acid and vitamin B_{12} which could be of great significance.

Dietary folic acid consists primarily of conjugates of reduced formyl or methyl folates. Current concepts of the events occurring during folate absorption are outlined as follows [45].

1. Digestion of δ-glutamyl peptide conjugates of folate by intestinal enzymes (δ-glutamyl carboxypeptidase, "conjugase") releases monoglutamic folate. This site of hydrolysis, whether intraluminally, as the brush border, or within the intestinal cell, is undetermined.

2. Absorption of monoglutamic folate takes place largely in the proximal small intestine by a process that may be specific and active. However, conflicting evidence leaves this mechanism unclear.

3. Metabolic conversion of monoglutamic folate to reduced methyl folate (largely N^5-methyl-THF) appears to occur within the intestinal cell prior to folate release into the portal circulation. Conversion may be quantitative for reduced folate but only partial

for unreduced pteroylglutamic acid because of the substrate specificity of dihydrofolate reductase and the methylation system.

Inhibitors of intestinal deconjugation present in some foodstuffs may affect the availability of polyglutamic folate in the diet [46]. Recently, it has been shown that approximately 70% of the total folates in bovine milk were methyl folate with three or fewer glutamates [47]. The average "free" folate of this same milk was 60 μg/liter which would indicate the usual classification of milk as a low-folate food may be inappropriate. Although 70% of milk folate existed as methyl folate, the extent of utilization of this form of folate by man is unknown. In a recent study, Tamura and Stokstad [46] have shown that the availability of pure folate compounds varied from 25 to 100%, depending on the diet to which they were added. These studies show clearly that very little is actually known concerning the levels of the various folic acid cofactors in human foods or the extent to which the various cofactor forms can be used to meet the folic acid requirement of man.

In tables of food composition, folate pattern is ignored, and only values based on assay with a single microorganism are presented. For example, Hardinge and Crooks [48], in a compilation of available data, reported that fruit juices contain less than 4 μg folate per serving. This range is in agreement with many reports published between 1950 and 1968 [49-54] and indicates that fruit juice contributes an insignificant proportion of the recommended dietary allowance for folate. However, more recent reports [55, 56], using improved assay procedures designed to measure total folate, indicate that orange juice contains at least 10 times as much folate as reported earlier. If this applies to all juices, then they would be significant sources of dietary folate, especially for hospitalized patients on soft or liquid diets. Unfortunately, this information is not currently available.

B. Measurement of Folic Acid Cofactors

The assay of folates is important for identifying the cause of megaloblastic anemia, which can be caused by a deficiency of folic acid or vitamin B_{12}, or both. Therapeutic uses of folic acid to treat megaloblastic anemia caused by vitamin B_{12} deficiency can lead to very dangerous consequences because folate therapy allows for temporary remission of the anemia, but does not alleviate the neurologic symptoms associated with vitamin B_{12} deficiency.

Microbiological methods depend on the metabolic requirements of certain microorganisms: *Lactobacillus casei*, *Streptococcus faecalis*, *Pediococcus cerevisiae*, and *Leuconostoc citrovorum*. Each microorganism has its own spectrum of folate that it can metabolize. The increase in number of the microorganisms, as measured by the change in optical density of the culture, is directly related to the amount of folate present in a given medium. The results obtained are dependent on the test organism used on account of the different metabolic spectra. It is essential to know which test organisms has been used before assessing and comparing results.

Certain forms of folates are available only to certain microorganisms. These include S. faecalis: monoglutamates other than N^5-methyl folate; L. casei: monoglutamates, triglutamates, and N^5-methyl folate; L. citrovorum: N^5-formyl-THF only. Certain other forms of folate are not available to these test organisms. These are the polyglutamates of folic acid conjugates which have more than three glutamic acid residues linked to the pteridine ring. They can be rendered available to assay after enzymatic action by incubation with extracts of pancreas or kidney which contain "conjugases" to split off the glutamic acid residues.

The biochemical method usually used to evaluate folic acid status is the urinary excretion of formiminoglutamic acid (FIGLU). This method depends on the excretion of an intermediate product of histidine metabolism when a deficiency of folic acid occurs or if the enzyme formiminotransferase is blocked at the level of FIGLU and an accumulation of this product occurs. This is followed by an increased excretion of FIGLU in the urine. An oral dose of histidine is administered (usually 15 g) and the urine is collected over the following 8 hr when the maximum excretion occurs. The amount of FIGLU is then assayed by electrophoresis [57-60], chromatography [61], enzyme microbiologic [62-64], or enzyme and spectrophotometry [65-67].

In nonpregnant patients with folate deficiency an increase in the amount of FIGLU is detected in the urine. The value of the test in pregnancy is very controversial. An increase in FIGLU excretion in early pregnancy with progressive fall in level toward term and a second rise in the puerperium were found by Berry and associates [68] and others [69-71]. Approximately one-third of normal women showed this rise in level in early pregnancy and the puerperium [72]. Several investigators [57, 60, 73, 74] feel that an increase in FIGLU excretion in pregnancy is diagnostic of folate deficiency. There is no adequate explanation of this discrepancy of results and opinions.

Chanarin and associates [75] have found the test to be unreliable in pregnancy and in megaloblastic anemia of pregnancy for the following reasons:

1. There is a delayed absorption of histidine from the intestinal lumen in pregnancy as suggested by low, prolonged serum levels after an oral dose of histidine [75, 76].

2. An increase in urinary excretion of histidine occurs in pregnancy because of an increased glomerular filtration and a reduced tubular reabsorption [77].
3. There is an active transfer of histidine to the fetus as shown by higher cord blood levels [78, 79].
4. There may also be an increased utilization of histidine for protein synthesis with a diversion of histidine from the FIGLU pathway.

Folate does not exist as a single chemical entity but can occur as a variety of structurally related compounds differing in (a) state of oxidation of the pteridine ring; (b) number of glutamate residues; and (c) the presence or absence of a single carbon substitution on the 5- or 10-nitrogen, or both, and the level of oxidation of this moiety. Theoretically, there are over 100 possible naturally occurring folates. These compounds are too closely related to allow separate identification and quantification by chemical methods. Partial definition of folate cofactor patterns in biological materials has been effected by utilizing the specificities of several folate-requiring microorganisms. These microbiological assays have several obvious limitations insofar as total quantitation of the folate pattern is concerned. These limitations include:

1. The inability to identify individually formyl THF, methylene THF, and free THF which are all metabolically active cofactors with separate functions.
2. The inability to identify individually folates with from one to seven glutamic acid residues. The significance of the number of glutamic acid residues in relation to cofactor activity is currently unknown.
3. The inhibition of cofactor activity by folate antagonists which render the assay inappropriate for measuring tissue folate levels in the individual treated with chemotherapeutic agents that are folate antagonists.
4. The uncontrollable influences such as mutation of organisms, stability of assay medium, and variability of inoculum potency.

Conventional ion exchange chromatographic procedures offer a partial solution since most folate cofactors can be separated from one another. However, this technique is very time-consuming and often allows for nonenzymatic interconversion of folates. Therefore, the number of analyses that can be performed is limited. Furthermore, such column procedures lack a desirable degree of reproducibility.

We have begun to apply the techniques of high-pressure liquid chromatography to the separation and quantitation of folate cofactor patterns in

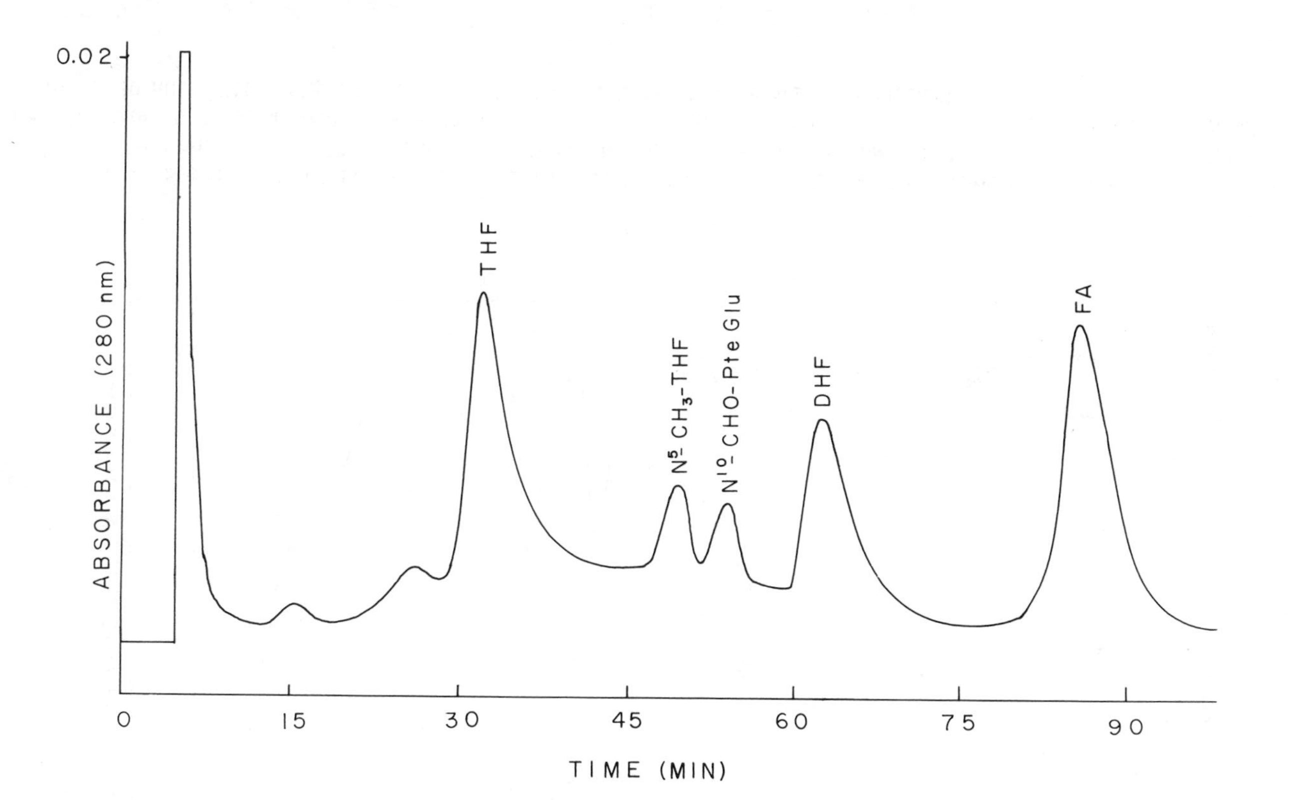

FIG. 2. Separation of a mixture of reagent-grade folate cofactor standards. Instrument: Varian Aerograph LCS-1000. THF, tetrahydrofolic acid; N^5-CH_3-THF, N^5-methyltetrahydrofolic acid; N^{10}-CHO-PteGlu, N^{10}-formylpteroylglutamate; DHF, dihydrofolic acid; FA, folic acid.

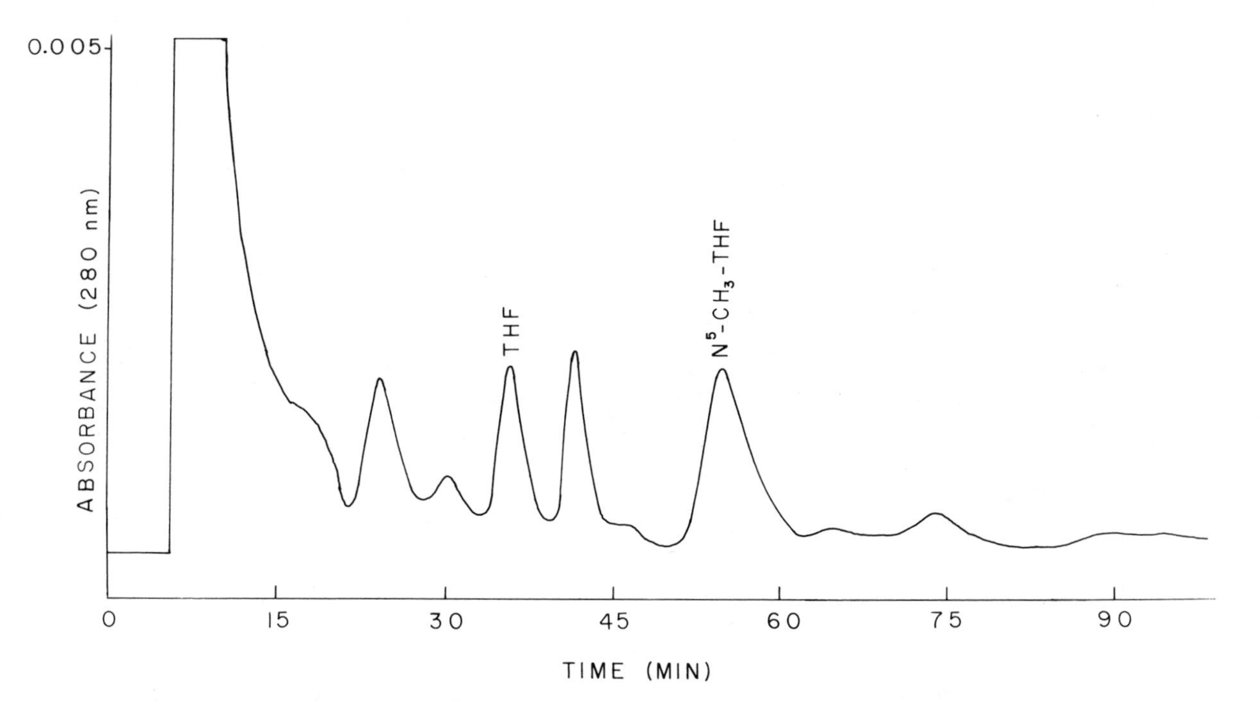

FIG. 3. Separation of folate cofactors in rat liver. Instrument: Varian Aerograph LCS-1000. UV detector 280 nm. Column: 1 mm × 300 cm; packed with pellicular anion exchange resin (Pellionex). Flow rates: 7.6 ml/hr. Eluents: 0.01 M potassium phosphate, pH 7.5; 0.01 M potassium phosphate in 1.0 M KCl, pH 7.5. Initial volume: 40 ml. THF, tetrahydrofolic acid; N^5-CH_3-THF, N^5-methyltetrahydrofolic acid.

biological materials [80]. The following describes our progress to date. Reagent-grade folate cofactors, purchased from Sigma Chemical Company, were used as standards. The compounds were dissolved in 0.01 M potassium phosphate buffer, pH 7.5, and separated on a 1-mm × 300-cm pellicular anion exchange column (Pellionex) with an ionic concentration gradient made by initiating the separation with 40 ml of 0.01 M potassium phosphate buffer (pH 7.5) to which was added 0.01 M potassium phosphate buffer in 1.0 M KCl, pH 7.5. The dilute and concentrated buffers were pumped at 7.6 and 3.5 ml/hr, respectively. The UV output of the column was monitored at 280 nm and the column temperature was 40°C. The separations of the standard folates obtained by this procedure are shown in Fig. 2. The order of elution was tetrahydrofolic acid, N^5-methyltetrahydrofolic acid, N^{10}-formyltetrahydrofolic acid, dihydrofolic acid, and folic acid. The total elution time was approximately 90 min. Identification of the peaks was accomplished by cochromatography with known pure compounds, adding compounds singly to the mixture being chromatographed and identifying the peak increasing in size.

Liver extracts for folate cofactor analysis were made according to the method of Bird et al. [81] and are briefly presented here. A 1:2 (w/v) dilution of fresh liver was homogenized in potassium phosphate buffer (0.01 M, pH 6.0) containing 0.2 M mercaptoethanol to maintain a reducing environment. The homogenates were placed in a boiling water bath for 30 sec, rapidly cooled on ice, and filtered through Millipore filters (0.45 nm). Ten microliters of the clear filtrate was applied to the column and the folate cofactors were eluted as described above. The chromatogram (Fig. 3) shows that the predominant folates in rat liver were N^5-methyltetrahydrofolate and tetrahydrofolate. There was a complete absence of folic acid per se. Although the identity of the remaining folates is somewhat uncertain at present, Fig. 3 illustrates the possibilities with respect to the folic acid analyses that are now available. In 90 min it is possible to determine the folic acid cofactor patterns in liver by rather simple, rapid procedures and with these procedures the study of absorption, metabolism, and biological function of folic acid will be greatly facilitated.

C. Other Water-Soluble Vitamins

High-pressure liquid chromatography has also been applied to the separation of other water-soluble vitamins, nicotinic acid, thiamine, ascorbic acid, folic acid, pyridoxine, pyridoxal, and pyridoxamine [82], vitamins B_2 and B_{12} [83], and riboflavin [84]. The effectiveness of the procedures was evaluated in determining the vitamins in the presence of their degradation products and other vitamins. These procedures were finally applied to the determination of the vitamins in several commercial multivitamin products. The applications thus far have been toward separating the vitamins from mixtures of vitamins; no work has been done to evaluate the levels of these vitamins in tissues, such as liver or blood, in order to assess the status of animals or man with respect to these vitamins.

D. Fat-Soluble Vitamins

High-pressure liquid chromatography has also been applied to the quantitative analyses of fat-soluble vitamins [83, 85, 86]. The currently accepted methods of analyses fall into three general categories: biological, physical, and chemical. The biological methods are the most sensitive; they were used for samples of small vitamin content where a direct correlation of vitamin content to animal growth responses was desirable. Unfortunately, the biological measurement has poor precision, is time-consuming, and is not practical for analyses of a large number of samples. Although the physical and chemical methods are faster and more reproducible than the biological methods, saponification and column chromatography were sometimes necessary to separate interfering substances from the sample. The saponification and column chromatography clean-up procedures often increased the possibility of oxidation and were inconvenient and time-consuming.

High-pressure liquid chromatography is especially useful for the analysis of vitamins A, D, and K which have large UV absorption at 254 nm. Using a spectrophotometer set at this wavelength, less than 50 ng of these compounds [86] can be detected. Vitamin E, on the other hand, a relatively poor UV absorber, had a minimal detectable level of approximately 300 ng. Reversed phase partition chromatography with water, alcohol solvents, and nonpolar column packings has been used effectively for the quantitative analyses of the fat-soluble vitamins. Retention times and peak area reproducibility were $\pm 2\%$ or less when internal standards and electronic integration were used.

IV. SUMMARY AND CONCLUSIONS

As shown throughout this chapter, the introduction of the technology of high-pressure liquid chromatography has extended the analytical era of nutrition. High-pressure liquid chromatography is a technique which can improve our understanding of the interaction and metabolism of nutrients which required the development of analytical procedures for their quantitative determination.

In the area of diet purines, uric acid metabolism, and gout, high-pressure liquid chromatography has made it possible to determine individual purines in foods and to evaluate the significance of these purines in diets. Until now the purines in foods have been analyzed and quantitated from the UV absorption of perchloric acid extracts, which did not distinguish among individual purine components. It is clear from the studies reported here that individual purine components play important definitive roles in purine metabolism and that an evaluation of these components in foods is important. With respect to the metabolism of purines it is now possible to evaluate purine excretion patterns in animals and man fed various diets. This information will enhance our knowledge of the metabolism and excretion of purine nucleic acids, compounds that may have far greater medical importance than is currently recognized. It also creates the possibility of evaluating nucleic acids in many new and unconventional proteins so that the protein contribution of these novel protein sources will not be compromised by their generally high nucleic acid content. High-pressure liquid chromatography has enhanced our understanding of uric acid metabolism in biochemical terms and the abnormalities associated with uric acid overproduction and gout in humans.

The second major area which has been stimulated by the development of high-pressure liquid chromatography is the analysis of both water- and fat-soluble vitamins and related compounds. At the present time, relatively little is known concerning the absorption, metabolism and excretion of folic acid because no reliable, rapid methods existed for evaluating the folic acid cofactor patterns in various biological tissues. The advent of high-pressure liquid chromatography has now made it possible to assay biological materials and diets for various folic acid cofactor patterns. This will enhance our understanding of the absorption of folic acid, the metabolism and biochemical steps regulated by folic acid, and the abnormalities of metabolism which occur when folic acid becomes deficient. Other water-soluble folic acid cofactors have been separated and quantitated by high-pressure liquid chromatography but most of these separations and quantitations have been on multivitamin preparations of vitamins where the concentrations of these cofactors are extremely high relative to biological material and in most instances are devoid of interfering substances. The real contribution of high-pressure liquid chromatography will come when it becomes possible to assay biological materials, including blood and liver, for the various water-soluble vitamins, in a simple, precise, and reliable way. Data from our laboratory are the only ones available concerning assays in biological materials and deal exclusively with folic acid.

Fat-soluble vitamins have also been assayed in multivitamin preparations; high-pressure liquid chromatography has made simplified methods available which eliminated the need for exhaustive saponification and complicated preliminary clean-up of samples before analysis. In an era when food labeling and nutrient content of foods are becoming very important, high-pressure liquid chromatography, although in its infancy, has tremendous potential in quantitating and evaluating micronutrients in foods and biological materials, and will immensely enhance our understanding of nutrition of single cells, clusters of cells, organs, and ultimately the whole organism of man itself.

ACKNOWLEDGMENTS

The studies reported in this chapter were supported by Grant AM 16726 from the U.S. Public Health Service, local institutional grant from the American Cancer Institute, and Hatch 2850 from the Agricultural Experiment Station. The author acknowledges the following colleagues for materials: Nevin S. Scrimshaw, Massachusetts Institute of Technology (single cell proteins), Angelita Fan, University of California, Davis, Sacramento Medical Center (low and high purine diets) and E. L. R. Stokstad, University of California, Berkeley (N^{10}-formylpteroyl glutamate).

REFERENCES

1. C. D. Scott and N. E. Lee, J. Chromatogr., 42, 263 (1968).
2. J. E. Seegmiller, L. Laster, and R. R. Howell, N. Engl. J. Med., 268, 712 (1963).
3. W. N. Kelley, M. L. Greene, F. M. Rosenbloom, J. F. Henderson, and J. E. Seegmiller, Ann. Intern. Med., 70, 155 (1969).
4. A. B. Gutman and T. F. Yu, N. Engl. J. Med., 273, 252 (1965).
5. M. A. Oeryzlo, Arthritis Rheum., 8, 799 (1965).
6. E. J. Bien, T. F. Yu, J. D. Benedict, A. B. Gutman, and D. Stetten, J. Clin. Invest., 32, 778 (1953).
7. C. A. Nugent and F. H. Tyler, J. Clin. Invest., 38, 1890 (1959).
8. C. A. Nugent, Arthritis Rheym., 8, 671 (1965).
9. H. B. Lewis and E. A. Doisy, J. Biol. Chem., 36, 1 (1918).
10. L. Poka, M. N. Csoka, G. Czirbusz, E. Foldi, and A. Torok, Nutr. Dieta (Basel), 9, 161 (1967).
11. W. C. Rose, J. S. Dimmitt, and H. L. Bartlett, J. Biol. Chem., 48, 575 (1921).
12. H. F. Host, J. Biol. Chem., 38, 17 (1918).
13. G. W. Raiziss, H. Dubin, and A. I. Ringer, J. Biol. Chem., 19, 473 (1914).
14. A. E. Taylor and W. C. Rose, J. Biol. Chem., 18, 519 (1914).

15. J. S. Leopold, A. Bernhard, and H. G. Jacobi, Am. J. Dis. Child., 29, 191 (1925).

16. C. I. Waslien, D. H. Calloway, and S. Margen, Am. J. Clin. Nutr., 21, 892 (1968).

17. J. Bowering, D. H. Calloway, S. Margen, and N. A. Kaufmann, J. Nutr., 100, 249 (1969).

18. C. J. Smyth, in Arthritis and Allied Conditions (J. L. Hollander, ed.), Lea & Febiger, Philadelphia, 1970, pp. 859-897.

19. J. C. Edozien, U. U. Udo, V. R. Young, and N. S. Scrimshaw, Nature, 228, 180 (1970).

20. J. B. Wyngaarden, in The Metabolic Basis of Inherited Disease, 3rd ed. (J. B. Stanbury, J. B. Wyngaarden, and D. S. Fredrickson, eds.), McGraw-Hill, New York, 1965, pp. 667-728.

21. J. E. Seegmiller, in Duncan's Diseases of Metabolism, 6th ed. (P. K. Bondy, ed.), Saunders, Philadelphia, 1969, pp. 516-599.

22. L. H. Smith, in Textbook of Medicine, 13th ed. (P. B. Besson and W. McDermott, eds.), Saunders, Philadelphia, 1971, pp. 1682-1697.

23. A. J. Clifford, J. A. Riumallo, V. R. Young, and N. S. Scrimshaw, J. Nutr., 106, 428 (1976).

24. A. J. Clifford and D. L. Story, J. Nutr., 106, 435 (1976).

25. N. Zöllner and A. Griebsch, Adv. Exp. Med. Biol., 41B, 433 (1974).

26. T. S. Shenoy and A. J. Clifford, Biochim. Biophys. Acta, 411, 133 (1975).

27. G. B. Elion, T. J. Taylor, and G. H. Hitchings, Sixth Int. Congr. Biochem. Meeting, 4, 305 (1964).

28. F. Kruger and H. Schmidt, Z. Physiol. Chem., 14, 1 (1905).

29. F. J. Stare and G. W. Thorn, J. AMA, 127, 1120 (1945).

30. P. R. Brown, High Pressure Liquid Chromatography, Academic Press, New York, 1973.

31. J. W. Drysdale and H. N. Munro, Biochim. Biophys. Acta, 138, 616 (1967).

32. A. V. Hoffbrand, Arch. Dis. Child., 45, 441 (1970).

33. B. A. Cooper, G. S. D. Cantlie, and L. Brunton, Am. J. Clin. Nutr., 23, 848 (1970).

34. E. R. Eichner and R. S. Hillman, Am. J. Med., 50, 218 (1971).

35. C. E. Butterworth, Jr., R. Santini, Jr., and W. B. Frommyer, Jr., J. Clin. Invest., 42, 1929 (1963).

36. R. Santini, Jr., F. M. Berger, G. Berdasco, T. W. Sheehy, J. Aviles, and I. Davila, J. Am. Diet. Assoc., 41, 562 (1962).

37. V. Herbert, Arch. Intern. Med., 110, 649 (1962).

38. V. Herbert and J. Bertini, in The Vitamins, Vol. 7 (P. Gregory and W. N. Pearson, eds.), Academic Press, New York, 1967, p. 243.

39. R. Zalusky, V. Herbert, and W. B. Casele, Arch. Intern. Med., 109, 545 (1962).

40. V. Herbert and R. Zalusky, J. Clin. Invest., 41, 1263 (1962).

41. I. Chanarin, M. C. Bennett, and V. Berry, J. Clin. Pathol., 15, 269 (1962).

42. R. Zalusky and V. Herbert, Lancet, 1, 108 (1962).

43. R. A. Joske, Gut, 4, 231 (1963).

44. L. E. Rosenberg, in The Metabolic Basis of Inherited Disease, 3rd ed. (J. B. Stanbury, J. B. Wyngaarden, and D. S. Fredrickson, eds.), McGraw-Hill, New York, 1972, p. 440.

45. I. H. Rosenberg and H. A. Godwin, Prog. Gastroenterol., 60, 445 (1971).

46. T. Tamura and E. L. R. Stokstad, Fed. Proc., 32, 927 (1973).

47. F. L. Dong and S. M. Oace, Fed. Proc., 32, 927 (1973).

48. M. G. Hardinge and H. Crooks, J. Am. Diet. Assoc., 38, 24 (1961).

49. L. J. Teply, P. H. Derse, C. H. Krieger, and C. A. Elvehjem, J. Agric. Food Chem., 1, 1204 (1953).

50. M. Burger, L. W. Hein, L. J. Teply, P. H. Derse, and C. H. Dreiger, J. Agric. Food Chem., 4, 418 (1956).

51. W. A. Krehl and G. R. Cowgill, Food Res., 15, 179 (1950).

52. E. W. Toepfer, E. G. Zook, M. L. Orr, and L. R. Richardson, USDA Agriculture Handbook No. 29, 1951.

53. V. Herbert, Am. J. Clin. Nutr., 12, 17 (1963).

54. A. D. F. Hurdle, D. Barton, and I. H. Searles, Am. J. Clin. Nutr., 21, 1202 (1968).

55. S. Butterfield and D. H. Calloway, J. Am. Diet. Assoc., 60, 310 (1972).

56. R. R. Streiff, Am. J. Clin. Nutr., 24, 1390 (1971).

57. E. D. Hibbard, J. Obstet. Gynaecol. Br. Commonw., 69, 739 (1962).

58. J. P. Knowles, T. A. J. Prankerd, and R. G. Westall, Lancet, 2, 347 (1960).

59. J. Kohn, D. L. Mollin, and L. M. Rosenbach, J. Clin. Pathol., 14, 345 (1961).

60. F. J. W. Lewis and G. R. Moore, Lancet, 1, 305 (1962).

61. H. P. Broquist, J. Am. Chem. Soc., 78, 6205 (1956).

62. H. P. Broquist and A. L. Luhby, Proc. Soc. Exp. Biol., 100, 349 (1959).

63. M. Silverman, R. C. Gardiner, and P. T. Condit, J. Natl. Cancer Inst., 20, 71 (1958).

64. G. H. Spray and L. J. Witts, Lancet, 2, 702 (1959).

65. I. Chanarin and M. C. Bennett, Br. Med. J., 1, 27 (1962).

66. A. L. Luhby, J. M. Cooperman, and D. N. Teller, Am. J. Clin. Nutr., 7, 397 (1959).

67. H. Tabor and L. Wyngaarden, J. Clin. Invest., 37, 824 (1958).

68. V. Berry, I. Chanarin, M. A. Booth, and D. Rothman, Br. Med. J., 2, 1103 (1963).

69. J. Metz, K. Stevens, T. Edelstein, V. Brandt, and N. Baumslag, S. Afr. J. Med. Sci., 28, 74 (1963).

70. H. A. Hansen, On the Diagnosis of Folic Acid Deficiency, Almquist & Wiksell, Stockholm, 1964.

71. T. Edelstein, K. Stevens, V. Brandt, and J. Metz, J. Obstet. Gynaecol. Br. Commonw., 73, 197 (1966).

72. D. Rothman, Am. J. Obstet. Gynecol., 108, 149 (1970).

73. A. L. Luhby, R. Feldman, L. J. Salerno, and J. M. Cooperman, Am. J. Obstet. Gynecol., 51, 553 (1946).

74. M. L. Stone, A. Luhby, R. Feldman, M. Gordon, and J. M. Cooperman, Am. J. Obstet. Gynecol., 99, 638 (1967).

75. I. Chanarin, D. Rothman, and E. J. Watson-Williams, Lancet, 1, 1068 (1963).

76. E. W. Page, Am. J. Obstet. Gynecol., 51, 553 (1946).

77. E. W. Page, M. B. Glendening, W. Dignam, and H. A. Harper, Am. J. Obstet. Gynecol., 68, 110 (1954).

78. I. Chanarin, D. Rothman, S. Ardeman, and A. McLean, Clin. Sci., 28, 377 (1965).

79. M. B. Glendening, A. J. Margolis, and E. W. Page, Am. J. Obstet. Gynecol., 81, 591 (1961).

80. A. J. Clifford and C. K. Clifford, Xth International Congress of Nutrition, Tokyo, Japan, August 1975 (in press).

81. O. D. Bird, V. M. McGlohon, and J. W. Vaitkus, Anal. Biochem., 12, 18 (1965).

82. R. C. Williams, D. R. Baker, and J. A. Schmit, J. Chromatogr. Sci., 11, 618 (1973).

83. J. A. Schmit, R. A. Henry, R. C. Williams, and J. F. Dieckman, J. Chromatogr. Sci., 9, 645 (1975).

84. D. Wittmer and W. G. Haney, Jr., J. Pharm. Sci., 63, 588 (1974).

85. D. F. Tomkins and R. J. Tscherne, Anal. Chem., 46, 1602 (1974).

86. R. C. Williams, J. A. Schmit, and R. A. Henry, J. Chromatogr. Sci., 10, 494 (1972).

Chapter 2

POLYELECTROLYTE EFFECTS IN GEL CHROMATOGRAPHY

Bengt Stenlund

Department for Development of Chemical Products
The Finnish Pulp and Paper Research Institute
Helsinki, Finland

I. INTRODUCTION

Although gel chromatography, or gel permeation chromatography, is very well known as a method of fractionation, the literature contains few discussions on the gel chromatography of polyelectrolytes and the influence of charged groups. This circumstance prompted the author to hesitate momentarily before beginning to write this chapter. In addition, this author's investigations had been limited to only one polyelectrolyte, namely sulfonated lignin or lignosulfonates.

However, lignosulfonates are separable into well-characterized fractions. This is evident from Fig. 1, which is a chromatogram obtained on refractionation of a mixture of three calcium lignosulfonate fractions on a Sephadex G-50 column using 0.25 M calcium chloride solution as eluent. These fractions were derived from fractionation of a lignosulfonate sample with broad molecular weight distribution (MWD), on a Sephadex G-50 column with a diameter of 20 cm and a length of 140 cm. In this case the eluent was water.

Lignosulfonates may be isolated from spent liquors derived in bisulfite cookings of wood. The composition of the lignosulfonates is determined by

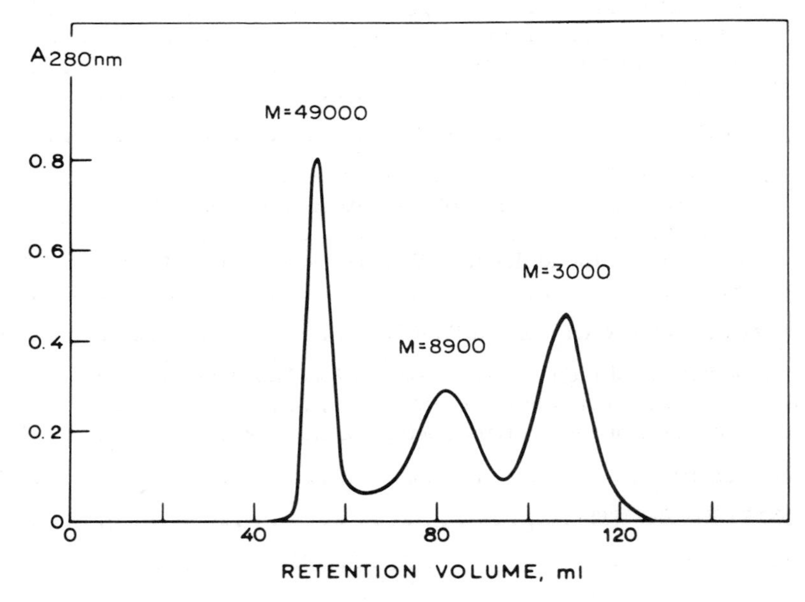

FIG. 1. Gel chromatography of three calcium lignosulfonate fractions using 0.25 M calcium chloride solution as eluent. Column: Sephadex G-50, diameter 1 cm, length 140 cm.

the cooking schedule. Lignosulfonates have been studied at The Finnish Pulp and Paper Research Institute since the mid-1950s.

The aim of this chapter is to direct the attention of those interested in fractionation to the very marked influence exercised by charge effects upon the gel chromatographic fractionation of ionic polymers. Occasionally, these effects outmaneuver normal fractionation according to hydrodynamic volume. The author also wishes to stress that in some cases, his statements are personal opinions, intended to provoke discussion, rather than proven facts.

II. THE FRACTIONATION MECHANISM

A. Ion Exclusion

If we consider a gel or a resin which is completely free from charged groups, it would be equally possible for both ionic and nonionic species to penetrate into the interior of the gel or resin. Most gels and resins, however, contain certain amounts of charged groups, thus giving rise to an ion exclusion effect. Ion exclusion refers to that phenomenon in which the diffusion of an ionic species into the interior of a gel or resin is restricted by electrostatic repulsion. Ion exclusion was first studied by Wheaton and Bauman [1]. They developed methods for the separation of ionic species from nonionic using ion exchange resins. For example, if a solution of sodium chloride and glucose were eluted from a column containing cation exchange resin loaded with sodium, sodium chloride would be prevented from entering into the resin because of repulsive forces. However, glucose would enter into the resin and, therefore, be retarded. Thus, sodium chloride would be eluted before glucose and the two species are separated.

Flodin [2] has pointed out that Sephadex gels contain small numbers of fixed, negatively charged groups (presumably carboxylate groups) which, for example, influence the fractionation of a number of amino acids when the elution is carried out with distilled water. According to Flodin, these groups represent 20 to 30 μEq/g of dry gel.

If a mixture being fractionated on a Sephadex gel consists of compounds with negatively charged groups, a repulsion will occur between the ionic solutes and the gel, with the consequence that some of the components of the mixture will be prevented from diffusing into the internal volume of the gel. Gelotte [3] has demonstrated that this interference is of importance in the fractionation of organic acids. The acids are excluded from the internal volume of the gel as a result of repulsive forces, and are thus eluted earlier than are nonionic substances of similar hydrodynamic volume.

The first to make quantitative estimations of the influence of the ion exclusion effect on gel-chromatographic fractionation of ionic solutes on Sephadex columns were Neddermeyer and Rogers [4]. By means of a specially developed technique, they determined the number of acid groups of the gel. The ion exchange capacities of Sephadex G-10 and G-25 were found to be 4.1 ± 2.8 and 45 ± 23 μEq/g of dry gel, respectively. These capacities represent about 1% of those usually found with ion exchange resins.

Eaker and Porath [5] have reported that oleate anions which remain from the manufacturing process of Sephadex G-10 and G-15 also contribute to the level of charged groups. The sodium oleate was reportedly extracted by washing the gel with 1 M pyridine.

In a study by Neddermeyer and Rogers of the effect of sample concentration upon the elution behavior of ionic solutes, 50-μl samples of sodium chloride solutions of varying concentrations were injected into a Sephadex G-10 column and eluted with water. The results obtained are presented in Fig. 2. It is seen that the chromatograms are superimposed on the leading edges, but differ in the locations of the peak maxima and trailing edges. The appearance volume was constant, and equaled the void volume of the column.

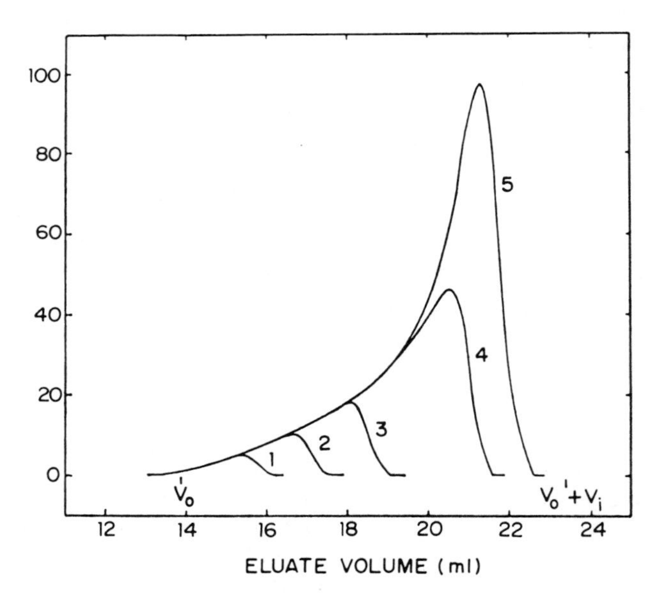

FIG. 2. Variation with sample concentration of chromatograms for 50-μl samples of NaCl on a 0.61 × 126 cm column of Sephadex G-10. 1, 0.010 M; 2, 0.025 M; 3, 0.050 M; 4, 0.10 M; 5, 0.20 M.

For confirmation of the results obtained by Neddermeyer and Rogers with a polyelectrolyte, Forss and Stenlund made a corresponding study, but with slight modifications [6]. A calcium lignosulfonate preparation, with a molecular weight of about 2500, and an equivalent weight of about 300, was eluted with water from Sephadex G-25 (curve 1 in Fig. 3). The column dimensions were 2 cm × 190 cm, and the sample concentration and sample volume were 50 g/liter and 10 ml, respectively. The fractions representing the high peak (to the right of the mark) were recovered by evaporation and refractionated. This procedure was repeated four times. The sample size naturally decreased on refractionation.

It is obvious that the diminution in sample size led to a stronger effect of ion exclusion. The proportion of the extended front part of the

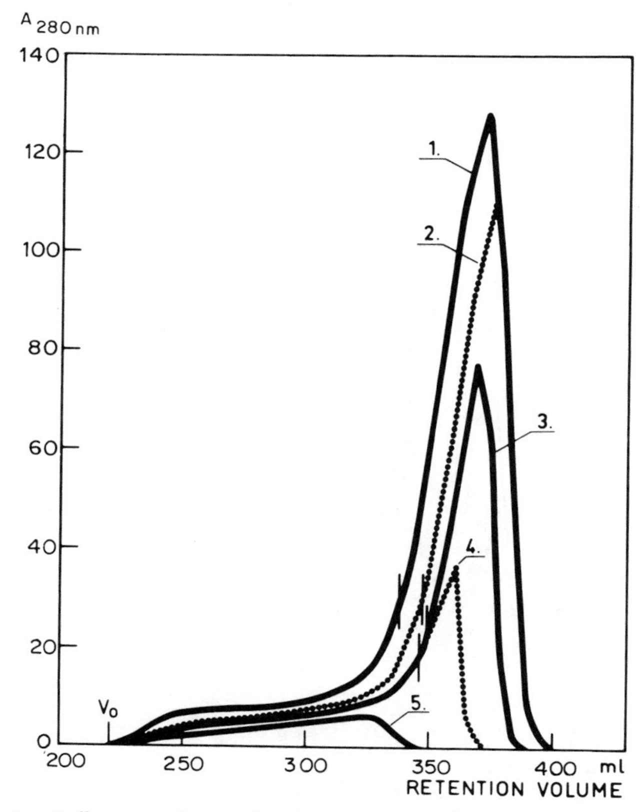

FIG. 3. Influence of sample size on ion exclusion when a low-MW calcium lignosulfonate fraction is eluted from Sephadex G-25. Flow rate 16 ml/hr (5 ml/hr · cm^2 [6]. (Courtesy of the Journal of Polymer Science.)

chromatogram increased with decreasing sample size, and the high peak vanished completely upon fractionation of the smallest sample (curve 5). On comparison with Fig. 2, it is discernible that the influence of the ion exclusion effect is similar for the elution of simple electrolytes and polyelectrolytes with water.

Gelotte [3] has demonstrated that the ion exclusion effect is suppressed if the eluent contains a simple electrolyte, or if the electrolyte content of the sample exceeds a certain limit. This effect was investigated quantitatively by Neddermeyer and Rogers [4]. Figure 4 illustrates the chromatograms obtained on the elution of sodium sulfate and sodium chloride samples of different molarity. When the sample had a low content of simple electrolytes, the separation achieved was very poor. When the electrolyte content of the sample was increased, the separation was improved, but some overlapping remained. On the elution of similar samples with 0.01 M solution

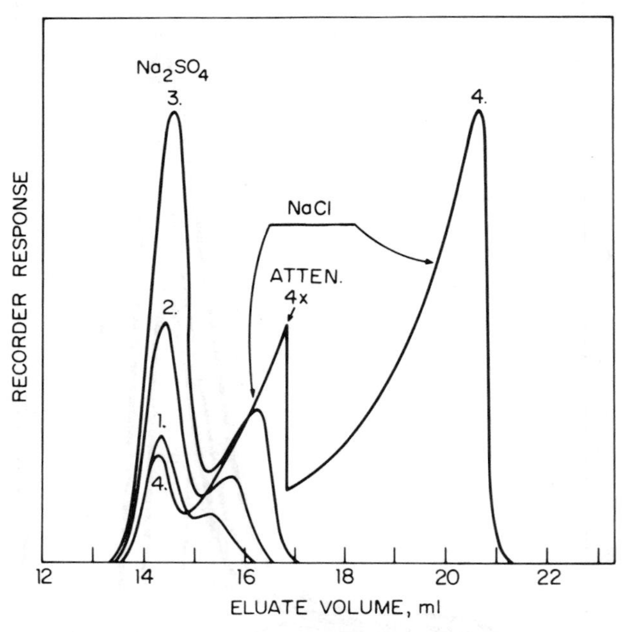

FIG. 4. Elution of 20-μl samples of sodium sulfate and sodium chloride mixtures with deionized water from a 0.61-cm × 124-cm Sephadex G-10 column [4]. Samples: 1, 0.0125 M Na_2SO_4 and 0.0125 M NaCl; 2, 0.025 M Na_2SO_4 and 0.025 M NaCl; 3, 0.050 M Na_2SO_4 and 0.050 M NaCl; 4, 0.010 Na_2SO_4 and 0.010 M NaCl. (Reprinted with permission of Analytical Chemistry. Copyright by the American Chemical Society.)

of sodium chloride, the separation was very much improved, as is evident from Fig. 5. The retention volumes of the two components of the sample solution were independent of the amount of salt. The peak area and peak height also increased linearly with increase in concentration of the sample.

In their discussion of the results, Neddermeyer and Rogers [4] pointed out that, on elution with water, the appearance volumes of the simple electrolytes equaled or were slightly less than the void volume of the column, thereby indicating that when the salt concentration of the samples if low, the salts are almost excluded from the internal volume of the gel. This is in accordance with the observations made by the present author.

Neddermeyer and Rogers also stressed that, in the front part of the solute zone, the charged sites in the gel restrict the diffusion of ions into the gel. By contrast, they enhance the diffusion out of the gel in the rear part of the solute zone. This is evidently the course of the extended leading edge and the sharp trailing edge obtained upon elution of ionic solutes with water from Sephadex gels. Since salts with greater hydrodynamic volumes are exposed to fewer of the fixed charges in the gel, their elution behavior is less influenced by the ion exclusion effect.

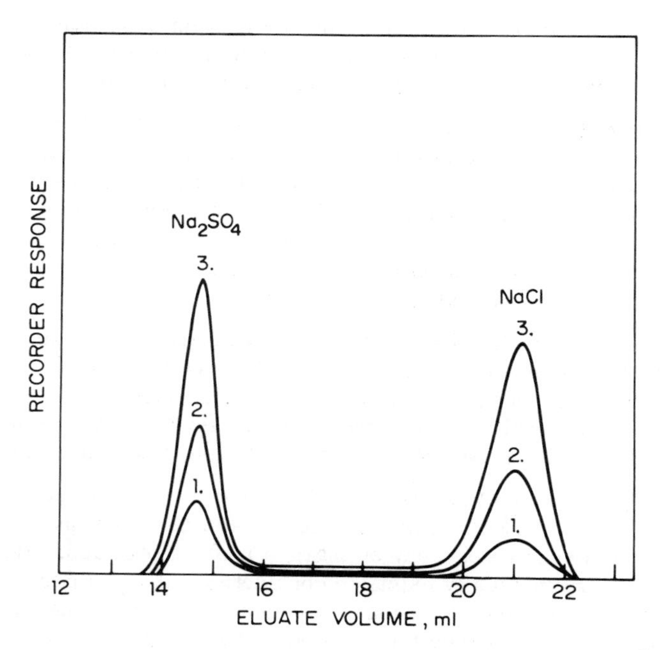

FIG. 5. Elution of the same samples as in Fig. 4, with 10^{-2} M sodium chloride solution [4]. (Reprinted with permission of Analytical Chemistry. Copyright by the American Chemical Society.)

The suppression of the ion exclusion effect on the elution of ionic solutes with an electrolyte solution is attributable to screening of the charges of the solute and the fixed charges of the gel. According to Neddermeyer and Rogers [4], the following factors need to be considered on elution with electrolyte solution:

1. For completely ionized salts, a concentration of 10^{-2} M is adequate to overcome the ion exclusion effect.
2. To reduce the number of possible cation-anion combinations that might result from ion exchanges, the electrolyte should have its cation or anion in common with that of the ionic solute.
3. One important consideration is that the hydrodynamic volume of the electrolyte in the eluent should be sufficiently small to permit full penetration of the internal volume of the gel.

B. Ion Inclusion

When a solution contains two or more different ionic solutes, and one of them is barred from regions (in a gel or membrane) penetrable by the other, a Donnan "equilibrium" will tend to be established [7-9].

A certain type of Donnan effect studied by Teorell [10] is evidently also operative in gel-chromatographic systems. Teorell observed a Donnan equilibrium resulting from a "diffusion effect" in membranes permeable to all the ions present, but to different extents. The ionic solute with the lowest permeability will dictate the behavior of the other ionic solutes to such an extent that it is possible to induce the transport of ionic solutes against their concentration gradients.

Flodin [11] has also observed that the separation of simple electrolytes from high-molecular-weight (high-MW) polyelectrolytes is improved by the increase in concentration of the low-molecular-weight (low-MW) solutes in the internal volume of the gel as a result of a "Donnan diffusion effect." This leads to retarded elution of the low-MW ionic solutes.

From their study of the elution behavior of dextran sulfate on Sephadex G-25 columns, Nichol et al. [12] conclude that a Gibbs-Donnan equilibrium operates in gel chromatography of charged macromolecules. Since a Donnan equilibrium usually exists in conjunction with semipermeable membranes, the authors also state that the occurrence of a Donnan effect provides experimental evidence for the model of gel chromatography based on liquid-liquid partition, with the gel matrix acting as a semipermeable membrane between the mobile and stationary phases.

The influence of the so-called Donnan diffusion effect in gel chromatography is the reverse of the ion exclusion effect. For the sake of clarity, Stenlund has suggested the term "ion inclusion effect" to describe the influence of the retarding Donnan diffusion effect [13].

In their paper concerning electrolyte effects in aqueous gel chromatography of inorganic salts, Neddermeyer and Rogers [14] rationalized the phenomena observed by invoking the existence of a Donnan diffusion or ion-retarding effect, involving an internal volume of the gel penetrable to some, but not all, ionic solutes in solution. On the elution of a linear polyphosphate or a cyclic metaphosphate with electrolyte solution, two positive peaks occurred if the sample salt differed from the eluent salt, one peak resulting from the sample salt, and the other from the eluent salt. The peak of the eluent salt was attributed to the ion-retarding effect, since the sample salt and the eluent salt were able to penetrate into the gel phase to different degrees.

In the system studied by Neddermeyer and Rogers, both the sample salt and the eluent salt were able to penetrate the internal volume. In their study of the influence of the ion inclusion effect on the gel-chromatographic fractionation of a polyelectrolyte, Forss and Stenlund [6] simplified the interpretation of the results by using a polyelectrolyte completely excluded from the gel phase as eluent electrolyte. Thus, all the peaks observed in the chromatograms represented ionic solutes of the sample. A schematic depiction of this system is given in Fig. 6. This type of Donnan effect (the high-MW, excluded polyelectrolyte influences the penetration of low-MW solutes) has a resemblance to that observed when a polyelectrolyte in a salt solution is barred from a pure salt solution by a membrane permeable only to the salt. In this case, the salt activity at equilibrium will be higher in the compartment that contains the pure salt solution than in that containing both polyelectrolyte and salt [15].

The investigation [6] mentioned above was carried out on a Sephadex G-25 column, 2 cm in diameter and 140 cm in length. The polyelectrolyte was obtained via fractionation of a polydisperse sample of lithium lignosulfonates. It was estimated that the gel-chromatographic molecular weight was about 5000. The eluents used were distilled water, 0.15 M lithium chloride solution, an aqueous solution containing 3.9 g high-MW lithium lignosulfonates per liter (a fractionation test showed that these lignosulfonates were completely excluded from Sephadex G-25), and, finally, an electrolyte solution of high-MW lithium lignosulfonates 0.15 M with respect to lithium chloride. The concentration of the low-MW lignosulfonate fraction (MW = 5000) in the sample solution was also 3.9 g/liter. Small amounts of a fraction of high-MW (excluded) lithium lignosulfonates and lithium chloride were added to indicate V_0 (void volume) and V_i' (a crude estimation of the saturated volume), respectively.

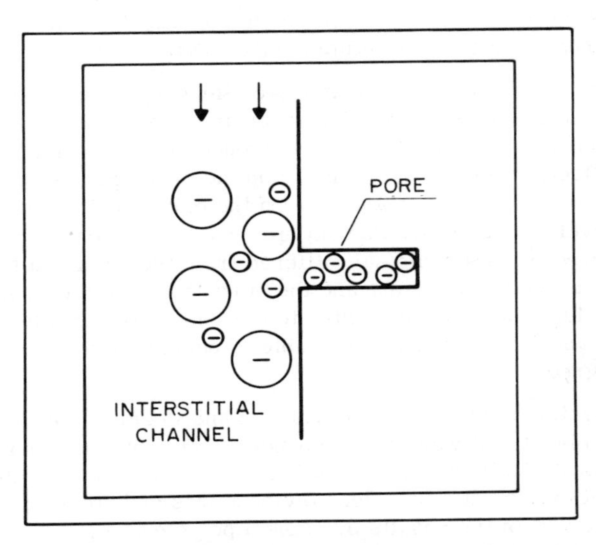

FIG. 6. Ion inclusion of a permeable polyelectrolyte component into a pore in a gel due to the presence of an excluded polyelectrolyte component [6]. (Courtesy of the Journal of Polymer Science.)

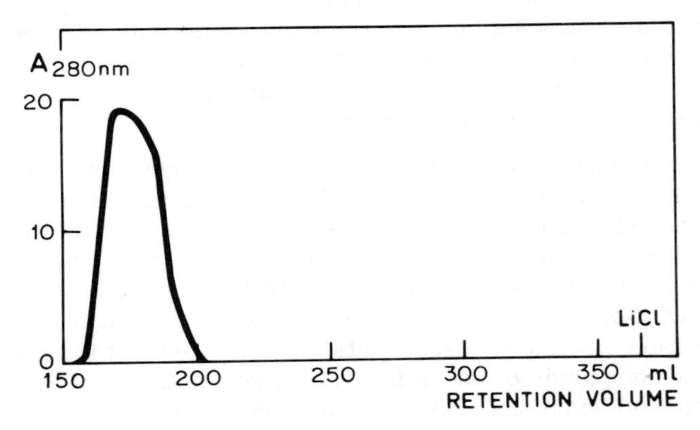

FIG. 7. Elution of a lithium lignosulfonate fraction from a Sephadex G-25 column, with water. Flow rate 20 ml/hr (6.4 ml/hr · cm^2) [6]. (Courtesy of the Journal of Polymer Science.)

Figure 7 illustrates the chromatogram obtained on elution of the sample solution with distilled water. It is evident that the high-MW reference lignosulfonates and the low-MW lignosulfonate fraction (MW = 5000) emerged from the column as one single peak on elution with distilled water. Thus, the low-MW lignosulfonate did not diffuse into the gel phase.

On the addition of high-MW lignosulfonates to the eluent, a further component is introduced into the interstitial volume. When the fractionation experiment is started, the interstitial volume contains two lignosulfonate components, one high MW and one low MW. Since the high-MW component is completely excluded from the gel phase, a Donnan potential acts on the low-MW, permeable, component, which, as opposed to the case in which pure water is the eluent, begins to diffuse into the internal volume of the gel. This is clearly evident in Fig. 8. The first peak in the chromatogram of Fig. 8 represents the high-MW reference lignosulfonates, with a retention volume equal to the void volume of the column. The second peak corresponds to the low-MW lignosulfonate fraction in the sample. It is obvious that an ion inclusion effect promoted the diffusion into the internal volume of the gel, and thus retarded the elution of the low-MW lignosulfonate. Means for improvement of the separation of different ionic solutes in this way are discussed in a later section.

It is well known that charge interactions in polyelectrolytes are swamped out by a simple electrolyte [16]. This also applies in the

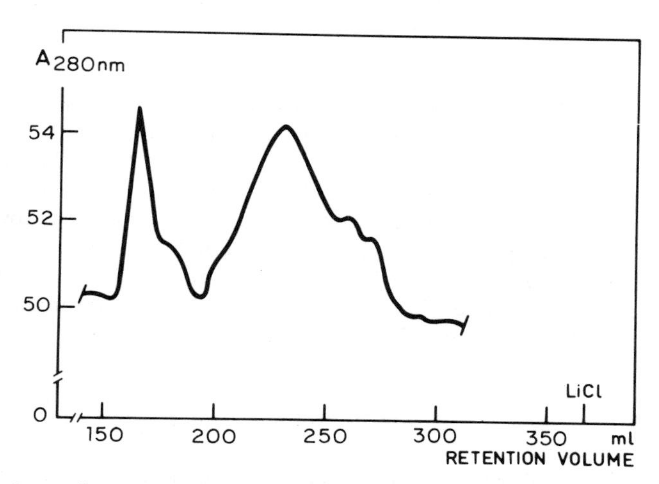

FIG. 8. Influence of the ion inclusion effect on elution of a low-MW lithium lignosulfonate fraction from Sephadex G-25 with an aqueous solution of high-MW lignosulfonates [6]. (Courtesy of the Journal of Polymer Science.)

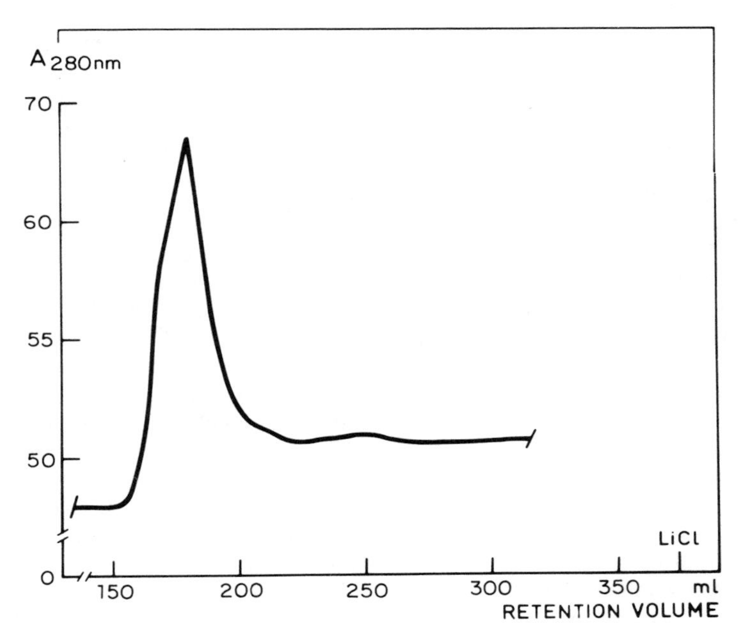

FIG. 9. Suppression of the ion inclusion effect on elution of a low-MW lithium lignosulfonate fraction with a solution of high-MW lithium lignosulfonates in 0.15 M lithium chloride [6]. (Courtesy of the Journal of Polymer Science.)

gel-chromatographic fractionation of polyelectrolytes, as can be seen in Fig. 9, which represents a chromatogram obtained on the elution of the low-MW lithium lignosulfonate fraction with an electrolyte solution of high-MW lithium lignosulfonates, 0.15 M with respect to lithium chloride. Figure 9 illustrates that both the reference lignosulfonates and the low-MW fraction were eluted again as a single peak; accordingly, no ion inclusion effect was detectable on elution with an electrolyte solution containing high-MW lignosulfates. This result is in agreement with the statement made by Tanford [8] that the influence of Donnan effects is suppressed in the presence of a simple electrolyte.

C. Excluded Volume Effect

On the elution of both ionic and nonionic solutes with solutions that contain high-MW polymers or polyelectrolytes, an excluded volume effect may influence the fractionation. The high-MW polymers excluded from the interior of the gel increase the activity of the permeable solutes, so that their concentration in the internal volume becomes greater than that in the void volume.

Hellsing [17] has studied this effect by the elution of human serum albumin from Sephadex G-200 with eluents containing nonionic polymers, such as dextran and polyethylene glycol, and found that the k_D value of the albumin increased significantly when the polymer concentration of the eluent was increased. Giddings and Dahlgren [18] have also investigated this effect, and suggest that the addition of different amounts of high-MW polymers to the eluent could provide a means for the control of retention volume, and for retention programming in gel chromatography. Vink [15] has also observed that an excluded volume effect occurs in membrane systems with both polyelectrolytes and nonionic polymers.

For determination of whether a similar effect prevails when lignosulfonate fractions were eluted with eluents containing high-MW lignosulfonates, Stenlund [19] performed an experiment in which the polyelectrolyte in the eluent was exchanged for the nonionic polymer, Dextran T 20 (MW 20,000 according to the manufacturer, Pharmacia Fine Chemicals, Uppsala, Sweden). The lithium lignosulfonate sample was similar to those used in the experiments described in Sect. II B. The concentrations of the sample solution and of the eluent were equal, i.e., 4 g/liter. The sample solution contained markers for void volume and total liquid volume. The resulting chromatogram is shown in Fig. 10.

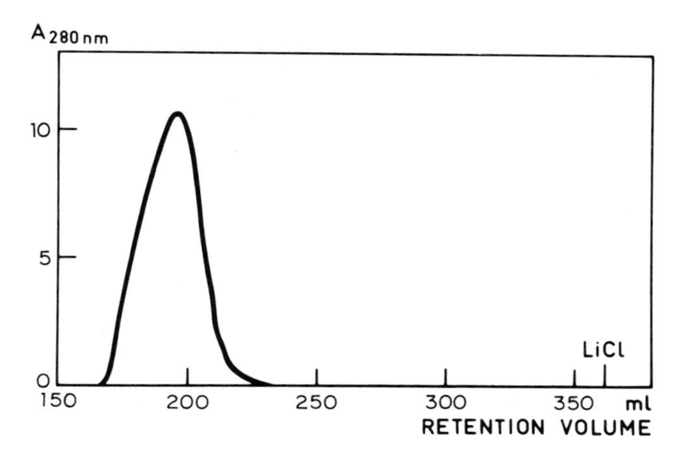

FIG. 10. Elution of a low-MW lithium lignosulfonate fraction from a Sephadex G-25 column with an aqueous dextran solution [19]. (Courtesy of Paperi ja Puu.)

It is obvious that the low-MW lithium lignosulfonate emerged together with the high-MW marker, and that no enhanced diffusion into the gel phase attributable to the presence of the dextran in the eluent could be observed. This accords with the results obtained by Giddings and Dahlgren [18]. When the polymer concentration of the eluent was low, the increase in k_D was insignificant. Consequently, it appears that the ion inclusion effect, and not the excluded volume effect, was responsible for the retarded elution of the low-MW lithium lignosulfonate fraction on elution with an aqueous solution of high-MW lignosulfonates (cf. Fig. 7).

D. Polyelectrolyte Swelling

Since gel-chromatographic fractionation is determined, to a great extent, by the hydrodynamic volume of the different components, the occurrence of polyelectrolyte expansion during a fractionation experiment may markedly affect the resulting chromatogram.

It has been well established that the reduced viscosity of polyelectrolytes in aqueous solution increases appreciably with decreasing polyelectrolyte concentration [20-22]. This increase arises from an increase in the hydrodynamic volume of macroions with decreasing concentration, presumably because of an increase in charge density.

During the course of a fractionation in a column, the sample is, to some extent, diluted. If the sample contains ionic solutes, the hydrodynamic volume increases; thus, the rate of migration also increases during the experiment. Furthermore, the charge density on the macroions also increases with decreasing concentration, a phenomenon which alters the influence of the ion inclusion effect in such a way that the permeable macroions are more severely retarded. Consequently, two polyelectrolyte effects, acting in opposite directions, influence the fractionation resulting from the elution of polyelectrolytes with water. An attempt at a more detailed description of these effects is presented in the following section.

In this context, one should be cautioned against the use of ionic strength gradients in the elution of polyelectrolytes. The studies mentioned above in regard to the dependence of reduced viscosity on polyelectrolyte concentration also indicated changes in the hydrodynamic volume of the macroions with varying background electrolyte concentration. Consequently, the elution behavior of a polyelectrolyte is not constant if the electrolyte concentration varies below a certain level. This level may be as high as 10^{-2} or even 10^{-1} M (see Sect. III A).

A good description of the extent of polyelectrolyte expansion as a function of ionic strength of the surrounding medium is found in a paper by Gupta and McCarthy [35], who have calculated the viscosity radii of different sodium lignosulfonate species in water and in sodium chloride solutions of varying concentrations. By this means, they estimated that the radius of a lignosulfonate species with a weight-average molecular weight of 59,000 was 67 Å in water, and only 37 Å in 0.1 M sodium chloride solution. The effective radius of a sodium lignosulfonate molecule may thus almost double on a change in its environment from 0.1 M sodium chloride to pure water.

It may thus be stated that polyelectrolyte swelling is considerable even if the macroion is compact and spherical, as is the case with lignosulfonates. With coiled polymers, the influence of polyelectrolyte swelling upon hydrodynamic volume and molecular configuration is even more drastic.

To avoid confusion that may arise from polyelectrolyte swelling during a fractionation experiment, the appropriate electrolyte level should be established. This can be done by comparing elution data obtained with eluents of varying ionic strength, or simply by measuring the viscosity of the sample in different concentrations of electrolyte.

E. Fractionation Model

In an endeavor to visualize the influence of the various polyelectrolyte effects described above, Forss and Stenlund [6] have prepared the following fractionation model for the gel-chromatographic fractionation of polyelectrolytes. It is assumed that two lignosulfonate zones AB and CD (Fig. 11), which overlap in zone CB, are present in a gel column. Lignosulfonate species LS_a has a hydrodynamic volume slightly exceeding that of lignosulfonate species LS_b.

(Note: The following three paragraphs are derived from Forss and Stenlund [6] with permission of the Journal of Polymer Science.)

1. Elution with Water

Owing to the ion inclusion effect, the molecular species LS_b in the intermediate zone CB is forced into the internal volume of the gel and the concentration of this species is hence greater in the internal than in the interstitial gel volume. The smaller molecular species LS_b consequently migrates slower in zone CB than in zone BD where there is no ion inclusion effect. However, the lignosulfonate concentration is lower in zone BD than in the intermediate zone CB. Consequently, the species LS_b expands and starts to move at a higher rate, i.e., at the same rate as species LS_a. The final result of these effects is that species LS_b is eluted immediately after species LS_a without any appreciable overlapping.

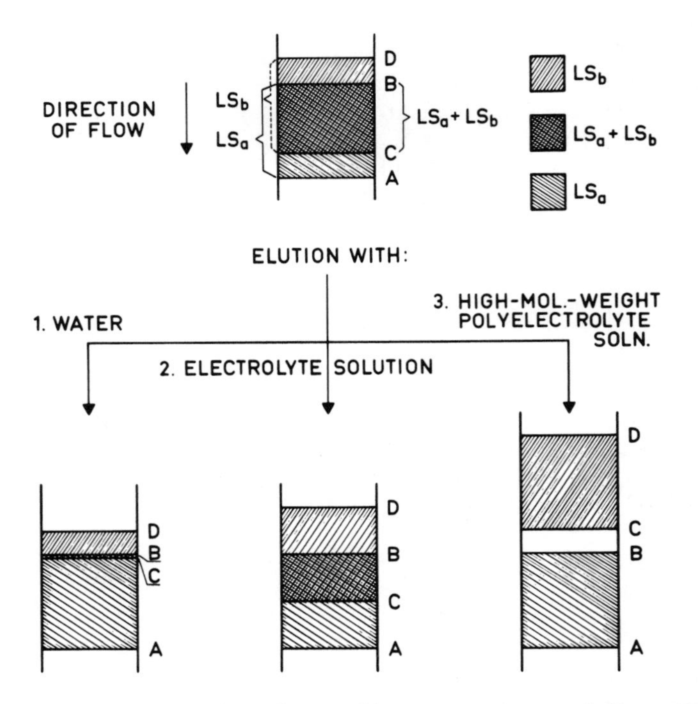

FIG. 11. Behavior of two lignosulfonate species on elution with water, an electrolyte solution, and a solution of high-MW lignosulfonates [6]. (Courtesy of the Journal of Polymer Science.)

2. Elution with Electrolyte Solution

If the lignosulfonates are eluted with an electrolyte solution, the ion inclusion effect is suppressed. If the difference in the migration rates of components LS_a and LS_b (Fig. 11) is small, both components will overlap on leaving the column. Since polyelectrolyte expansion does not occur in an electrolyte solution, the migration rate of LS_b does not increase in the zone BD. Broadening of zone CD is thus more pronounced when the eluent is an electrolyte solution than when it is pure water.

3. Elution with a Solution of High-Molecular-Weight Lignosulfonates

If the elution is performed with a solution of high-MW lignosulfonates that are excluded from the gel phase, polyelectrolyte expansion and the ion exclusion effect are suppressed because of the constant ionic strength produced by the charged groups of the lignosulfonates in the eluent. The ion inclusion is, therefore, the main charge effect in this case. The smaller molecular species LS_b does not increase its migration rate on leaving the intermediate zone CB, because there also exists an ion inclusion effect outside the intermediate zone because of the presence of the high-MW lignosulfonates in the eluent. Therefore, LS_b migrates slower than LS_a and these components thus separate and form sharp zones.

III. POLYELECTROLYTE EFFECTS

A. Influence of Sample Volume and Concentration

The preceding section was concerned with some specific effects that occur in the fractionation of polyelectrolytes. We will now discuss the direct influence of these effects on the appearance of actual chromatograms.

It is well known that, to a certain extent, the sample volume and concentration affect the fractionation of nonionic polymers [23, 24]. On the elution of a polyelectrolyte with an eluent of low ionic strength, however, the influence of sample volume and concentration is an order of magnitude greater.

The influence of sample concentration on the gel-chromatographic fractionation of a polyelectrolyte has been studied by Stenlund [25]. Equal volumes of three dilutions of a calcium lignosulfonate solution with a broad MWD were subjected to gel chromatography on a Sephadex G-50 column (2 cm × 140 cm). The elution was carried out with water. In this case, the chromatogram was recorded by continuous measurement of absorbance of the effluent at 340 nm. To facilitate comparison of the chromatograms, normalization to equal areas was performed. The chromatograms are illustrated in Fig. 12.

It is obvious that sample concentration exerts a marked effect on the course of fractionation. The amount of calcium lignosulfonates excluded by the gel is found to decrease significantly with increasing sample concentration; the lignosulfonates appear to have had their hydrodynamic volumes diminished. However, it is seen in Fig. 12 that the effect of sample concentration is relatively small when the sample concentration is decreased from 101 to 74.6 g/liter. The fractionation results differ markedly, however, when the concentration is decreased from 74.6 to 38.5 g/liter. It is thus impossible to compare fractionation results if considerable variations in polyelectrolyte concentration occur within a fairly low concentration range, with the elution being carried out with an eluent of low ionic strength.

A variation in sample volume at constant concentration affects the fractionation result in a manner similar to that of the variations in sample

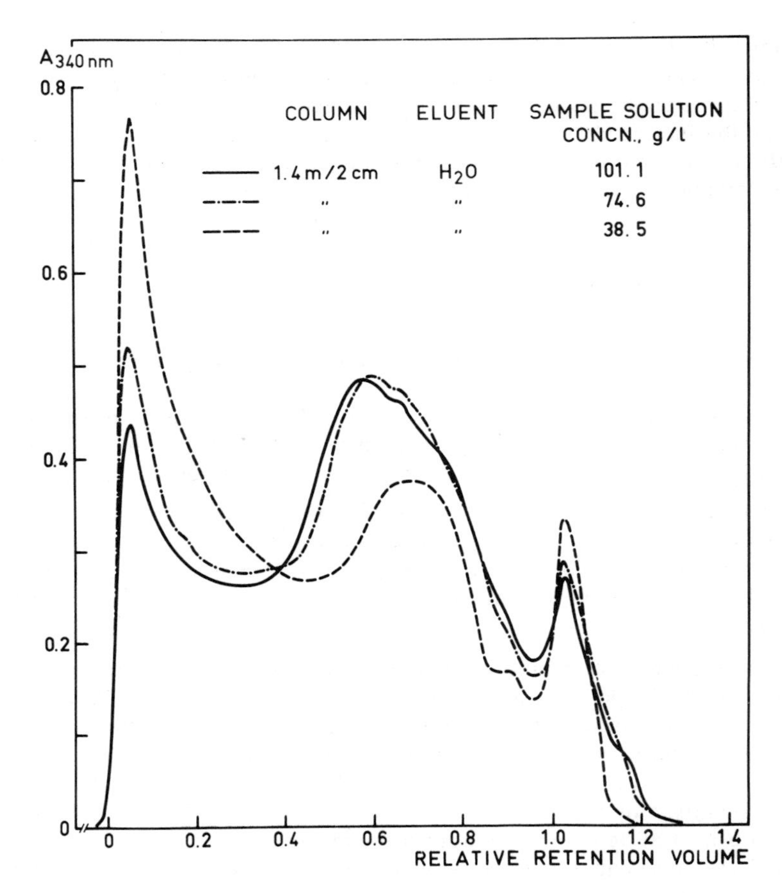

FIG. 12. Effect of solute concentration in the sample solution on the elution of calcium lignosulfonates with distilled water through a Sephadex G-50 column. Flow rate 15 ml/hr (4.78 ml/hr · cm^2) [25]. (Courtesy of Paperi ja Puu.)

concentration. According to Fisher [26], the optimum sample size in gel chromatography is equal to about 2% of the volume of the column. For instance, with a column volume of about 1.9 liters, a sample size of 35 ml can still be regarded as giving satisfactory resolution. Chromatograms obtained in a series of experiments are shown in Fig. 13. It is discernible that the influence of the sample volume is great when the sample volume is increased from 10 to 20 ml, and further to 30 ml. When the sample volume is changed from 30 to 40 ml, the difference in the result of fractionation is relatively slight. This result closely resembles that obtained when the

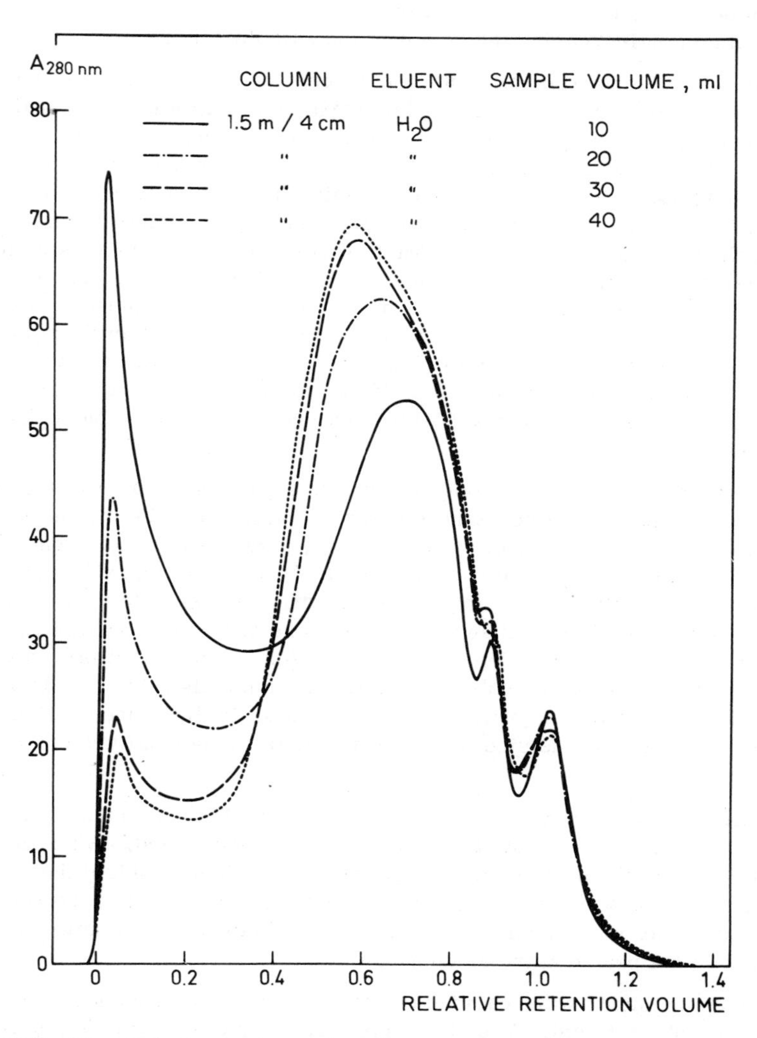

FIG. 13. Gel chromatography of samples of varying volume of a calcium lignosulfonate solution on a Sephadex G-50 column using distilled water as eluent. Elution rate 35 ml/hr (2.78 ml/hr · cm^2) [25]. (Courtesy of Paperi ja Puu.)

sample concentration was varied; this is obviously attributable to the concentration of the solute zone changing as a result of variations in the sample volume. This finds its explanation in the decrease in the dilution effect in the column on increasing the height of the sample zone.

The most likely explanation for the concentration and volume effect is, of course, to be found in polyelectrolyte expansion. As was mentioned in the preceding section, lignosulfonates behave as a swelling polyelectrolyte. Consequently, the hydrodynamic volume increases with decreasing macroion concentration. This brings about a diminution in the permeability of the lignosulfonates, and the exclusion of a major part of the lignosulfonate from the gel phase. The changes in the fractionation results are much smaller when the concentration varies within a relatively broad range, than when it does so within a narrow range. This may be because the lignosulfonates are in their most compact mode, and do not begin to expand until the concentration falls below a certain critical level. It is the opinion of the present author that the ion exclusion effect does not influence the fractionation of lignosulfonates with broad MWD on a preparative scale on Sephadex G-50 columns.

As was mentioned above, polyelectrolyte expansion is suppressed if the ionic strength is kept constant during the fractionation by means of a supporting simple electrolyte. Consequently, a series of experiments were performed with a view to determining the electrolyte concentration needed to eliminate the influence of sample concentration and sample volume [25]. It can be seen from Figs. 12 and 13 that the size of the peak representing excluded solutes is greatly dependent on the sample concentration and volume, and, accordingly, on polyelectrolyte expansion also. For as long as polyelectrolyte expansion prevails with a decrease in the concentration, a change also occurs in the size of the first peak of the normalized chromatogram.

When a nonionic polymer sample is fractionated, the amount of excluded matter is proportional to the amount of sample added. When the ionic strength of the supporting electrolyte is sufficient to shield the charged groups of the macroion, its behavior resembles that of the nonionic macromolecule, and a linear relationship is established between the size of the first peak and the amount of sample.

The experiments were carried out on a preparative Sephadex G-50 column (4 cm × 140 cm). The solute concentration of the sample solution was 96.2 g/liter, and the eluents were distilled water and 0.05 and 0.25 M calcium chloride solutions, respectively. The relative retention volumes chosen as limits for the first peak, which represents excluded macroions, were 0 and 0.1, as indicated in Fig. 14. This figure illustrates the relationship between the area under this peak, divided by the area of the whole chromatogram, and the volume of the sample solution. For comparative purposes, a series of elutions were carried out with distilled water as eluent. The relationship between the relative area of the first peak and the volume of sample solution was linear only up to a sample volume of 10 ml; at larger sample volumes, however, the plot ran parallel to the horizontal axis, implying that the lignosulfonates which should have been eluted in the fractions giving the void peak were subsequently eluted within the fractionation range. When the elutions were carried out with 0.05 M calcium chloride solution, the relationship was almost linear. On elution with 0.25 M

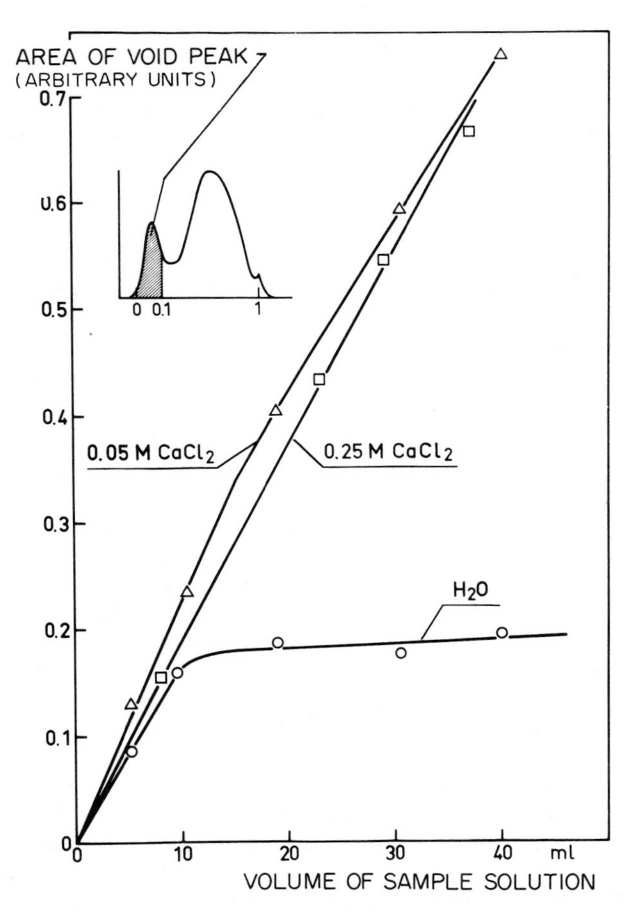

FIG. 14. Area of the peak representing excluded matter as a function of the volume of the sample of calcium lignosulfonate solution added to the column [25]. (Courtesy of Paperi ja Puu.)

calcium chloride solution, the relationship was linear throughout the entire range of sample volume, thereby showing that calcium lignosulfonates behave like nonionic polymers when the elution is done with 0.25 M calcium chloride solution. The corresponding behavior of lignosulfonates with a monovalent counterion is discussed in the following section.

B. Influence of the Valence of the Cation of the Sample Solution

As has been stated by Oosawa [27] the valence of the counterion considerably influences the state and behavior of the macroion. Polyvalent counterions are more strongly bound to the macroion than are monovalent counterions. Polyvalent counterions start to "condense" in the volume occupied by the macroion at a considerably lower concentration than do monovalent counterions.

It thus becomes evident that the number of free charges on the macroion is much lower in the presence of polyvalent than in the presence of monovalent counterions. As the polyelectrolyte effects in gel chromatography represent entire-charge effect, it is clear that the fractionation result strongly depends on the counterion of the sample solution if the eluent is of low ionic strength.

For elucidation of the effect of the counterion on the result of fractionation, let us give a description of some experiments performed with lignosulfonates associated with either monovalent or polyvalent counterions [25]. Samples of calcium and lithium lignosulfonates were prepared by ion exchange from a lignosulfonic acid solution. These samples were eluted with water from preparative Sephadex G-50 columns, which had been equilibrated with calcium and lithium chloride solutions, respectively, prior to being used. The chromatograms, normalized to equal area, are shown in Fig. 15. A very high proportion of the lithium lignosulfonates emerged as a narrow zone at relative retention volumes that ranged from 0.6 to 1.0, whereas most of the calcium lignosulfonates emerged at relative retention volumes that ranged between 0.35 and 1.0. The main part of the lithium lignosulfonates was accordingly retarded to a higher extent than was the major part of the calcium lignosulfonates. This is also evident from Fig. 16, which is a calibration plot obtained by means of molecular weight determinations derived from light-scattering experiments [25, 28]. The plot in Fig. 16 illustrates that the relative retention volume for calcium lignosulfonates with a weight-average molecular weight of 30,000 was 0.22, whereas the corresponding value for lithium lignosulfonates of similar molecular weight was 0.51.

The character of the counterion consequently exerts a marked effect on the fractionation of lignosulfonates on elution with distilled water. But

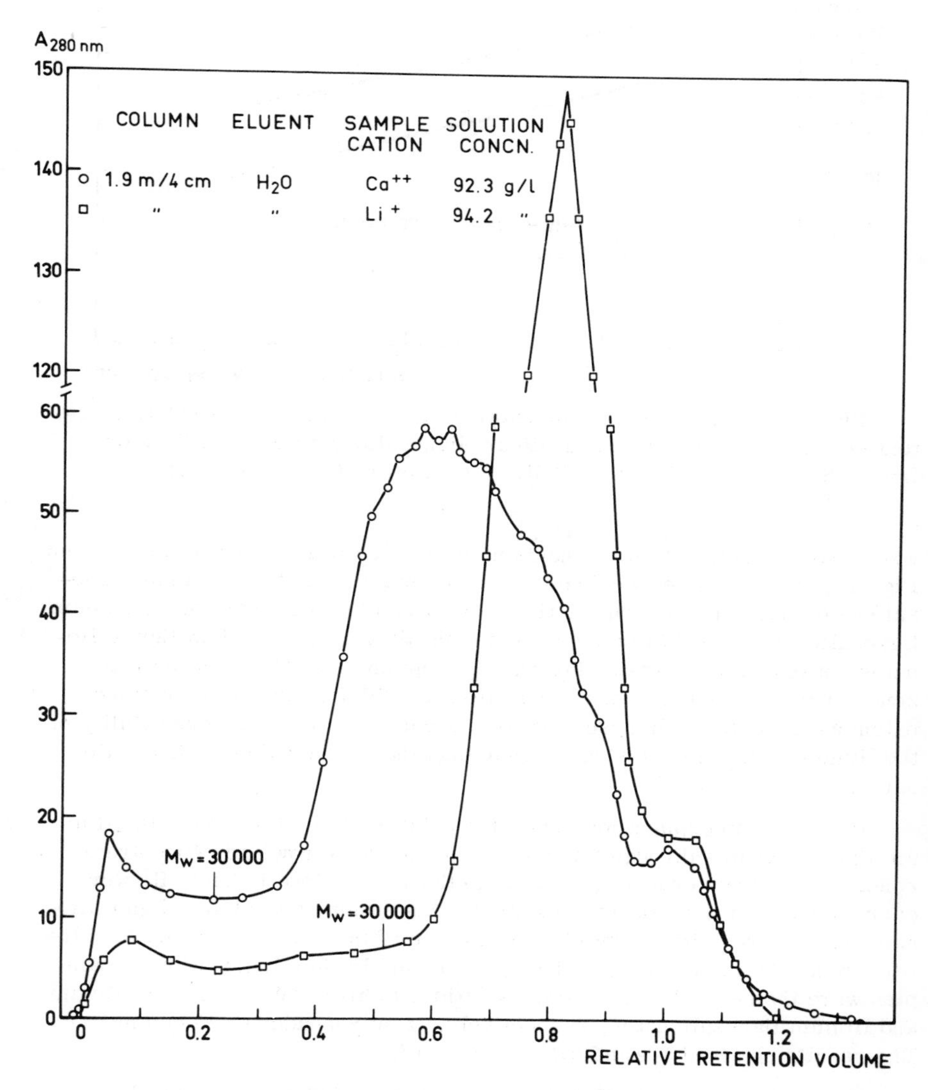

FIG. 15. Fractionation of calcium and lithium lignosulfonates on a Sephadex G-50 column using water as eluent. Elution rate 35 ml/hr [25]. (Courtesy of Paperi ja Puu.)

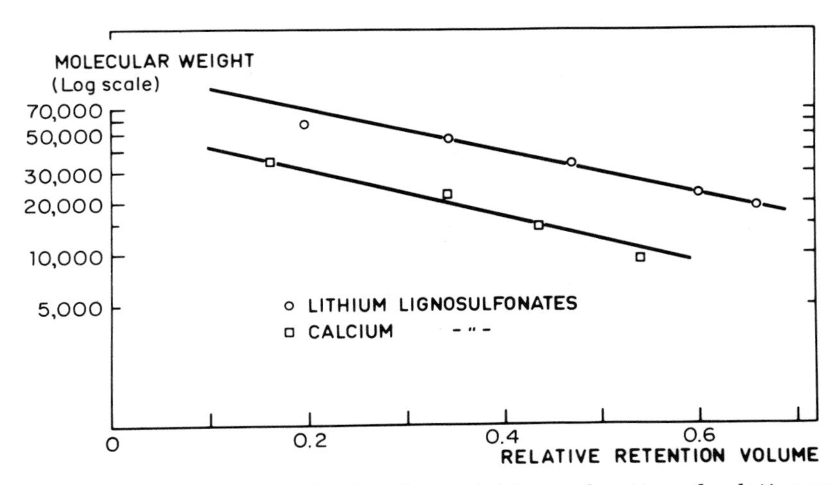

FIG. 16. Logarithm of molecular weight as a function of relative retention volume for calcium and lithium lignosulfonates eluted with water from a Sephadex G-50 column [25]. (Courtesy of Paperi ja Puu.)

what kind of fractionation mechanism determines this elution order? From Fig. 15, it is possible to calculate that the main part of the calcium lignosulfonates was diluted about 20 times, whereas the main part of the lithium lignosulfonates was diluted only about eight times. The lithium lignosulfonates consequently migrated through the column as a fairly concentrated zone. This means that these lignosulfonates did not expand to the same extent as the calcium lignosulfonates. Because of this, the permeability of the lithium lignosulfonates was higher than that of the calcium lignosulfonates.

However, an explanation must also exist for the migration of the lithium lignosulfonates through the column as such a narrow and concentrated zone. Model fractionations, similar to those described in Sect. II, were carried out to investigate this question. A solution of a calcium lignosulfonate fraction was divided into two parts; one of these was converted into the lithium salt by ion exchange. The calcium and lithium lignosulfonate samples were then eluted with aqueous solutions of high-MW, excluded calcium and lithium lignosulfonates, respectively, from Sephadex G-25 columns. The results obtained are indicated in Fig. 17.

It was shown in Sect. II B that the presence of high MW lignosulfonates in the eluent enhances the permeability of low-MW lignosulfonates. Figure 17 illustrates that the lithium lignosulfonate sample was eluted considerably later than the corresponding calcium lignosulfonate sample, obviously because counterion dissociation is more prevalent when the counterion is

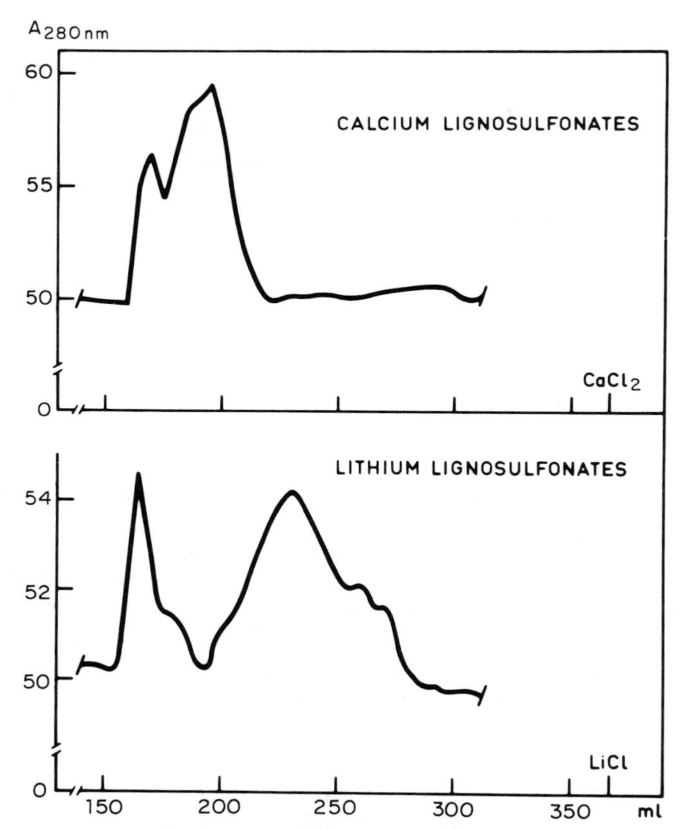

FIG. 17. Elution of a calcium lignosulfonate fraction and a lithium lignosulfonate fraction through Sephadex G-25 columns using aqueous solutions of high-MW lignosulfonates as eluents. High-MW excluded lignosulfonates were added to the sample solution as markers for the void volume [6]. (Courtesy of the Journal of Polymer Science.)

monovalent than when it is polyvalent. This has been demonstrated by Strauss and Leung [29] by means of dilatometric measurements with polystyrenesulfonates and polyvinylsulfonates. To check the validity of this for lignosulfonates, Stenlund [13] estimated the activity of free counterions in aqueous solutions of sodium and calcium lignosulfonates by the application of ion-specific electrodes. The results indicated that solutions of lignosulfonates with a monovalent counterion have a concentration of free (unbound) counterions that is significantly higher than that of solutions of lignosulfonates with divalent counterions. The macroions in a lithium lignosulfonate solution thus carry a larger number of free charges than do the macroions in an aqueous solution of calcium lignosulfonates.

For the elution of a sample of lithium lignosulfonates of broad MWD with water from a gel column, a stronger interaction prevails between high-MW lignosulfonates with low permeability and more permeable lignosulfonates than is the case when the sample contains calcium lignosulfonates. This interaction induces a fairly strong retardation of the permeable lithium lignosulfonate; this is the primary cause for lithium lignosulfonates being eluted considerably later than corresponding calcium lignosulfonates if the elution is done with water.

C. Influence of the Molecular Weight Distribution of the Sample Solution

The preceding sections have shown that, on the elution of a polyelectrolyte with an eluent of low ionic strength, the high-MW components usually influence the fractionation of the permeable components to a certain degree. If the proportion of high-MW compounds differs from one sample to another, it could be expected that the ion inclusion effect influences the fractionation result to varying degrees. For a more quantitative determination of this effect, Stenlund [30] has performed a series of fractionations with lignosulfonate samples comprising varying proportions of high-MW components. The variation in the MWD was achieved by variation of the digestion time of the acid calcium bisulfite cookings from which the lignosulfonates were isolated.

Samples of calcium lignosulfonates were eluted with water from preparative Sephadex G-50 columns, 190 cm long and 6 cm in diameter. The concentration of the sample solution was about 93 g/liter, and the volume of the sample was 70 ml. The results of these fractionations are indicated in Fig. 18. The chromatograms are distinguished by the digestion time at 130° C. The highest proportions of high-MW compounds were obtained after periods of digestion of 2 and 3 hr, whereas relatively low proportions of high-MW material were obtained after 0.5 and 8 hr.

For characterization of the fractionation, fractions were taken for the determination of molecular weight by application of the light-scattering method at four retention volumes within the fractionation range. These fractions are indicated by the Roman numerals I-IV in Fig. 18. Fractions were taken solely from the earlier part of the chromatogram because of the great difficulty experienced in applying the light-scattering method to lignosulfonate samples with molecular weights below 7000 [28]. The mean molecular weights of the lignosulfonate fractions I-IV have been plotted as a function of digestion time at 130° C in Fig. 19.

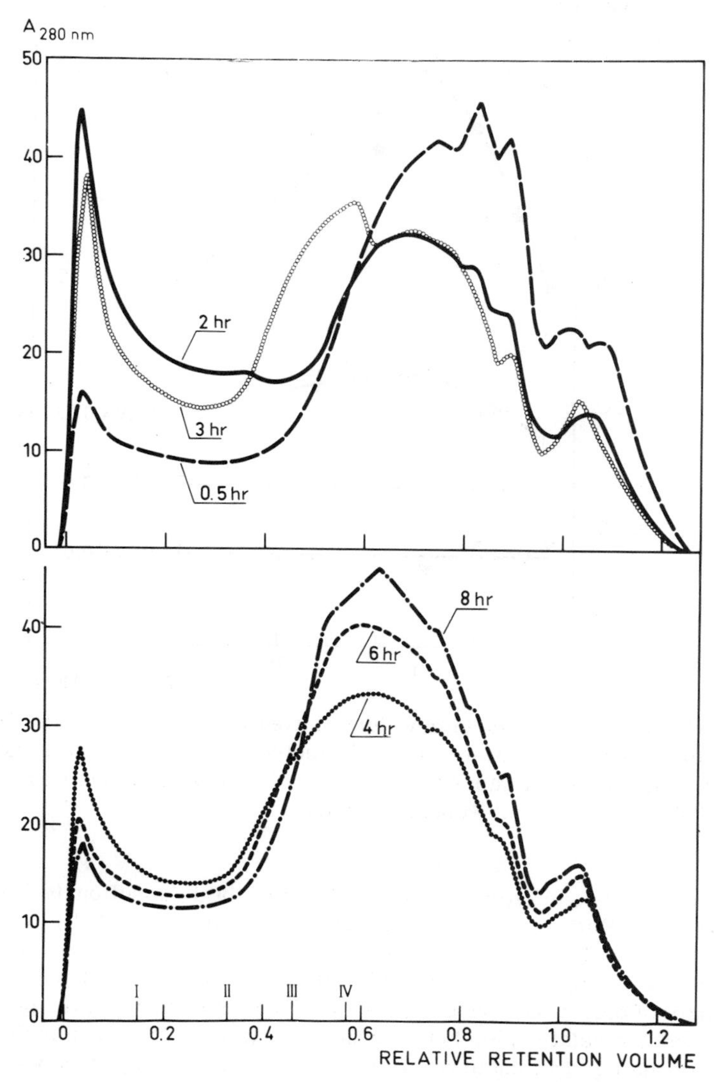

FIG. 18. Elution of lignosulfonates in spent sulfite liquor samples with water through Sephadex G-50 columns. Flow rate 80 ml/hr (2.8 ml/hr · cm^2) [30]. (Courtesy of Paperi ja Puu.)

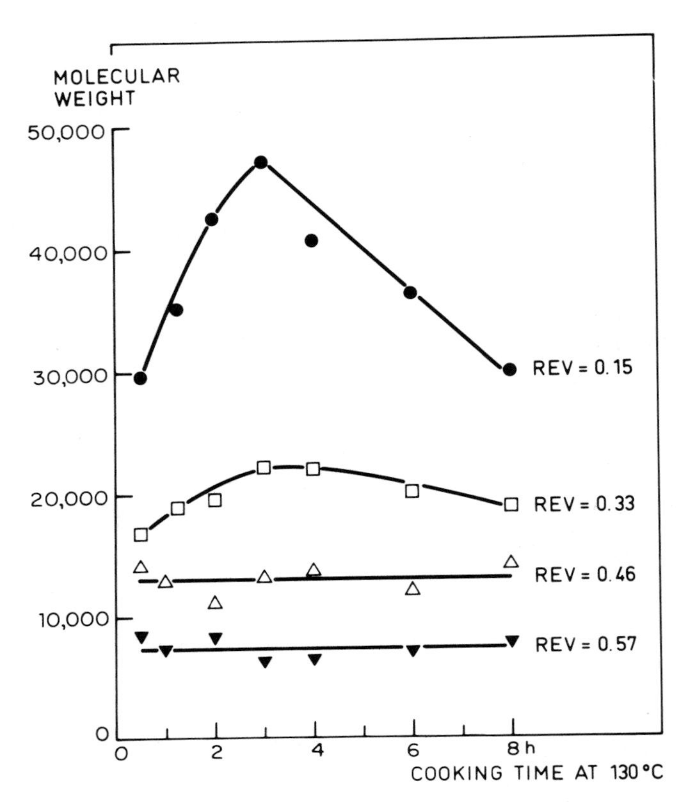

FIG. 19. Mean molecular weight plotted versus cooking time at 130° C for calcium lignosulfonate fractions collected during elution of lignosulfonates in spent sulfite liquors through Sephadex G-50 columns with distilled water [30]. (Courtesy of Paperi ja Puu.)

The weight-average molecular weights of the lignosulfonate fractions emerging at relative retention volumes of 0.15 (I) and 0.33 (II) diverged considerably with variations in length of the digestion period. However, the molecular weights of the two later fractions were almost independent of the length of digestion period. This is further evident from the plot that illustrates the relationship between the logarithm of the mean molecular weight of the lignosulfonate fractions and the relative retention volume in Fig. 20.

If a comparison is made between Figs. 19 and 20 and Fig. 18, it is seen that a similar trend exists in the variation of the molecular weights of the lignosulfonates that emerge at relative retention volumes of 0.15 and 0.33, and in the variation of the area of the first peak that corresponds to

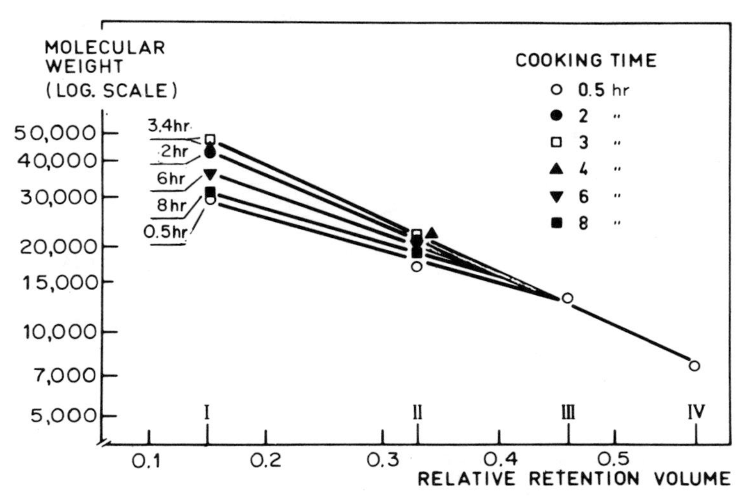

FIG. 20. Relationship between molecular weight and elution volume for calcium lignosulfonate fractions collected during the fractionation of lignosulfonates in spent sulfite liquors from sulfite cooks of different lengths. Eluent: water [30]. (Courtesy of Paperi ja Puu.)

high-MW excluded matter. Both the molecular weights and the peak areas increase in magnitude up to a digestion time of 2 to 3 hr. Subsequent to this, they begin to decrease, with the implication that the variation in the elution of the fractions with digestion time is dependent on the amount of high-MW lignosulfonates in the sample.

In order to confirm this, determinations were made of the molecular weight distributions of the lignosulfonate samples. The proportion of high-MW matter as a function of digestion time has been plotted in Fig. 21. It is observable from Figs. 19 and 21 that the molecular weight of the lignosulfonates that emerged at relative retention volumes of 0.15 (I) and 0.33 (II) rose with the increasing proportion of high-MW lignosulfonates in the samples. When the digestion period exceeded 3 hr, the molecular weights of the lignosulfonates that emerged at the relative retention volumes mentioned declined with the decrease in the proportion of high-MW lignosulfonates. The course of the fractionation thus obviously depends on the composition of the sample in such a way that when the sample contains high levels of high-MW matter, lignosulfonates with fairly low permeability are forced to enter into the gel phase because of a strong inclusion effect, and are, accordingly, eluted later than when the sample contains smaller amounts of high-MW components.

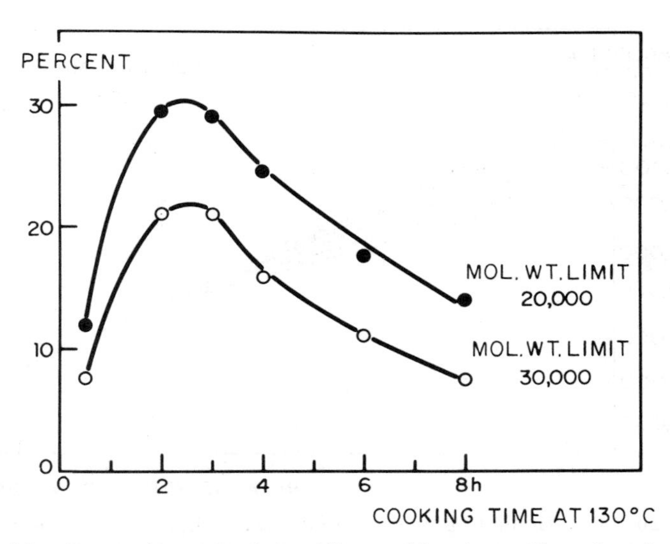

FIG. 21. Proportion of calcium lignosulfonates with molecular weights exceeding 20,000 and 30,000, respectively, as a function of cooking time at 130° C [30]. (Courtesy of Paperi ja Puu.)

It is interesting to note that in the fractionations studied this inclusion effect only affected the fractionation of lignosulfonates with molecular weights exceeding 12,000. This corresponds to an estimated k_D value of 0.47. However, the ion inclusion effect may at times be of longer range, as will become evident in the following section.

IV. OPTIMIZATION OF RESOLUTION

A. Influence of the Concentration of High-Molecular-Weight Polyelectrolytes in the Eluent

It becomes evident, from the experiments with eluents containing high-MW lignosulfonates as described in the preceding sections, that the retention volumes of lignosulfonate fractions can be altered. Accordingly, Forss and Stenlund [6] have investigated the practicability of improving the resolution of a sample containing a number of low-MW lignosulfonate fractions, through the use of eluents containing different concentrations of high-MW lignosulfonates that are excluded by the gel.

These experiments were performed with a Sephadex G-50 column, whose length and diameter were 140 cm and 1 cm, respectively. The sample contained some low-MW, neighboring lithium lignosulfonate fractions obtained from a preparative fractionation on a Sephadex G-50 column with water as the eluent. It was estimated that the gel-chromatographic mean molecular weight was about 6000, and that the concentration and volume of the sample solution were 4 g/liter and 0.5 ml, respectively. The interstitial volume of the column was 38 ml, as determined by means of high-MW lignosulfonates eluted with water, and the retention volume of a lithium chloride reference was 140 ml. Minor amounts of high-MW, excluded lithium lignosulfonates and lithium chloride were added to the sample solution as internal standards.

The lignosulfonate concentration of the effluent was monitored with a UV detector. The interference filter of the detector was removed to permit use of the relatively high light intensity required for the measurements of the elution curves in the presence of eluents of relatively high absorbances. Consequently, the measurements were not made at any specific wavelength; this approach was valid since all the fractions in the sample solution had similar spectra. For comparative purposes, one sample was eluted with distilled water (curve A in Fig. 22). The first peak to the left in Fig. 22 represents high-MW reference lignosulfonates. On elution with distilled water the low-MW fractions were spread over a range of relative retention volumes from 0.2 to 0.5. This was evidently a result of the influence of the ion exclusion effect. When the concentration of high-MW lignosulfonates in the eluent was 0.4 g/liter (curve B), the trailing edge was displaced toward higher retention volume, i.e., 0.7. The ion exclusion effect was thus partially suppressed by the charge effect of the high-MW lignosulfonates in the eluent. When the strength of the eluent was increased, peak formation became more evident (curve C), and as is observable from curve D, the favorable influence upon the resolution of the ion inclusion effect was greatest when the concentration of high-MW lignosulfonates in the eluent equaled 1.7 g/liter. It is obvious that the sample contained at least five different components. When the lignosulfonate concentration of the eluent was further increased, the ion exclusion effect became almost completely suppressed. Deterioration in the resolution was also apparent, since the ion inclusion effect became so marked that the components in the sample were eluted too close to each other.

It can be concluded that very good resolution is attainable if the eluent contains high-MW polyelectrolytes, and if the experimental conditions are optimized. As will be shown in the next section, good resolution is impracticable if the eluent consists of an electrolyte solution.

B. Influence of Electrolyte-Containing Eluents

It occurs to the author that some confusion might arise in a reader's mind by the rather extensive discussion of the fractionation behavior of

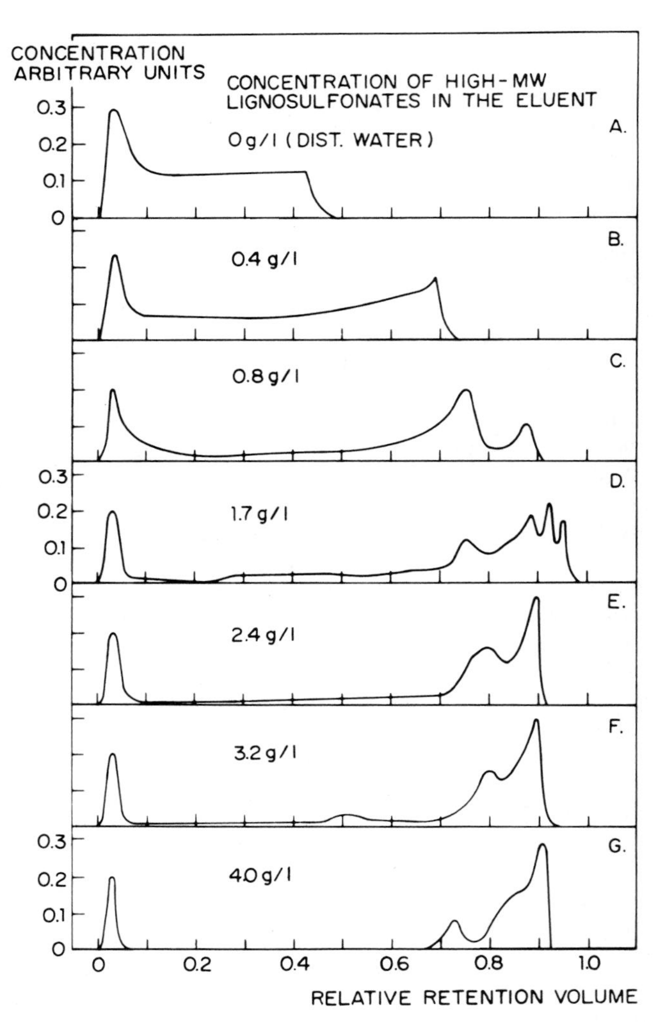

FIG. 22. Different degrees of resolution of lithium lignosulfonates on elution with water and solutions of high-MW lithium lignosulfonates of varying concentration. Flow rate 20 ml/hr (25 ml/hr · cm^2) [6]. (Courtesy of the Journal of Polymer Science.)

polyelectrolytes in aqueous solutions, with little being said about fractionation in electrolyte solutions. Of course, it is well established that polyelectrolyte effects are suppressed if a simple electrolyte is present within certain concentration ranges. It must be borne in mind that weighty reasons argue against the use of electrolyte solutions as the only possible eluents, an important one being the possible deterioration of resolution. Before we embark on a discussion of resolution, however, a brief discussion will be given about general fractionation behavior on elution with electrolyte solutions.

Skalka [31] has studied polyelectrolyte effects in the fractionation of heparin with Sephadex columns; he demonstrated that the electrolyte content of the eluent influenced fractionation as follows: When the ionic strength of the eluent was increased, heparin was retarded and eluted later. The peak became broadened, obviously because of the heterogeneity of the sample. One reason for the retarded elution was the suppression of charge effects in the gel (fixed charges); i.e., the ion exclusion effect was suppressed when the ionic strength of the eluent was increased. Another explanation involves a suggestion that, with changing ionic strength of the solution, the molecules of heparin change in shape and size, as well as in degree of solvation.

Cha [32] has carried out gel chromatography of polyacrylonitrile bearing some sulfonate groups (PAN-S) in pure dimethylformamide (DMF), and in DMF made 0.1 M in respect to lithium bromide. By addition of the electrolyte, the polyelectrolyte was transformed from a charged into an almost uncharged coil. On elution with the electrolyte solution, the retention volume markedly increased compared to elution with pure DMF. According to Cha, this effect was attributable to an impressive diminution in coil size as a result of charge neutralization.

In the opinion of Coppola and co-workers [33], a decrease in molecular dimensions is the normal effect to be expected, but it is difficult to explain the magnitude of the retardation solely by consideration of coil contraction. From their investigations of a similar system, these authors concluded that the striking difference in the elution behavior of PAN-S in pure DMF and DMF containing electrolyte arises primarily from an intermolecular effect, and not an intramolecular one. This means that PAN-S forms aggregates in pure DMF which are split on the addition of electrolyte, this being the cause of the retarded elution in electrolyte solution. This statement was based in part on the results of an investigation made by Chiang and Stauffer [34].

In his studies concerning gel-chromatographic fractionation of lignosulfonates, Stenlund has included many fractionations in electrolyte solutions. The influence of sample concentration and sample volume was thus eliminated when water was replaced by an electrolyte solution as eluent [25].

However, the results indicated that a fairly high electrolyte concentration was needed to achieve complete suppression of the polyelectrolyte effects. In the fractionation of calcium lignosulfonates, the strength of calcium chloride in the eluent should be about 0.25 M, and on the fractionation of lithium lignosulfonates, the concentration of lithium chloride in the eluent has to exceed 1 M in some applications [25]. In an unpublished work, Stenlund has observed that a 3 M lithium chloride solution may completely suppress the fairly strong polyelectrolyte effects that occur in lithium lignosulfonate solutions. The strong influence of the cation of the lignosulfonates is also suppressed upon elution with electrolyte solution [25].

Through the use of an electrolyte solution, accordingly, it becomes possible to make the fractionation results independent of experimental conditions. However, in some cases the peaks observed on elution with water are almost completely smoothed out on elution with electrolyte solution. This tendency is very obvious on the elution of lignosulfonates with electrolyte solution from Sephadex G-25, as can be seen in Fig. 23. One chromatogram was obtained on elution with water, and the other on elution with 0.5 M lithium chloride solution. Only a slight tendency toward peak formation was discernible when the elution was carried out with electrolyte solution, whereas the peaks seen in the other chromatogram are fairly sharp and regular.

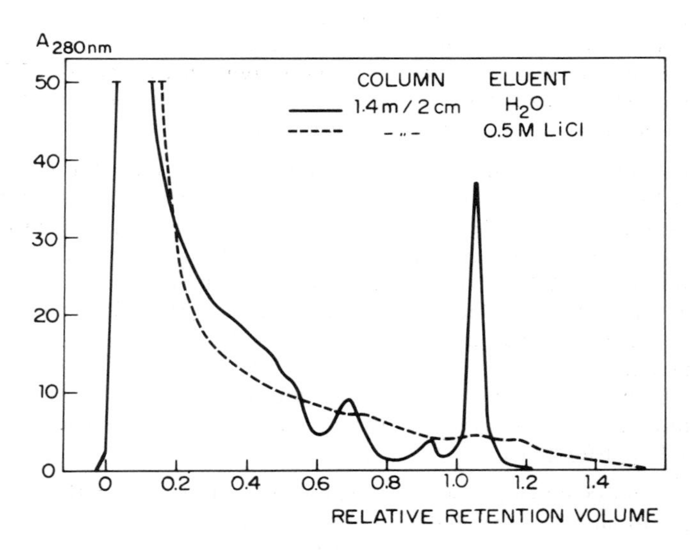

FIG. 23. Elution of lithium lignosulfonates from Sephadex G-25 with water and 0.5 M lithium chloride solution, respectively.

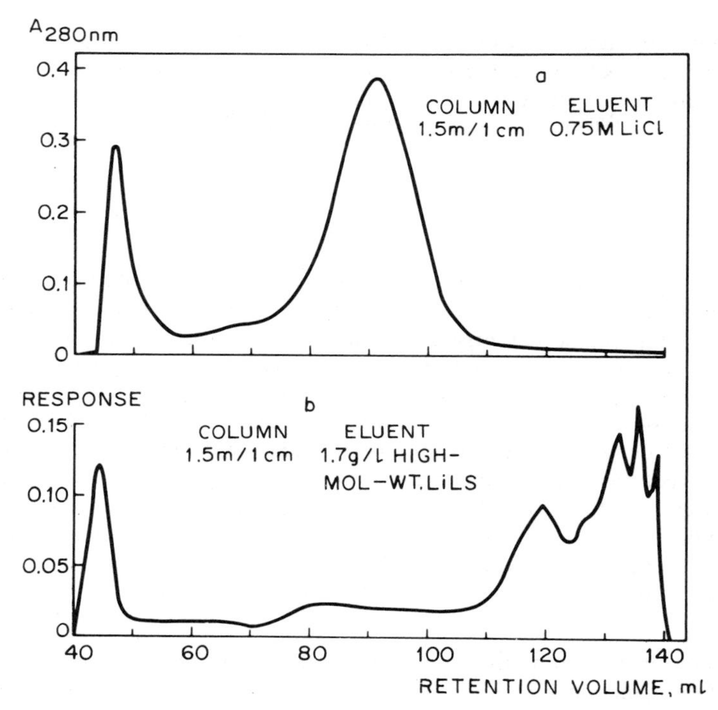

FIG. 24. Elution of a low-MW lithium lignosulfonate sample from Sephadex G-50. (a) 0.75 M lithium chloride solution and (b) a solution containing 1.7 g high-MW lithium lignosulfonates per liter. Flow rate 20 ml/hr [6]. (Courtesy of the Journal of Polymer Science.)

For study of the influence of electrolyte solution on resolution, Forss and Stenlund [6] have eluted a lithium lignosulfonate sample similar to the one used to obtain Fig. 22 (Sect. IV A) with 0.75 M lithium chloride solution. The result of this fractionation is comparable with the chromatogram representing optimal resolution on elution with high-MW lignosulfonates (curve D in Fig. 22) in Fig. 24. The peak to the left again represents the excluded high-MW reference lignosulfonates. On elution with electrolyte solution, the sample gives rise to only one peak, which is quite symmetric. On elution with a solution containing high-MW lignosulfonates, the improvement in resolution is quite significant. On the one hand, this effect may be due to the existence of different charge densities in different lignosulfonate compounds, and on the other, to the lignosulfonates being forced further into the internal volume, where the steric hindrance to diffusion of components with fairly small differences in hydrodynamic volume is more selective than is the case on elution with electrolyte solution.

In our opinion electrolyte solutions are quite satisfactory as eluents if one wishes to obtain a general picture of the sample, for example, if it is high or low MW. However, if the "fine structure" of the sample is in question, a solution of an excluded polyelectrolyte could be used as eluent.

V. SUMMARY

When polyelectrolytes (i.e., lignosulfonates) are eluted from Sephadex gels with an eluent of low ionic strength, the fractionation is affected by steric exclusion, polyelectrolyte expansion, ion exclusion, and ion inclusion effects. The last-mentioned effect arises from the interaction of charged sites in the high-MW macroions, which are more or less excluded from the gel phase with charged sites in more permeable macroions. The ion inclusion effect depends on the counterion in such a way that the fractionation of polyelectrolytes with a univalent counterion is affected to a greater extent than is the fractionation of polyelectrolytes with a divalent counterion. This is attributable to the existence of a higher density of free charges in the former case, since univalent counterions have a lesser tendency to form ion pairs with the charged sites of the macroion than have divalent counterions.

The ion inclusion effect may have a favorable influence on the resolution of components of relatively equal molar volume, but of different charge densities.

Polyelectrolyte effects are suppressed if the elution is carried out with an electrolyte solution of adequate ionic strength, so that the polyelectrolyte behaves as a nonionic polymer. However, since the separation effected by differences in charge densities is eliminated (swamped out), resolution may deteriorate; nevertheless, the fractionation order is independent of the composition of the sample, the concentration of sample solution, and other fractionation variables.

If one wishes to acquire a general picture of the composition of a certain polyelectrolyte, a solution of a simple electrolyte may be the most appropriate eluent. However, if polyelectrolyte components with small differences in hydrodynamic volume are to be separated, an eluent comprising a solution of a high-MW polyelectrolyte, completely excluded by the gel (e.g., dextran sulfate), can be used.

ACKNOWLEDGMENTS

The author expresses his sincere gratitude to Dr. K. Forss, Chairman of the Department for Development of Chemical Products of The Finnish Pulp and Paper Research Institute, for valuable advice and discussions during the investigations of polyelectrolyte effects in gel chromatography. He is also indebted to the staff of his own and other departments for assistance both with experimental work and preparation of the manuscript. The author also thanks the authors and publishers whose figures have been reproduced, and for their kind permission to include them in this work. Pehr-Erik Sågfors, M.Sc., is acknowledged for Fig. 1.

REFERENCES

1. R. M. Wheaton and W. C. Bauman, Ind. Eng. Chem., 45, 228 (1953).
2. P. Flodin, Dextran Gels and Their Application in Gel Filtration, Dissertation, Uppsala, 1962, p. 73.
3. B. Gelotte, J. Chromatogr., 3, 330 (1960).
4. P. A. Neddermeyer and L. B. Rogers, Anal. Chem., 40, 755 (1968).
5. D. Eaker and J. Porath, Sep. Sci., 2, 507 (1967).
6. K. G. Forss and B. G. Stenlund, J. Polymer Sci., Symposium, 42, 951 (1973).
7. F. G. Donnan, Chem. Rev., 1, 73 (1924).
8. C. Tanford, Physical Chemistry of Macromolecules, Wiley, New York, 1961, p. 225.
9. J. Th. G. Overbeek, Prog. Biophys. Biophys. Chem., 6, 57 (1956).
10. T. Teorell, Discuss. Faraday Soc., 21, 9 (1956).
11. P. Flodin, Dextran Gels and Their Application in Gel Filtration, Dissertation, Uppsala, 1962, p. 50.
12. L. W. Nichol, W. H. Sawyer, and D. J. Winzor, Biochem. J., 112, 259 (1969).
13. B. Stenlund, Paperi ja Puu (Paper and Timber), 52, 671 (1970).
14. P. A. Neddermeyer and L. B. Rogers, Anal. Chem., 41, 94 (1969).
15. H. Vink, Acta Chem. Scand., 17, 2524 (1963).
16. F. Oosawa, Polyelectrolytes, Marcel Dekker, New York, 1971.
17. K. Hellsing, J. Chromatogr., 36, 170 (1968).
18. J. C. Giddings and K. Dahlgren, Sep. Sci., 5, 717 (1970).
19. B. Stenlund, Paperi ja Puu (Paper and Timber), 52, 197 (1970).
20. J. L. Gardon and S. G. Mason, Can. J. Chem., 33, 1491 (1955).
21. M. Nagasawa and I. Kagawa, J. Polymer Sci., 25, 61 (1957).

22. A. Rezanowich and D. A. I. Goring, J. Colloid Sci., 15, 452 (1960).

23. J. Hazell, L. A. Prince, and H. E. Stapelfeldt, Am. Chem. Soc., Polymer Preprints, 8, 1303 (1967).

24. H. E. Adams, L. A. Farhat, and B. L. Johnson, Ind. Eng. Chem. Prod. Res. Dev., 5, 126 (1966).

25. B. Stenlund, Paperi ja Puu (Paper and Timber), 52, 121 (1970).

26. L. Fisher, An Introduction to Gel Chromatography, in Laboratory Techniques in Biochemistry and Molecular Biology, Vol. I (T. S. Work and E. Work, eds.), North-Holland, Amsterdam, 1969, Part II, p. 304.

27. F. Oosawa, Polyelectrolytes, Marcel Dekker, New York, 1971, p. 14.

28. K. G. Forss and B. Stenlund, Paperi ja Puu (Paper and Timber), 51, 93 (1969).

29. U. P. Strauss and Y. P. Leung, J. Am. Chem. Soc., 87, 1476 (1965).

30. B. Stenlund, Paperi ja Puu (Paper and Timber), 52, 333 (1970).

31. M. Skalka, J. Chromatogr., 33, 456 (1968).

32. C. Y. Cha, J. Polymer Sci., B7, 343 (1969).

33. G. Coppola, P. Fabbri, P. Bice, and U. Bianchi, J. Polymer Sci., 16, 2829 (1972).

34. R. Chiang and J. C. Stauffer, J. Polymer Sci., A-2, 5, 101 (1967).

35. P. R. Gupta and J. L. McCarthy, Macromolec., 1, 236 (1968).

Chapter 3

CHEMICALLY BONDED PHASES IN CHROMATOGRAPHY

Imrich Sebestian and István Halász

Angewandte Physikalische Chemie
Universität Saarbrücken
West Germany

I. INTRODUCTION

The conventional active supports in chromatography are silica and alumina, independent of whether the mobile phase is a gas or a liquid. The hydroxyl groups on the surface of both solids determine their adsorptive properties and their selectivities as stationary phases. The chromatographic properties of both solids, but especially that of silica, is determined by the amount of physically adsorbed water on the surface, including their pores. The heat of the reversible adsorbed water on a highly dispersed silica (i.e., Aerosil) is 12.6 kcal/mole [1]. The high heat of the specific reversible adsorption can be interpreted only with the assumption of hydrogen bonding [1].

In liquid chromatography the amount of the adsorbed water on the surface of silica is an extremely sensitive function of the water content of the eluent. The sorption properties of the silica change if the water content of the n-heptane eluent increases from 50 to 70 ppm and the capacity ratios decrease up to 30%. The water-silica equilibrium is relatively slow; consequently, the retentions change continuously for a long period if the water content of the liquid organic eluents changes. Because of the great affinity of water to silica it can happen that although the water content of an apolar eluent is very low, this water will be picked up by the active support. The consequence is a heavily loaded column with a water content up to 50% by weight or more, depending on the specific pore volume of the silica. Similar problems arise if the gradient elution method is used, because some polar organic compounds are also adsorbed on silica or alumina through hydrogen bonding. To use active adsorbents is sometimes dangerous because of their catalytic activity, especially at high temperature.

If the hydroxyl groups on the surface of alumina or preferably on silica are chemically reacted with organic compounds, the surface properties of the active supports are more or less changed. Such chemically bonded stationary phases in gas or liquid chromatography are based mainly on silica support. The surface properties of the silica (i.e., specific surface area, total pore volume, pore distribution) codetermine the properties of the chemically bonded stationary phase as described in the literature [2-5].

If the particle size of the support is higher than 10 μm, the speed of mass transfer "in" a porous silica is relatively slow. To increase the speed of analysis porous layer beads (PLB) are used, i.e., an impenetrable core (e.g., glass) is coated with a thin layer of active support [6, 7]. The thickness of the active layer used is varied between 0.1 and 5 μm. Experiences in our laboratories and elsewhere [8] show that if the particle size of the support is below 10 μm the speed of analysis is more or less independent of whether PLB or porous supports are used. Further advantages (or disadvantages) of the PLB packing are caused by the smaller amount of active stationary phase per unit of column volume: (a) the capacity ratios $[k' = (t_R - t_0)/t_0]$ are smaller, consequently the time of analysis can be shortened; (b) the analysis can be made at lower temperatures; (c) the loadability (i.e., maximum sample size) is smaller. Chemically bonded phases can be made, of course, with porous layer beads also.

Irregardless of the surface properties of the inorganic support the chemically bonded stationary phases have the following advantages if compared with the conventional stationary (active solid or stationary liquid) phases:

1. The vapor pressure (precisely the sublimation pressure) of chemically bonded phases is negligible. At a given temperature the stationary phase decomposes and the column "bleeds." Another reason for bleeding, especially in liquid chromatography, is that the phase is chemically attacked by the eluent or by the samples (e.g., hydrolysis). Usually the baseline is as stable with columns packed with chemically bonded phases as with those

packed with active solid supports. As a consequence of the baseline stability, extremely small sample sizes can be separated or detected and the quantitative analysis is accurate. In prepscale work it is advantageous that the fractions are not contaminated with the stationary phase. This is also important if the chromatographic separation is coupled with other analytical methods, for example with mass spectroscopy. The excellent baseline stability in programmed temperature or programmed flow rate gas chromatography is advantageous with chemically bonded stationary phases. The same advantage is expected with liquid eluents if gradient elution or flow-programmed methods are applied, because none of these phases is dissolved in the mobile phase.

2. The mass transfer in (or on) the chemically bonded phases is extremely speedy as long as the "film thickness" is small, i.e., as long as a monomeric (and not a polymeric) layer of organic molecules is bonded on the surface of the inorganic support. Especially in gas chromatography the h versus u curve has a broad minimum and the slope of the curve (the constant C in the van Deemter equation) is on the order of magnitude of a few milliseconds. The consequence is an unusually high speed of gas-chromatographic analysis. With monomeric chemically bonded phases the relative peak broadening h is more or less independent of (a) the capacity ratio k'; (b) the temperature; (c) the quality of the sample in gas and in liquid chromatography; (d) the sample size in a broad range (i.e., in gas chromatography between 10^{-8} and 10^{-3} g in a column with 2 mm ID).

3. The selectivity of brush-type (i.e., monomeric) stationary phases can be varied in an unusually broad range with bristles containing functional groups (e.g., —CN, —COOH, aldehyde groups). In gas chromatography the chemically bonded phases without functional groups are more or less inefficient, probably because of the competition of the unreacted silanol and silioxane groups on the surface of the silica in the sorption mechanism. In liquid chromatography these groups are blocked by the polar eluent; consequently, excellent separations are achieved with paraffinic (e.g., octadecyl) bristles in the field of "reversed" phase chromatography.

II. TYPES OF CHEMICALLY BONDED PHASES

The structure of the organic component of the chemically bonded phases (CBP) can be monomeric or polymeric. This structure determines some properties of the stationary phase, as, for example, the speed of mass transfer and sometimes the selectivity.

In monomeric CBP the organic molecules form a more or less "unimolecular" layer and are bonded to the surface of the inorganic support by ionic or covalent bonding. Depending on the type of bonding between silica and organic molecules they can be ordered into three groups: (a) ≡Si—O—C≡ , (b) ≡Si—NH—C≡ , (c) ≡Si—C≡ . The thermal stability and the stability against hydrolysis increases from (a) to (c). Unfortunately, however, the difficulties of preparation of the phases increase in the same order.

The polymeric CBP are characterized by the formation of organic polymer (silicones) chemically attached to the surface of the silica. Furthermore, it is possible to coat an impenetrable [9] or porous support with any kind of organic polymer [10]. This coating procedure is similar to that used in gas chromatography. The organic polymer is dissolved in an appropriate solvent (e.g., polyamide in formic acid). The support is coated with this solution and the solvent is evaporated. These polymer-coated supports are of some interest in high-speed liquid chromatography.

A. Monomeric Stationary Phases

1. Ionic Bond

The cations of alumina silicates are exchangeable by strong organic bases by ion exchange. Bentones are produced by treating bentonites (e.g., Bentone 34) with quaternary organic ammonium cations (e.g., dioctadecyldimethylammonium cations) [11, 12]. Excellent separations of aromatics are achieved with such stationary phases in gas chromatography [11, 12], including the difficult separation of the xylene isomers [13, 14].

Kiselev et al. coated Chromaton with 15% (w/w) Benton 34 to separate aromatic isomers in rapid liquid chromatography [15]. Terphenyl, phenoxyphenol, and nitrophenol isomers were analyzed with n-hexane and ethanol eluents.

Substituted aromatics were separated on a porous layer bead stationary phase with a specific surface area of 15 m^2/g in liquid chromatography. In this way, glass beads were coated with a porous layer of molecular sieves. The cations of the molecular sieves were substituted by diethanolamine [16].

2. Covalent Bond Phases

a. ≡Si—O—C≡ Bond. The weak acidic silanol groups (≡Si—OH) on the surface of the silica can be esterified with alcohols at elevated temperatures [17-19]. Deuel and Wartmann [20] produced silica gel chloride in the first step by reacting silica with thionyl chloride:

$$\equiv Si{-}OH + SO_2Cl \longrightarrow \equiv SiCl + SO_2 + HCl$$

In the second step of the preparation of the alkoxy derivatives the chlorinated silica was reacted with alcohols:

$$\equiv Si{-}Cl + R{-}OH \longrightarrow \equiv Si{-}O{-}R + HCl$$

Rossi et al. [21] were the first to use esterified silica as stationary phase in gas chromatography. They separated C_1-C_4 hydrocarbons on a silica esterified with lauryl or benzyl alcohol. These separations were not outstanding. Excellent gas-chromatographic separations were achieved, however, by the introduction of functional groups at the end of the bristle (ω position to the silica surface) [22, 23]. Up to 30 theoretical plates per second were generated with such types of stationary phases, where the h values were independent of the capacity ratios up to k' = 138. The diethyl ether peaks were symmetrical in these experiments, showing the same h values as paraffins. The symmetry of the ether peak is the most sensitive indicator of whether or not the surface reaction was quantitative. The tailing of the ether peak indicates the presence of unreacted free silanol or siloxane groups not shielded by the organic bristles and therefore available for the sample molecules.

The capacity ratios and the relative retentions can be varied within a broad region by variation of the functional groups [23]. Using esterified silica as stationary phase in gas chromatography, chlorinated and nonchlorinated hydrocarbons in the C_4 range were separated [24]. Excellent relative retentions were achieved at room temperature for C_1-C_5 hydrocarbons with silica reacted with phenyl isocyanate or phenyl thiocyanates [25]. These phases also had monomeric structure.

In liquid chromatography porous layer-type brush phases with $\equiv Si{-}O{-}C\equiv$ bonding were used. Glass beads were coated with a porous layer of silica where the thickness of the porous layer was on the order of magnitude of 1 μm and the silica was reacted with alcohols. The specific surface area of these stationary phases (i.e., Corasil, Zipex, Perisorb) varied between 0.5 and 14 m^2/g. On Corasil (Waters Associates Inc., Milford, Mass.) esterified with polyethylene glycol 400, aromatic amines [26] were separated. On the same stationary phase and on silica esterified with oxydipropionitrile, drugs (metalozones) and benzodiazepines in urine were separated [27, 28]. It should be pointed out, again, that esterified silica (estersil) phases are not stable against hydrolysis. This is extremely disadvantageous in liquid chromatography, because even the water content of apolar solvents such as heptane is not negligible. It can happen that after a few days or weeks the estersil is hydrolyzed by the heptane eluent also, especially at temperatures around or above 50° C. During this period the relative retentions change.

b. $\equiv$Si—NH—C$\equiv$ Bond. Stationary phases with such bonding are stable against hydrolysis, at least in the pH range of 4 to 8. If pH > 8, the silica support itself is dissolved. These CBP are prepared by reacting chlorinated silica with primary or secondary organic amines [29]. Aromatic hydrocarbons can be easily separated on such stationary phases in liquid chromatography if nitrophenyl groups are introduced at the end of the bristles [30]. The spatial requirement of a $\equiv$Si—NH—C$\equiv$ bonded bristle is not small enough to allow application in the field of gas chromatography (i.e., incomplete coverage of the silica).

c. $\equiv$Si—C$\equiv$ Bond. It is a well-known fact that the temperature stability and the stability against hydrolysis of the $\equiv$Si—C$\equiv$ bond is excellent [31]. Although the preparation of stationary phases with this type of bonding is much more sophisticated than that of $\equiv$Si—O—C$\equiv$ and $\equiv$Si—NH—C$\equiv$ bonding, the $\equiv$Si—C$\equiv$ bonded stationary phases are very important in liquid as well as in gas chromatography.

3. Deactivation of the Support

Inert supports are needed in gas partition chromatography. Unfortunately, however, there are always active groups on the surface of the conventional inorganic supports. Howard and Martin silanized diatomaceous earth with dimethylchlorosilane in the gas phase [32] to separate fatty acids by gas-chromatographic methods. Other authors describe the deactivation of the support with the same silanizing agent (dissolved in toluene) in the liquid phase [33-36]. Kiselev and Stcherbakova achieved good results with trimethylchlorosilane [37]; Bohemen et al. silanized with hexamethyldisilazane dissolved in petrol ether [38]. In this reaction NH_3 is produced; consequently, the reaction is more efficient than those with chlorosilanes. Kirkland compared the efficiency of the different deactivation methods. The support was reacted with dimethylchlorosilane, trimethylchlorosilane, and hexamethyldisilazane [39]. No basic differences were found although the best peak symmetry was achieved with dimethyldichlorosilane. Similar results were published by Kabot and Ettre [40]. Of course, any highly reactive silanizing agent can be used to deactivate a support.

A deactivated support can hardly be used as a stationary phase itself. Usually it is coated with a high-boiling partitioning liquid. To prepare an efficient CBP for chromatography organic bristles other than the methyl group have to be anchored onto the surface of the inorganic support. The chlorine atoms of a chlorinated silica can be substituted through organic groups via metal organic reagents [40-42], i.e., Grignard reagents [43]. With that Grignard reaction or with a Wurtz-Fittig synthesis on Porasil C (a silica with a specific surface area of about 70 m^2/g and an average pore diameter of 500 Å), naphthyl groups were introduced. On this phase condensed aromatics and their nitro derivatives were separated in the gas phase. The capacity ratios of the samples were a function of the sample size [43]. This and the tailing of the peaks demonstrate the inhomogeneity of the surface (unreacted silanol groups).

For liquid chromatography cation exchangers can be produced by sulfonating the chemically bonded aryl groups on the surface of silica. The sulfonation is done either by oleum or by chlorosulfonic acid [43-47].

Reversed (i.e., apolar) stationary phases are produced for liquid chromatography by treating silica with mono-, di-, or trichloroalkyl silanes. The length of the alkyl groups usually varies between C_8 and C_{18}. Such phases are commercially available from different manufacturers. If trichloro alkyl silanes are used for steric reasons, it is possible that not all three chlorine atoms react simultaneously with the silanol groups on the surface of the silica. Precaution has to be taken to exclude traces of water in the solution, otherwise unwanted polymerization may occur. The unreacted chlorine atoms on the bristle have to be removed (e.g., by alcoholysis [48]) after the silanization reaction. Fewer problems are involved if mono- or dichloro alkyl silanes are applied.

There are two basic methods for introducing the wanted functional groups (if possible, the ω position) onto the surface. If silane derivatives with the wanted functional groups are available, and these functional groups do not compete (e.g., with the chlorine atoms) in the silanization reaction, the stationary phase can be prepared as described above.

The second method is to introduce the wanted functional groups into the bristles of a reversed phase [49]. To exclude the influence of unreacted silanol groups the reversed phase is resilanized with hexamethyldisilazane. Afterwards one or more hydrogens of the alkyl bristle are substituted by halogen (e.g., bromine) atoms. These reactive bromines can be substituted by amines, alcohols, etc., or by their derivatives [49]. Consequently, any functional group can be introduced to the $\equiv$Si—C$\equiv$ bonded bristle, including ion exchanger groups. Stationary phases with this type of bond can be used in gas [49, 50] as well as in high-speed liquid chromatography [49].

B. Polymeric Stationary Phases

Such types of stationary phases are prepared by hydrolysis and the subsequent polymerization of trichloro alkyl silanes on inorganic (i.e., silica) surfaces. Through control and/or repetition of these reactions the thickness of the polymer layer can be varied. The advantage of the polymeric CBP is that all unreacted active silanol groups are shielded by the "thick" organic layer. The disadvantage is that the speed of mass transfer in the "thick" polymer layer is much slower than that on the "surface" of the bristles of the monomeric bonded stationary phases.

Abel et al. bonded up to 14% (body weight) silicones on the surface of silica by the polymerization of hexadecyltrichlorosilane [51]. (It should be pointed out, however, that monomeric bonded reversed phases with octadecyl bristles were produced with a carbon content up to 20% by weight on the surface, 350 m^2/g, of silica [52].) The properties (i.e., the relative retentions) of these polymeric CBP were in gas chromatography very similar to those achieved with silicon-coated column packings. The use of such polymeric CBP in liquid chromatography was also demonstrated [53].

Aue, Hastings, and their associates polymerized different alkyl chlorosilanes on the surface of Chromosorb. The chain length of the alkyl group was varied between C_1 and C_4 [54-56]. The kind of chemical bond between the inorganic support and the organic polymer, if one exists, is unknown. Optimum stationary phases were prepared by the polymerization of butyltrichlorosilanes. Alcohols with higher molecular weights and alkanes were separated by gas-chromatographic methods on these stationary phases.

Kirkland polymerized different silicon layers on the surface of Zipax [57, 58]. Zipax (E.I. du Pont de Nemours and Co., Wilmington, Del.) is a silica-coated glass bead type of stationary phase with a specific surface area of about 1 m^2/g. Up to 1% (by weight) silicon was polymerized on the surface of these porous layer beads. Using substituted alkyl silanes different functional groups (e.g., nitril groups) were introduced into this polymer. The relative retentions of different samples varied with these functional groups. In high-speed liquid chromatography, for instance, sulfonamides and thiolhydroxomates, in gas chromatography chlorinated aromatics were separated in columns packed with this stationary phase.

Polymeric reversed CBP with octadecyl chains were also produced to separate substituted ureas with water/methanol (65/35 v/v) eluent [26]. Schmit et al. [60] separated condensed aromatics, exhaust gases, vitamins, and other materials, with this stationary phase using the gradient elution method. The concentration of the methanol in water was changed from 20 to 100%. The separation of the derivatives of anthraquinones and of chlorinated biphenyl with the same method is also described by these authors [60].

Novotny et al. polymerized di- and trichlorosilanes substituted first with chloromethyl phenyl groups and then with functional groups on the surface of Porasil C and Corasil [61].

An important application of polymeric CBP is the enrichment of trace compounds from the gas or liquid phase. The concentrated sample can be displaced from the column and analyzed with gas- or liquid-chromatographic methods. Water pollution, exhaust gases, and insecticide impurities were analyzed by this method [56].

III. CONCLUSIONS

The differences between monomeric and polymeric CBP have been discussed in detail. The advantages and disadvantages of both types of CBP were compared, including the conventional types of stationary phases (i.e., active solids or liquid-coated supports). With monomeric phases the speed of analysis in gas and in liquid chromatography is high. Through the introduction of functional groups the selectivity of monomeric and polymeric phases can be varied.

However, it should be pointed out that most of the separations with CBP described up to now can be performed with conventional stationary phases in liquid chromatography as well. It seems that the biggest advantage of CBP in liquid chromatography is their rapid equilibration with polar eluent components, especially water. (All organic solvents contain water.) This primarily determines the retentions with a given pair of stationary phase and eluent. Unfortunately, the silica/water equilibrium is slow, and therefore, it sometimes takes hours or days until the relative retentions are constant.

The gradient elution method in liquid chromatography assumes the use of eluents with different polarities, as well as different water content. With silica as the stationary phase the analysis is made usually within a few minutes. The reequilibration of the silica, returning to the less polar eluent again, needs hours because of the slow equilibrium as discussed above. Therefore, CBP are preferred if the gradient elution method is necessary.

Last but not least it should be pointed out that the introduction of the reversed phases enlarged the scope of high-speed liquid chromatography. The same will probably be true for the chemically bonded ion exchange phases in the near future.*

REFERENCES

1. W. Stöber, Kolloid Z., 145, 17 (1956).
2. D. C. Locke, J. Chromatogr. Sci., 11, 120 (1971).
3. R. E. Majors, Am. Lab., 4 27 (1972).
4. I. Halász and I. Sebestian, J. Chromatogr. Sci., 12, 161 (1974).
5. K. Unger, P. Ringe, J. Schick-Kalb, and B. Straube, Z. Anal. Chem., 164, 267 (1973).
6. I. Halász and Cs. Horvath, Anal. Chem., 36, 1178 (1964).
7. I. Halász and P. Walkling, J. Chromatogr. Sci., 7, 129 (1969).
8. J. Kirkland, Private communication, 1972.

---

*This manuscript was originally written in 1974.

9. F. M. Rabel, Anal. Chem., 45, 957 (1973).

10. H. Wiedemann, Ph.D. Thesis, Saarbrücken, 1973.

11. D. White, Nature, 179, 1075 (1957).

12. D. White and C. T. Cowan, Trans. Faraday Soc., 54, 557 (1958).

13. M. A. Hugnes and A. L. Roberts, Nature, 184, 1796 (1959).

14. M. Taramasso and F. Veniale, Contrib. Mineral Petrol. (Berlin), 21, 53 (1969).

15. A. V. Kiselev, N. P. Lebedeva, J. J. Frolov, and Ya. J. Yashin, Chromatographia, 5, 341 (1972).

16. H. M. McNair and C. D. Chandler, in Advances in Chromatography (A. Zlatkis, ed.), Houston, Texas, 1973, p. 357.

17. R. K. Iler, U.S. Patent No. 2,654,149 (1953).

18. R. K. Iler, The Colloid Chemistry of Silica and Silicates, Cornell Univ. Press, Ithaca, New York, 1955.

19. W. Stöber, G. Bauer, and K. Thomas, Liebigs Ann. Chem., 604, 104 (1957).

20. H. Deuel and J. Wartmann, Helv. Chim. Acta, 42, 1160 (1959).

21. C. Rossi, S. Munari, C. Cengari, and G. F. Tealdo, Chim. Ind. (Milano), 42, 724 (1960).

22. I. Halász and I. Sebestian, Angew. Chem., Int. Ed., 8, 453 (1969).

23. I. Sebestian, Doktorarbeit, Universität Frankfurt/M., 1969.

24. J. N. Little, W. A. Dark, P. W. Farlinger, and K. J. Bombaugh, J. Chromatogr. Sci., 8, 647 (1970).

25. J. Asshauer and I. Halász, Liebigs Ann. Chem., 758, 202 (1972).

26. J. J. Kirkland, Anal. Chem., 43, 36A (1971).

27. O. N. Hinsvark, W. Zazulak, and A. J. Cohen, J. Chromatogr. Sci., 8, 379 (1970).

28. C. G. Scott and P. Bommer, J. Chromatogr. Sci., 8, 446 (1970).

29. O.-E. Brust, Doktorarbeit, Universität Saarbrücken, 1972.

30. I. Sebestian, O.-E. Brust, and I. Halász, in Gas Chromatography 1972 (S. G. Perry, ed.), Appl. Science Publ., London, 1972, p. 281.

31. W. Noll, Chemie u. Technologie der Silicone, Verlag Chemie, Weinheim, 1968.

32. G. A. Howard and A. J. P. Martin, Biochem. J., 46, 532 (1950).

33. A. Kwantes and G. W. A. Rijnders, in Gas Chromatography 1958 (D. H. Desty, ed.), Butterworth, London, 1958, p. 125.

34. C. C. Sweeley and E. A. Moscatelli, J. Lipid Res., 1, 40 (1959).

35. W. L. Holmes and E. Stack, Biochim. Biophys. Acta, 56, 163 (1962).

36. E. C. Horning, K. C. Maddock, K. J. Anthony, and W. J. A. Vandenheuvel, Anal. Chem., 35, 526 (1963).

37. A. V. Kiselev and K. D. Stcherbakova, Gas Chromatographie, Akademie Verlag, Berlin, 1962.

38. J. Bohemen, S. H. Langer, R. H. Perrett, and J. H. Purnell, J. Chem. Soc., 1960, 2444.

39. J. J. Kirkland, in Gas Chromatography (L. Fowler, ed.), Academic Press, New York, 1963, p. 77.

40. F. J. Kabot and L. S. Ettre, J. Gas Chromatogr., 2, 21 (1964).

40a. J. Wartmann and H. Deuel, Chimia, 12, 82 (1958).

41. J. Wartmann and H. Deuel, Helv. Chim. Acta, 42, 1166 (1959).

42. H. P. Boehm and M. Schneider, Z. Anorg. Allg. Chem., 304, 326 (1959).

43. D. C. Locke, J. T. Schmermund, and B. Banner, Anal. Chem., 44, 90 (1972).

44. H. J. Wartmann, Doktorarbeit, Zurich, 1958.

45. K. Unger and K. Berg, Z. Naturforsch., 24b, 454 (1969).

46. K. Unger, W. Thomas, and P. Adrian, Kolloid Z.u.Z. Polymere, 251, 45 (1973).

47. K. Unger and D. Nyamah, Chromatographia, 7, 63 (1974).

48. R. H. Stehl, U.S. Patent No. 3,664,967 (1972).

49. I. Sebestian and I. Halász, Chromatographia, 7, 371 (1974).

50. W. Werner, Diplomarbeit, Universität Saarbrücken, 1973.

51. E. W. Abel, F. H. Pollard, P. C. Uden, and G. Nickless, J. Chromatogr., 22, 23 (1966).

52. K. Karch, Ph.D. Thesis, Universität Saarbrücken, 1974.

53. H. N. M. Stewart and S. G. Perry, J. Chromatogr., 37, 97 (1968).

54. W. A. Aue and C. R. Hastings, J. Chromatogr., 42, 319 (1969).

55. C. R. Hastings, W. A. Aue, and J. M. Augl, J. Chromatogr., 53, 497 (1970).

56. W. A. Aue and P. M. Teli, J. Chromatogr., 62, 15 (1971).

57. J. J. Kirkland and J. J. De Stefano, J. Chromatogr. Sci., 8, 309 (1970).

58. J. J. Kirkland, J. Chromatogr. Sci., 9, 206 (1971).

59. J. A. Schmit, R. A. Henry, R. C. Williams, and J. F. Dieckman, J. Chromatogr. Sci., 9, 645 (1971).

60. S. H. Byrne, J. A. Schmit, and P. E. Johnson, J. Chromatogr. Sci., 9, 592 (1971).

61. M. Novotny, S. L. Bektesh, K. B. Denson, K. Grohmann, and W. Parr, in Advances in Chromatography (A. Zlatkis, ed.), Houston, Texas, 1973, p. 347.

Chapter 4

PHYSICOCHEMICAL MEASUREMENTS USING CHROMATOGRAPHY

David C. Locke

Department of Chemistry
Queens College, CUNY
Flushing, New York

I. INTRODUCTION

Should there be only one area in which chromatographic theory and practice are in fact related, it is the direct application of chromatography to physicochemical measurements. Users of chromatography are aware of the obvious and widespread application of chromatography as an analytical tool, but in addition one can quantitatively relate the observed chromatographic retention and spreading parameters to physical and chemical characteristics of the system. Several of the latter applications are the subject of this chapter. As suggested and demonstrated by Martin and Synge [1], the first chromatographic physical measurements were of liquid-liquid partition coefficients, which were in agreement with static values. Next, Glueckauf [2] in 1945 showed how to determine adsorption isotherms from liquid chromatograms. In gas-liquid chromatography (GLC), the first systematic thermodynamic measurements were probably those of Hoare and Purnell at Cambridge [3], Phillips and co-workers at Oxford [4], and Porter et al. at Shell Oil [5]. Surely, the theory and apparatus are rather more sophisticated nowadays, but even a cursory perusal of the literature indicates increasing interest in this direct use of chromatography by chemical engineers and physical chemists. Application of such information in practical situations requires rapid, reliable methods such as chromatographic ones. By and large, the engineer's interest is the most important testament to the importance of chromatography for physical measurements.

In Table 1 are listed the general types of physical and chemical information available. GLC and liquid-liquid chromatography (LLC) have the widest application in the study of solution thermodynamics. Measurement of solubilities in ordinary solvents as well as in more exotic ones such as liquid crystals is a related area. A correction to the retention volume for vapor phase nonideality is required when exact GLC measurements of solute activity coefficients in the stationary phase are desired. More reliable solution data result, and in addition the correction offers an experimental route to interaction second virial coefficients for the solute vapor-carrier gas mixture. Another type of stationary phase is comprised of polymeric materials, whose sorption propensities for lower-molecular-weight solvents can readily be determined by gas chromatography (GC). Separation of polymers by liquid phase gel permeation chromatography is a powerful technique for polymer characterization. Addition to one or the other phase of substances which interact with some sample components, greatly increases the selectivity of the chromatographic system for analytical purposes. If the cause of this is the formation of a well-defined complex, the stability constant and related parameters can be calculated from chromatographic data. Deleterious as well as beneficial effects result from interfacial adsorption phenomena. Both phase interfacial and solid support adsorption are complicating factors which require correction. In gas-solid chromatography (GSC) and liquid-solid chromatography (LSC), on the other hand, this is the basis for retention which in turn can be studied fundamentally. GSC and LSC provide a variety of surface information: surface areas, heats, and entropies of adsorption, and sorption isotherms. Direct as well as indirect chromatographic methods have been developed for studying the kinetics of reactions occurring on catalytically active surfaces in the column. A rapid and precise route to solute diffusion coefficients in the mobile phase comes from study of peak broadening as a solute band moves through an empty column. Diffusion coefficients in the stationary phase can also be deduced from peak-broadening measurements with packed columns. Determination of miscellaneous physical properties results through empirical correlations between molecular structure and such properties as boiling points and vapor pressures. Yet another physical property, the stationary solvent molecular weight, can be estimated through application of solution theory to chromatography. Whether or not one regards ancillary chromatographic equipment as part of the overall

TABLE 1

Categories of Physicochemical Information Available from Chromatography

Solution thermodynamics
Vapor phase intermolecular interactions
Polymer studies
Complex formation equilibria in solution
Interfacial phenomena
Reaction kinetics
Diffusion coefficients
Miscellaneous physical properties
Use of ancillary chromatographic equipment

chromatographic process, several types of physical information are available from it. Examples are included here because these components are usually an integral part of the chromatograph and are not ordinarily used independently. Listing these varied areas of application should clearly indicate why chromatography is of interest to those using and requiring physical data.

Over the past 15 years the subject has been reviewed a number of times. Various annual reviews of analytical [6] and physical [7-11] chemistry contain some references to this application of GC but are not meant to be comprehensive. Early reviews were written by Hardy and Pollard [12], Purnell [13], and others [14-16]. Young [17], Conder [18], and Korol [19] have brought the field more up to date, especially in the area of solution thermodynamics. A short book on physical measurements using chromatography was published in Russian [20], and one is forthcoming from R. J. Laub and R. L. Pecsok (John Wiley & Co.).

II. INSTRUMENTAL REQUIREMENTS

The chromatographic apparatus required here is basically similar to analytical GC or LC, although for accurate measurements a few refinements are necessary. In particular, it is required that the retention parameber be (a) independent of operating conditions, (b) reproducible, (c) accurately measurable, and (d) quantitatively accounted for by a known mechanism; i.e., the retention volume should reflect, or be correctable to reflect, only solute-bulk stationary phase interaction, or whatever the studied interaction is. Again, in some cases the correction procedure itself leads to useful information, such as adsorption constants or second virial coefficients. In addition, the desired data themselves require control of the system temperature, pressure, and in some cases composition to specified limits.

A. Gas-Liquid Chromatography

The experimental retention volume corrected for the column pressure drop is

$$V_R^0 = t_r F_c J_3^2 \tag{1}$$

where t_r is the solute retention time in the column, F_c is the flow rate at the outlet of the column, and J_3^2 is the Martin and James gas compressibility factor for an ideal carrier gas. One generally measures a distance d_r on the recorder chart which moves at chart speed c'; the column outlet flow rate is measured with a soap bubble flow meter ordinarily at ambient temperature T_a rather than column temperature T; the atmospheric pressure in the flow meter P_a must be corrected for the vapor pressure of soap solution (water) at T_a, p_{H_2O}; and J_3^2 is calculated from the column inlet and outlet pressures, p_i and p_o. Thus,

$$V_R^0 = \frac{d_r}{c'} F_a \frac{T}{T_a} \left(1 - \frac{p_{H_2O}}{P_a}\right) \frac{3}{2} \frac{(p_i/p_o)^2 - 1}{(p_i/p_o)^3 - 1} \tag{2}$$

Based on this equation, Goedert and Guiochon [21-23] made an intensive theoretical study of the sources of error in the measurement of GC retention times, considering errors caused by pressure and temperature fluctuations, thermal gradients, and retention distance measurement on the recorder chart paper. They concluded that to achieve a precision of ±0.01% in retention time, one must stabilize the column pressure drop to ±0.006%, the temperature to ±0.001% (about 0.002° C), and measure the retention distance to ±0.005%. Since temperature gradients of 0.1° C/cm in the column oven and temperature fluctuations of 0.5° to 1° C produce errors in t_r of 0.5 to 2%, unmodified gas chromatographs are generally unsuitable for exact thermodynamic measurements. At the other extreme, several ultrasophisticated systems ("gas chromatographs" would be demeaning as a description) have recently been described [23-26]. These have such features as digital computer control of sample injection, digital data acquisition, temperature control to ±0.005° C, and dedicated computer handling of the data. The most precise of these [26] is reported to allow measurement of retention time reproducible to ±0.02%.

Such precisions are truly remarkable, but in terms of physical measurements, except perhaps differential measurements, are somewhat specious for two reasons. In the first place, one is interested in the net retention time, that of the solute t_r less that of a nonsorbed species t_m. Being close to the point of injection, the "dead time" t_m is always a less precise measurement. Thus, Goedert and Guiochon [23] find, with their apparatus, precision in retention time of benzene on a graphitized carbon of ±0.06%, whereas that of nonsorbed CH_4 is only ±0.3%. This is the reason for computer-controlled sample introduction [24, 26]. An additional problem is choosing a suitable compound to measure the interstitial volume of the column. With thermal conductivity detectors, an air peak is readily seen, and one is generally justified in assuming that the distribution coefficient K^0 for air or a fixed gas is zero. With flame ionization detectors, which are often required to ensure effectively zero sample size, air peaks are not seen so methane is often substituted. However, the assumption of zero solubility of CH_4 in the stationary phase is not always true. Several methods have been devised for locating the dead time with an ionization detector, including extrapolation of the retention times of n-paraffins to that of CH_4 [27], switching carrier gases and using vacancy chromatography [28], and temporarily changing the fuel and air flows to the detector to render it somewhat sensitive to N_2 [29]. The latter two references clearly demonstrate the unreliability of using the retention time of methane for the purpose. This problem in liquid chromatography has not been studied in any detail, although a solvent front is often seen with both the refractive index and UV monitors because of a refractive index change or sometimes because of the passage of a small air bubble. This peak is used to determine t_m.

In the second place, one requires for solution thermodynamic measurements by GLC the exact weight w_3 of stationary phase present in the column. According to Janak et al. [30], uncertainty in w_3 is the largest contribution to error in measuring the specific retention volume, producing an uncertainty of ±0.2% for a packed column and ±0.3% for a capillary column. Although one can accurately weigh stationary liquid phase and solid support, one cannot always be certain that all the liquid is ultimately sorbed onto the support. It is at best tedious to transfer quantitatively the coated support into the column. Most important, it is difficult to account quantitatively for losses of stationary phase during conditioning and operation of the column. In any case (Sec. IV) there is some question as to whether all the liquid functions as bulk solvent, in particular, that liquid deposited in very thin layers on some regions of the solid support, and that filling the very narrow pores of the support. What can be done is to prepare the packing and fill the column as accurately as possible, and after completing the measurements to remove the packing material and analyze a sample gravimetrically by either Soxhlet extraction or pyrolysis. The latter technique requires careful humidity equilibration of the packing before and after pyrolysis. Overall, the uncertainty in w_3 is probably never better than ±10 mg at best, or for a 1-m packed column about ±0.2%.

While the "super GC" apparatus clearly offer great precision, many excellent and reliable physicochemical measurements have been made with rather more modest equipment. It is probably true that most if not all commercial GC apparatus are not amenable to this application without modification in the pressure and temperature controls. A more or less isolated and air-conditioned room is clearly desirable for maintenance of thermal, pressure, and operator equilibrium. Attempts should be made to maintain column temperature to ±0.02° C, with a gradient of 0.2° C or less. Two-stage tank regulators with additional precision pressure controllers (such as Negretti and Zambra or equivalent) are entirely adequate. For high-pressure work or precise second virial coefficient measurements, or both, column outlet pressure control is also required. Flow rate can be measured at the column inlet end of the flow system using calibrated rotameter devices or more precisely with a special soap bubble flow meter [31], or at the outlet of the column with the bubble meter; in any case flow should be measured frequently. Mercury barometers or accurate, calibrated pressure gauges for higher pressures are essential for absolute pressure measurement, and a differential manometer is required to determine the pressure drop. The sample injection system should be immediately upstream of the column; all dead volumes should be held to a minimum. A sensitive detector is required for measurements at effectively infinite dilution. Obviously the stationary phase, gases, and liquid mobile phases should be pure and of high quality, and again the weight of the stationary phase must usually be accurately known. Automatic data acquisition is clearly desirable, if optional. A good example of a gas chromatograph meeting all these requirements for solution and gas phase measurements was described in detail by Cruickshank et al. [31].

For measurements at finite concentrations, or at supercritical pressures, specialized equipment is required [32, 33] which must be assembled from components, since no whole commercial units are available.

B. Liquid Chromatography

The requirements for LC have not as yet been clearly established [34]. Temperature control to ±0.01° C is not a major problem with a liquid bath, but flow control can be. High-pressure LC pumps are either of the constant- (average) flow (e.g., reciprocating piston displacement pump) or constant-pressure (e.g., pneumatic amplifier pump) type [35]. If a small obstruction develops, caused, for example, by a change in viscosity in the mobile phase accompanying a local temperature fluctuation, the pressure will increase to maintain constant flow, or the flow rate will drop with a constant-pressure pump. Constant-flow pumps are probably more desirable for this application since current theory [34] indicates column pressure to be of secondary importance at most. However, constant flow means constant average flow, since with simplex reciprocating-type pumps, the actual flow profile is more or less sinusoidal. Duplexed or triplexed pumps, which are commercially available, go far toward producing a true, steady flow. In addition it is possible to incorporate various pulsation-damping devices into the system based on RC filter principles. Continuous displacement pumps using gas pressure provide steady flow but may present problems with gas dissolution at the pump end and subsequent gas evolution at the low-pressure detector end, producing noise. Piston-drive continuous displacement pumps are generally quite expensive for high pressures, and have a finite capacity.

The biggest apparatus problem in LC is the lack of a sensitive, universal detector. At present the only two detectors in widespread use are the UV monitor, which requires absorbing solutes in UV-transparent solvents, and the refractive index detector, which is generally insensitive and requires close temperature and flow control for stable operation.

Overall, several of the commercially available liquid chromatographs have sample injection systems and flow stability of sufficient quality for physical measurements, and the investigator is probably just as well off using one of these with the required modifications, as assembling an LC from components.

III. LIMITATIONS

A. Gas-Liquid Chromatography

1. Stationary Phase Volatility

For highest accuracy, GLC is best restricted to volatile solutes and nonvolatile (vapor pressure less than 0.1 mm Hg at column temperature) solvents. There has been some successful work done with volatile stationary phases and presaturated carrier gases [36-38], but because there are a number of corrections required here, lower accuracies are to be expected; more thorough studies are needed. This requirement of stationary phase nonvolatility restricts both the types of stationary phase amenable to study and the temperature range one can use.

2. Solid Support Interactions

The solid supports ordinarily used for GLC are diatomaceous earth products having siliceous surfaces which are able to adsorb polar vapors. When coated with nonpolar stationary phases, exposed patches as well as the liquid-solid interface are accessible to polar solute vapors, producing the tailed peaks indicative of a mixed retention mechanism. Treatment with silanizing reagents such as hexamethyldisilizane is generally recommended to reduce surface activity. Although powdered Teflon® has the lowest surface energy of any support material, it has been shown to be not totally inert in this regard [39].

3. Peak Asymmetry

Peaks should be ostensibly symmetric, although even at infinite dilution they are not truly so because of the generally anti-Langmuir form of the sorption isotherm [40]. Peak skew values outside the range of 0.8 to 1.2 could be the result of excess sample size, adsorption at a solid or liquid interface (which are usually flow rate or sample size dependent and can thus be tested for), or poor sample introduction. A tailed peak for a system supposedly at infinite dilution is to be regarded as a danger signal and requires further investigation.

4. Effect of Spreading the Stationary Liquid on a Solid Support

By inference the effect on the bulk solvent properties of spreading a liquid on the solid support has been regarded as trivial except for lightly loaded columns. Agreement between static and chromatographic measurements has been within experimental error in many systems. Using static methods, Ashworth and Everett [41] found virtually identical sorption isotherms when the liquid was spread on Celite or studied in bulk form; however, the rate of attainment of equilibrium was much faster in the former case. Similarly, Freeguard and Stock [42] found the same increase in the rate of equilibration and no effect on the isotherm with Celite solid support, but found adsorption of halocarbons to be noticeable at low liquid loadings on Firebrick. Pecsok and Gump [43] also found adsorptive effects with alcohols on Chromosorb P. In any case, the effect on bulk liquid properties has not heretofore been considered a problem.

Recent work by Serpinet [44-46], while not disputing these observations, has made careful retention volume measurements of soluble and insoluble compounds over a range of temperatures bracketing the normal melting point, on columns of various liquid loadings. A dramatic effect on the bulk properties of the liquid phase caused by the solid support was clear. Three effects are apparently at work.

1. Nonuniform coating of the support leaves substantial areas of the adsorptive solid support surface exposed.
2. The adsorptive properties of the support have an orientating effect on the liquid phase, producing a situation in the solution one could envision as a solute-stationary liquid-support ternary system of properties different from the desired solute-solvent binary.
3. Solute vapors dissolved in stationary liquid located in fine pores of capillary dimension have abnormally low vapor pressure and are thus retained longer than they would be otherwise, an effect called Kelvin retention [46].

Again, the precise significance of Serpinet's work in the present context is not absolutely clear since in many systems accurately studied both by chromatographic and static means, such effects (except adsorption) have not been perceptible [31]. In any case, columns with at least 15 to 20% liquid loading on deactivated supports should be used. Effects such as these are probably more important in solute-solvent systems of diverse polarity. These are subject to a variety of complications, such as that discussed in the next section. Consequently, the Kelvin retention may not provide a contribution perceptible above the increased error in measurements corrected for other effects.

5. Liquid Interfacial Adsorption

Polar solutes are known [47] (Sec. IX) to adsorb in some cases on the gas-liquid interface with both polar and nonpolar stationary phases. The contribution of this phenomenon to retention is most important at lower loadings, where the liquid surface area/volume ratio is highest. Where the possibility of such adsorption exists, a systematic study of the effect of variation of liquid loading and sample size on retention should be made. Methods exist for correction, as discussed below; however, they not only add greatly to the experimental burden, but the corrective procedures detract appreciably from the ultimate accuracy of the results.

6. Finite Sample Concentration

A principal advantage of the GLC method is that data can be obtained directly at infinite dilution. This condition, which is met when dissolved solute molecules interact only with solvent molecules, being too far dispersed to interact with each other, requires the use of very small samples. For example, it has been calculated [31] that with a nonpolar column of 20% loading, for limiting activity coefficients in the range of 0.5 to 1.5, samples less than about a micromole are required, i.e., less than about 0.2 μl of liquid sample. This is no problem for the usual GC thermal conductivity detectors, which can see even less than this volume of vaporized sample. In some cases where linearity of the sorption isotherm obtains at still smaller sample sizes, ionization detectors are required which present the problem of determining the true void volume of the column. Another aspect of larger samples is the local temperature changes accompanying sorption and desorption of the moving solute band [31, 48]. Chromatographic methods for studying solutions at finite concentrations are discussed (Sec. V.B.).

7. Chromatographic Nonequilibrium

Chromatography is intrinsically a nonequilibrium process [49]. Thus, sharp, square wave solute input distributions become the normally nearly Gaussian, spread-out peaks. Even so, at least on an empirical basis, calculations using the peak maximum solute retention volume produce thermodynamic quantities generally in good agreement with statically determined (i.e., equilibrium) values. Theoretical calculations [50], however, indicate a deviation of the peak maximum retention volume from the chromatographic hypothetical equilibrium value as stationary phase mass transfer limitations become significant. When very precise measurements are made, this effect is indeed observable [31, 51, 52]. It has thus been recommended [52, 53] that for highest accuracy, peak maximum retention volumes be measured over a range of flow rates on the rising part of the van Deemter plot and extrapolated to zero flow rate. This should in principle produce an effective equilibrium value, even if at the expense of increasing the experimental task. Whether it is necessary in many situations is another matter, since the effect will be imperceptible when precisions in derived activity coefficients of the order of 1% are adequate. In any case, the peak maximum retention volume may not be the best retention parameter to use. Indeed, the so-called first statistical moment or center of gravity of the peak depends only on the equilibrium distribution coefficient and the longitudinal solute diffusion coefficient, and is independent of mass transfer effects [54]. The problem is to find it experimentally, if possible without the aid of a computer [55]. J. R. Conder (private communication) informs me that Hicks has shown [56] that this retention volume can in fact be found by drawing tangents at the points of inflection on both sides of the peak, extrapolating these to the baseline, and finding the midpoint. This value should a flow-independent retention volume.

B. Liquid Chromatography

Many of the same limitations presumably exist in LC although none has been well studied. The theory of LLC [34] is not as well worked out as that of GLC. The problem, of course, is not stationary phase volatility but in LLC is rather the matter of mutual phase miscibility [34, 57]. Clearly, if one is interested in solution properties with a pure solvent, even partial phase miscibility will produce a whole new system.

Interfacial and support adsorption exist, as discussed below, and although presumably analogous methods for handling these phenomena will eventually be developed they do not exist at present. Similarly, sample size effects and chromatographic nonideality have not been studied.

Again, the most severe immediate limitation in LC for any application is the lack of a sensitive, universal detector. It is far more difficult sensitively to detect a liquid in a liquid than a vapor in a gas.

IV. ADVANTAGES OF CHROMATOGRAPHY

While chromatography has limitations for this application, it has several unique advantages over conventional techniques. Generally, chromatographic methods are more rapid than static ones, since the rate of equilibration is greater. The results are generally as accurate as those of conventional methods if no corrections are required for interfacial adsorption or carrier gas solubility. For infinite dilution measurements, GLC is probably more accurate than gravimetric methods because no extrapolations are required. The apparatus need not be overly sophisticated, and the one chromatograph gives several quite different types of physical and chemical information. In a single experiment (i.e., one injection of a multicomponent sample) one can make several simultaneous measurements. Because chromatography is a separation technique, absolutely pure samples are not always required, and in any case only a few milligrams are needed, so rare, expensive, or research compounds can be characterized. Likewise only a few grams of solvents are required for the stationary phase in GLC or LLC (larger quantitites of pure eluent are required), or a few grams of an adsorbent. GC experiments are conducted in an inert atmosphere, and LC can also be performed in an entirely closed system.

V. SOLUTION THERMODYNAMICS

A. Gas-Liquid Chromatography at Infinite Dilution

1. Limiting Activity Coefficients

According to chromatographic theory, the net retention volume V_N is related to the solute distribution coefficient K^0 according to

$$V_N = V_R^0 - V_2 = K^0 V_3 \tag{3}$$

where the retention volume V_R^0 and the gas hold-up volume V_2 are both corrected for the column pressure drop as in Eq. (2),

$$K^0 = \frac{\text{solute concentration in stationary phase}}{\text{solute concentration in mobile phase}} \tag{4}$$

and V_3 is the volume of stationary liquid phase in the column. It turns out that K^0 corresponds to a zero mean column pressure, equilibrium value.

From simple Raoult's law considerations, it can be shown (cf. Young [17]) that

$$V_N = \frac{RTn_3}{\gamma_{13}^{\infty} p_1^0} \tag{5}$$

where n_3 is the number of moles of stationary liquid phase present in the column, and γ_{13}^{∞} and p_1^0 are, respectively, the solute infinite dilution activity coefficient in the stationary phase and its saturation vapor pressure at the column temperature T. Unless otherwise noted, hereafter subscript 1 refers to the solute, 2 to the carrier gas, and 3 to the stationary liquid solvent.

Alternatively, the specific retention volume V_g^0 is often used [50], which is the net retention volume per gram of stationary liquid, corrected to 0° C, according to

$$V_g^0 = \frac{273.2 V_N}{T w_3} \tag{6}$$

from which it can be shown that

$$V_g^0 = \frac{273.2R}{\gamma_{13}^{\infty} p_1^0 M_3} \tag{7}$$

The considerations leading to these equations ignore the nonideality of the (a) solute vapors, (b) solute vapor-carrier gas mixture, and (c) carrier gas. In addition, the use of these equations to determine values of γ_{13}^{∞} implies the absence of both stationary phase mass transfer resistance (i.e., ideal chromatography) and of carrier gas solubility in the stationary phase. The conditions described above must also be met, i.e., effectively zero sample size, absence of competing retention mechanisms (solute adsorption at an interface), and no alteration of the physical properties of the bulk stationary solvent resulting from its being spread on an inert support.

Some authors have made partial corrections by substituting the solute fugacity in Eq. (5) or (7) for its vapor pressure. This procedure ignores interactions between carrier gas and solute vapor, and does not properly correct the measured retention volume for the column pressure drop. Desty et al. [58] and Everett and Stoddart [53] were the first to consider, if incompletely, the carrier gas-solute interactions. The correction turns out to be small, of the order of 1%, and in many cases may be obscured by experimental error. Thus, the fugacity correction is often justified.

By far the most sophisticated and thorough approach to the study of GLC solutions and mobile phase mixtures has been made by Cruickshank and associates at the University of Bristol [52, 59]. All of the problems mentioned above have been considered. Although the exact theory does not lead to a simple, explicit equation for nonideal carrier gases, approximations to it produce tractable methods for the accurate analysis of experimental data.

In its most general form [52, 59], the net retention volume is related to properties of the system according to

$$\ln V_N + f(a, b, c) = \ln K^0 V_3 + \beta' p_o J_3^4 + \xi' p_o^2 J_3^5 + \cdots \tag{8}$$

where

$$f(a, b, c) = \ln \frac{[1 + bp_o(J_3^4 - J_2^3) + ap_o(J_3^4 - J_2^3) - b^2 p_o^2 (J_4^6 - J_3^5) + cp_o^2 (2J_4^6 - J_3^5)]}{1 + bp_o + cp_o^2}$$

$$\cong \ln\left(\frac{1 + bp_o J_4^5 + cp_o^2 J_4^6}{1 + bp_o + cp_o^2}\right)$$

$$\cong \ln\left(\frac{1 + bp_o J_2^3}{1 + bp_o}\right) \tag{9}$$

for fixed gases. This term correctly accounts for the compressibility of a nonideal carrier gas and enables one to calculate a truly compressibility-corrected net retention volume from the measured retention time. In this equation, p_o is the carrier gas pressure at the column outlet; $b = B_{22}/RT$, where B_{22} is the second virial coefficient of the pure carrier gas; a is related to the pressure variation of the carrier gas viscosity and is approximately 0.175b; c is a second-order gas imperfection term, $c = (C_{222} - B_{22}^2)/(RT)^2$, where C_{222} is the third virial coefficient of the carrier gas, accounting for three-body collisions among carrier gas molecules at high pressures; and

$$J_n^m = \frac{n(p_i/p_o)^m - 1}{m(p_i/p_o)^n - 1} \tag{10}$$

K^0V_3 is the zero-pressure value of the net retention volume, and is related to the infinite dilution solute activity coefficient by

$$K^0V_3 = \frac{RTn_3}{p_1^0 \gamma_{13}^{\infty}} \exp[\alpha(p_1^0)] \tag{11}$$

where V_3, n_3, and γ_{13}^{∞} have the same meaning as above, and

$$\alpha(p_1^0) = \frac{v_1^0 - B_{11}}{RT} p_1^0 + \frac{C_{111} - B_{11}^2}{(RT)^2} (p_1^0)^2 + \cdots \tag{12}$$

which accounts for the imperfection of the solute vapors. B_{11} and C_{111} are, respectively, the second and third virial coefficients of the solute vapor, p_1^0 is its vapor pressure, and v_1^0 is its molar volume.

The carrier gas-solute interaction is accounted for by the remaining two terms on the right-hand side of Eq. (8), in which the effect of carrier gas solubility is taken into account. Here,

$$\beta' = \frac{2B_{12} - v_{13}^{\infty}}{RT} + \lambda \left[1 - \left(\frac{\partial \ln \gamma_1^{\infty}}{\partial x_2}\right)_0\right] \tag{13}$$

and

$$\zeta' = \frac{3C_{122} - 4B_{12}B_{22}}{2(RT)^2} + \phi \left[1 - \left(\frac{\partial \ln \gamma_1^{\infty}}{\partial x_2}\right)_0\right] + \frac{\lambda^2}{2}\left[1 - \left(\frac{\partial^2 \ln \gamma_1^{\infty}}{\partial x_2^2}\right)_0\right] + \frac{\kappa' v_1^{\infty}}{RT} \tag{14}$$

v_{13}^{∞} is the infinite dilution partial molar volume of solute in the pure solvent; γ_1^{∞} is the limiting solute activity coefficient in the hypothetical mixture of solvent and dissolved carrier gas at the zero-pressure state; x_2 is the mole fraction of dissolved carrier gas, and λ and ϕ are coefficients in the carrier gas solubility-pressure series

$$x_2 = \lambda P + \phi P^2 + \psi P^3 + \cdots \tag{15}$$

and are zero when the carrier gas is insoluble. The last term in Eq. (14) is a stationary phase compressibility contribution.

Although it is apparent that Eqs. (5) and (7) are oversimplifications (although still frequently used), Eq. (8) is rather overcomplicated, to say the least, for many systems of interest. At low column pressures, it is clearly unnecessary to use all of Eq. (8) to determine solute activity coefficients. For $p_o < 15$ atm, and $p_i < 2$ atm, which situation Cruickshank et al. [51] designate medium-pressure GLC, f(a, b, c) contributes less than -0.001 to $\ln V_N$ for most common carrier gases. At mean column pressures of less than about 20 atm, the last term in Eq. (8) is undetectable. Thus, to a very good approximation, in most practical GLC situations the following equation is an adequate description:

$$\ln V_N = \ln K^0 V_3 + \beta' p_o J_3^4 \tag{16}$$

and, for an effectively insoluble carrier gas,

$$\ln V_N = \ln K^0 V_3 + \beta p_o J_3^4 \tag{17}$$

where β differs from β' in that the carrier gas solubility contribution in Eq. (13) is zero. Generally, v_{13}^{∞} is taken to be v_1^0 because of ignorance of the former in most cases. Equation (17) differs from the equation presented by Desty et al. [58] in the use of J_3^4 rather than J_2^3.

One or the other of these equations can be used directly if the required virial coefficients are accurately and independently known, which is usually not the case, or are calculable from corresponding states theory, which to within experimental error can often be done, according to Conder and Langer [60]. Using the procedure of Cruickshank et al. gives both γ_{13}^{∞} and B_{12}: fully corrected values of γ_{13}^{∞} are obtained by measuring V_N over a range of p_o at fixed small $p_i - p_o$. The zero-pressure intercept of the resulting ln V_N versus $p_o J_3^4$ plot is ln V_N^0, from which γ_{13}^{∞} is calculated. The slope of this plot provides β' or β and thus for an effectively insoluble carrier gas a value for B_{12}, the interaction second virial coefficient for solute vapor in carrier gas.

An indication of the wide range of solute-solvent systems that have been studied by GLC is apparent in Table 2. This list is more or less complete (my apologies to overlooked authors); some work is obviously more reliable than others, but the point of this chapter is not to comment on the data themselves but rather on the use of GLC for this purpose. In some cases, as noted, only specific retention volumes are given, but from these a fugacity-corrected γ_{13}^{∞} can be calculated. The type of correction, if any, is given. "No" correction means Eq. (5) or (7) was used; "fugacity" means the activity coefficient was calculated from these equations using the pure solute vapor fugacity rather than the vapor pressure, calculated from

$$\ln f_1^0 = \ln p_1^0 + \frac{p_1^0 B_{11}}{RT} \tag{18}$$

"Everett" indicates the use of an Everett and Stoddart type of correction [53]. "Yes," or fully corrected, means Eq. (8) (or some part of it) or one equivalent to it was used. In some cases, the infinite dilution value was extrapolated from higher concentrations, as indicated by "extrap."

What is the precision of GLC measurement of γ_{13}^{∞}? Cruickshank et al. [52, 59] estimate the experimental error in γ_{13}^{∞} to be of the order of $\pm 0.1\%$ to $\pm 1\%$. Martire et al. [101] estimate an uncertainty of $\pm 0.7\%$ in their activity coefficients. Similar estimates are given by others. In general, I think one can anticipate 1 to 2% precision when no corrections for interfacial adsorption are required, i.e., in the simplest cases. Better precisions refer to a single operator with a single, high-precision chromatograph, over a relatively short period of time with a new column. When

TABLE 2

Summary of Limiting Activity Coefficients Measured by Gas-Liquid Chromatography

Stationary phase	Solutes	Temperature range (° C)	Corrections	Ref.	Comments
Squalane	Five hydrocarbons	80-105	No	5	Give K
	15 hydrocarbons	80-135	No	36	
	16 hydrocarbons and nitro compounds	80-139	No	61	
	n-Butane-n-nonane	25-75	No	62	Give V_g^0
	Chlorides of Sn, Ti, Nb, and Ta	100-200	No	63	Give V_g^0
	27 hydrocarbons	20	Fugacity	64	
	57 organics	80	No	65	Give V_g^0
	CCl_4, $CHCl_3$, CH_2Cl_2	0-50	Yes	66	Static
	39 hydrocarbons and halo-carbons	53-94	No	67	
	Five C_6-C_8 hydrocarbons	30	Yes	68	Used $V_{N_{initial}}$
	n-Hexane	30	Yes	69	Static
	Six alkanes, benzene	25-60	Yes	31	
	35 C_5-C_{12} alkanes	80-120	No	70	Give V_g^0
	Four polar compounds	30-50	Yes	43	Static

	Five aromatic hydrocarbons	70	Yes	71	
	Benzene, n-octane, isooctane	50, 65	Yes	72	
	Six C_5-C_7 hydrocarbons	78	No	73	
	27 polar organics	80	Fugacity	74	
	n-Hexane	30	Yes	75	Extrap
	Five alcohols	50-70	Yes	76	
	n-Pentane, n-hexane, benzene	30-55	No	30	Give V_g^0
	10 chloroalkanes	30-60	Everett	77	Elution, frontal, static
	10 hydrocarbons	80	No	78	
	Benzene, n-hexane	60	Yes	32	Extrap
	Six chloroalkanes	0-55	Yes	79	Static
	Seven organics	50	No	80	Give V_g^0; static
	n-Pentane, n-hexane, benzene	30	Everett	81	
	Four aromatic hydrocarbons	50-80	Fugacity	82	
	C_5-C_9 n-alkanes	60-160	No	83	
	Six hydrocarbons	30	Yes	84	Static
Hexadecane	18 organics	25-40	No	36	
	15 hydrocarbons	30-60	No	36	

TABLE 2 (Continued)

Stationary phase	Solutes	Temperature range (° C)	Corrections	Ref.	Comments
Hexadecane (Continued)	Three fluorocarbons	30-80	No	85	
	39 organics	40-60	No	86	Give V_g^0
	n-Hexane	20-60	Yes	31	
	15 hydrocarbons	40-90	No	87	
	Benzene	20-50	Yes	88	
	n-Butane-n-heptane	30	Yes	89	
	n-Hexane, cyclohexane	50	Yes	32	
	n-Hexane, benzene	50	Yes	81	
	H_2, CH_4, ethane	22-202	Yes	90	Give Henry's constants; static
	n-Hexane	20-60	No	91	Extrap
	n-Hexane	30	Yes	92	Extrap
Heptadecane	Hydrocarbons, alcohols, amines	22-80	No	93	
	n-Hexane, 2-propanol	50	No	94	Extrap
	n-Hexane, 2-propanol, acetone	40-80	No	91	Extrap

Octadecane	$SnCl_4$ and $TiCl_4$	100-200	No	63	Give V_g^0
	Benzene, n-octane, isooctane	40, 50	Yes	72	
	Benzene	32-50	Yes	88	
	C_4-C_8 hydrocarbons	35	Yes	95	
	27 polar organics	80	Fugacity	74	
	Five C_1-C_3 hydrocarbons	35-200	No	96	Give Henry's constants
	Benzene, right fluorobenzenes	32-50	Yes	51	
	Hexene-1, heptene-1, octene-1, dienes	35-60	Yes	97	
	16 hydrocarbons and chloro compounds	30-60	No	98	Give V_g^0
	Seven hydrocarbons	60	No	99	Give V_g^0
n-$C_{20}H_{42}$	30 hydrocarbons and chloro-alkanes	53-94	No	67	
	n-Pentane, n-hexane, n-heptane	40	Yes	89	
	Five C_1-C_3 hydrocarbons	35-200	No	96	Give Henry's constants
n-$C_{22}H_{46}$	Benzene	50-68	Yes	88	
	Five C_1-C_3 hydrocarbons	35-200	No	96	Give Henry's constants

TABLE 2 (Continued)

Stationary phase	Solutes	Temperature range (° C)	Corrections	Ref.	Comments
n-$C_{22}H_{46}$ (Continued)	n-Butane, n-pentane, n-hexane	50, 60	Yes	89	
	Three hydrocarbons, hexafluorobenzene	35	Yes	100	
n-$C_{24}H_{50}$	15 hydrocarbons	60-105	No	36	
	Eight organics	20-65	Yes	100	
	Benzene	50-65	Yes	88	
	Hexene-1, heptene-1	60	Yes	98	
	23 hydrocarbons	76-88	Fugacity	101	
	11 haloalkanes	76-88	Fugacity	101a	
	18 polar organics	60	Fugacity	102	
n-$C_{28}H_{58}$	35 C_5-C_{12} hydrocarbons	80-120	No	70	Give V_g^0
	Benzene	50-75	Yes	88	
	n-Pentane, n-hexane	70	Yes	89	
n-$C_{30}H_{62}$	23 hydrocarbons	76-88	Fugacity	101	
	11 haloalkanes	76-88	Fugacity	101a	
	Methanol	30-70	No	91	Extrap

n-$C_{32}H_{66}$	35 C_5-C_{12} alkanes	80-120	No	70	Give V_g^0
	Three C_6-C_8 alkanes, hexafluorobenzene	75	Yes	100	
	Benzene	50, 75	Yes	88	
n-$C_{35}H_{72}$	15 hydrocarbons	80-105	No	36	
n-$C_{36}H_{74}$	35 C_5-C_{12} hydrocarbons	80-120	No	70	Give V_g^0
	23 hydrocarbons	76-88	Fugacity	101	
	11 haloalkanes	76-88	Fugacity	101a	
n-$C_{10}H_{22}$, satd. with methane	Ethane, propane, butane	-30 to +20	Fugacity	103	
n-$C_{16}H_{34}$, satd. with methane	Propane	20	Fugacity	103	
Hexadecene-1	39 organics	40-60	No	86	Give V_g^0
Squalene	27 polar organics	80	Fugacity	74	
Bicyclohexyl, diphenylmethane	H_2, CH_4, ethane	22-202	Fugacity	90	Give Henry's constants; static
Five n-alkylbenzenes	C_5-C_8 hydrocarbons	32-50	Yes	104	

TABLE 2 (Continued)

Stationary phase	Solutes	Temperature range (° C)	Corrections	Ref.	Comments
Benzyldiphenyl	32 aromatics and halo-aromatics	100	No	105	
	39 hydrocarbons and halo-alkanes	53-94	No	67	
	27 polar organics	80	Fugacity	74	
	Xylenes	105	No	30	Give V_g^0
Diphenyl, m-terphenyl, o-terphenyl	Benzene	90	Fugacity	106	
1-Methylnaphthalene	27 hydrocarbons	20	Fugacity	64	
Apiezon M	Various organics	100	Yes	107	Used manufacturer's MW
1,2,4-Trichlorobenzene	12 hydrocarbons	30	No	36	
1-Chloronaphthalene	27 hydrocarbons	20	Fugacity	64	
1-Bromo- and 1-chlorohexadecane	39 organics	40-60	No	86	Give V_g^0

1,3,5-Trichloro-benzene	n-Hexane	30	Fugacity	108	Extrap
1-Chlorooctadecane	Seven hydrocarbons	60	No	99	Give V_g^0
1-Hexadecanol	39 organics	60-80	No	86	Give V_g^0
	39 hydrocarbons and chloro-alkanes	53-94	No	109	Give V_g^0
1-Dodecanol	57 organics	80	No	109	Give V_g^0
1- and 2-Dodecanol	10 alcohols	56, 80	No	110	Give V_g^0
1-Octadecanol	Seven hydrocarbons	60	No	99	Give V_g^0
Glycerol	C_1-C_4 alcohols	62	Fugacity	111	
	Benzene	50	Yes	59	
Diglycerol	Six alcohols	111	No	61	
	14 organics	80	No	65	Give V_g^0
Diethyleneglycol	Benzene	50-90	Fugacity	108	Extrap
Diethylene glycol, triethylene glycol, tetraethylene glycol	14 C_6-C_{12} hydrocarbons	25-70	Yes	112	
Carbowax 400	11 C_6-C_{10} hydrocarbons	58-88	No	73	

TABLE 2 (Continued)

Stationary phase	Solutes	Temperature range (° C)	Corrections	Ref.	Comments
Carbowax 750	10 hydrocarbons	80	No	78	
Furfuryl alcohol, tetrahydrofurfuryl alcohol	11 hydrocarbons	27	No	113	
Furfuraldehyde	Five C_4 hydrocarbons	-20 to +40	No	114	
Di-n-octyl ketone	Hydrocarbons, alcohols	50-80	No	93	
	2-Propanol, n-hexane	50	No	94	Extrap
	2-Propanol, n-hexane, acetone	50	No	91	Extrap
Di-n-nonyl ketone	18 polar organics	60	Fugacity	102	
Di-n-octyl ether	12 haloalkanes	26-77	No	115	
	Hydrocarbons, alcohols, amines	22-50	No	93	
	16 hydrocarbons and haloalkanes	30-60	No	98	Give V_g^0
	2-Propanol, n-hexane	50	No	94	Extrap
	2-Propanol, n-hexane, acetone	50	No	91	Extrap

Di-n-octyl thioether	16 hydrocarbons and haloalkanes	30-60	No	98	Give V_g^0
Tri-n-hexylamine, di-n-octylmethylamine	15 hydrocarbons and haloalkanes	30-60	No	116	Give V_g^0
Furfurylamine	11 hydrocarbons	27	No	113	
Triethanolamine	10 hydrocarbons	80	No	78	
a-Naphthylamine	Two hydrocarbons	80	No	117	
	n-Nonane	65.5	Everett	81	
Aniline	27 hydrocarbons	20	Fugacity	64	
7,8-Benzoquinoline	32 aromatics and haloaromatics	100	No	105	
Isoquinoline, quinoline, 2-methylquinoline	27 hydrocarbons	20	Fugacity	64	
N,N-Dimethylmyristamide	39 hydrocarbons and haloalkanes	53-94	No	67	
Nine amides	Eight C_5-C_6 hydrocarbons	24-100	No	118	
Palmitonitrile	39 organics	40-60	No	86	Give V_g^0

TABLE 2 (Continued)

Stationary phase	Solutes	Temperature range (° C)	Corrections	Ref.	Comments
Oxydipropionitrile	10 hydrocarbons	80	No	78	
	C_1-C_4 alcohols	60	Fugacity	101	
	Seven organics	50	No	80	Give V_g^0; static
	Six aromatic hydrocarbons	60	Yes	119	
Thiodipropionitrile	Benzene, cyclohexane	30	Yes	120, 69	Static
	15 organics	25	Fugacity	111	
	Five alkylbenzenes	60	Yes	121	
	Six aromatic hydrocarbons	60	Yes	119	
Dinonylphthalate	18 halocarbons	19-98	No	115	
	16 hydrocarbons and nitro compounds	80-111	No	61	
	Seven hydrocarbons	20-40	Yes	41	Static
	Benzene, cyclohexane	48-110	Fugacity	122	
	Eight hydrocarbons	30	Everett	53	
	CCl_4, ethanol	20-45	Everett	123	

Dinonylphthalate (Continued)	57 organics	80	No	65	Give V_g^0
	Four hydrocarbons, ethanol	40-80	No	124	
	12 hydrocarbons	20	Fugacity	125a	
	CCl_4, $CHCl_3$, CH_2Cl_2	0-50	Yes	66	Static
	39 hydrocarbons and halo-alkanes	53-94	No	67	
	Six alkenes	20-50	Yes	31	
	Benzene	50	Yes	72	
	n-Heptane	30	Yes	75	Extrap
	10 chlorocarbons	30-60	Yes	77	Extrap, static
	Six chloroalkanes	0-55	Yes	79	Static
	12 hydrocarbons	60	Yes	125	
	Six hydrocarbons	30	Yes	84	Static
Dibutylphthalate	57 organics	80	No	65	Give V_g^0
Diisodecylphthalate	16 organics	75-135	No	5	Give K values
Di-n-propyltetra-chlorophthalate	34 hydrocarbons and halo-alkanes	100	No	126	
Di-n-butyltetra-chlorophthalate	39 hydrocarbons and halo-alkanes	53-94	No	67	

TABLE 2 (Continued)

Stationary phase	Solutes	Temperature range (° C)	Corrections	Ref.	Comments
Nine alkyl phthalates	Six aromatic hydrocarbons	120-140	No	127, 128	
Five alkyl phthalates	Five alkyl benzenes	60	Yes	121	
10 alkyl phthalates	Six aromatic hydrocarbons	60	Yes	119	
Four dialkyltetra-chlorophthalates	36 hydrocarbons and halo-alkanes	100, 110	No	129	
n-Butyl maleate	Five alkyl benzenes	60	Yes	121	
Four n-butyl esters of aliphatic diacids	Six aromatic hydrocarbons	60	Yes	119	
Three n-butyl esters of aliphatic diacids	Six aromatic hydrocarbons	40-60	Fugacity	130	
Di-(2-ethylhexyl)-sebacate	Four hydrocarbons	55	No	131	Solubility
Tricresylphosphate	Nine organics	80	No	65	Give V_g^0
Tritolylphosphate	11 chloroalkanes	40-146	No	132	

Tris(2,4,-dimethylphenyl phosphate	10 hydrocarbons	80	No	78	
Cobalt stearate	Eight organics	156	No	133	
Carbon tetrabromide	Six aromatic hydrocarbons	94-124	No	134	
1.77 N $AgNO_3$ in ethylene glycol	16 olefins	30	No	135	
1,3,5-Trinitrobenzene in dinonylphthalate	12 hydrocarbons	60	Yes	125	
16 Polar binaries	Six hydrocarbons	25, 34	No	136	
Di-ω-H-octafluoropentyl ester of tetranitrodiphenic acid; tetrameric bis(1,1,9-trihydrohexafluorononyl) phosphonitrilate; Apiezon L; and polyethylene glycolsuccinate	60 organics	100, 150	No	137	Give V_g^0 based on initial retention time
Solutions of uranyl nitrate, tributyl phosphate + dodecane, and dodecane + tributyl phosphate	13 organics, used as solvent extraction diluents	25-120	No	138	

TABLE 2 (Continued)

Stationary phase	Solutes	Temperature range (° C)	Corrections	Ref.	Comments
Dibutylphosphoric acid	11 organics	25-80	No	139	
Diphenylmethyl phosphate	11 organics	25-80	No	140	
17 polar solvents	Seven hydrocarbons	25, 34	No	141	
19 polar solvents	Nine hydrocarbons	25-70	No	142	Give Henry's constants
Water	CH_2Cl_2, $CHCl_3$, CCl_4, methanol	25	No	115	
	18 organics	20-40	No	109	
	Eight polar organics	25	No	143	
	Chlorohydrocarbons	12.5	No	38	
2,3,4-Trimethylpentane; 2-pentanone	18 organics	20-40	No	109	
1-Propanol	Benzene	40	Yes	37	Did not correct for changing stationary phase composition down the column
	Toluene, 1,2-dichloroethane	65	Yes	37	

Toluene	1,2,-Dichloroethane, 1-propanol	65	Yes	37	Did not correct for changing stationary phase composition down the column
1,2-Dichloroethane	1-Propanol, toluene	30-50	Yes	37	Did not correct for changing stationary phase composition down the column
Cholesterylmyristate	Seven hydrocarbons	70-89	Fugacity	144	
p-Azoxyanisole; 4,4'-dihexoxyazoxybenzene	Nine hydrocarbons	91-141	No	145	Give V_g^0
4,4'-Dihexoxyazoxybenzene	m-p- isomers of divinyl benzene and ethylvinyl benzene	96-155	Fugacity	146	
4,4'-Dihexoxyazoxybenzene; 4,4'-dimethoxyazoxybenzene	42 hydrocarbons and chlorinated hydrocarbons	115-150	Fugacity	147	Give plots of $\ln \gamma_f^{\infty}$ vs. $T^{-1} \infty$
4,4'-Diheptoxyazoxybenzene; p-(p-ethoxyphenylazo)phenyl undecylenate	Nine hydrocarbons	80-132	Fugacity	148	

TABLE 2 (Continued)

Stationary phase	Solutes	Temperature range (°C)	Corrections	Ref.	Comments
4,4'-Dihexoxyazoxybenzene	42 hydrocarbons and haloalkanes	88-126	Fugacity	149	
p-Azoxyanisole	42 hydrocarbons and haloalkanes	120-197	Fugacity	149	
4,4'-Dihexoxyazoxybenzene	n-Heptane	90.1	Yes	150	Static[a]

[a]See Chow and Martire [147].

complicating factors requiring correction, such as adsorption, exist, the precision is reduced to probably ±5% at best. Even so, GLC is clearly competitive with static methods from this point of view, and is very well capable of providing data sufficiently reliable for comparison with theoretical calculations, or for most practical applications. Some idea of how well GLC measurements agree with themselves and with statically determined values (McBain balance methods, generally) can be obtained from Tables 3 and 4. One is led to wonder from these tables whether the relatively complex experimental approach necessitated by the use of Eq. (8) or even Eq. (17) is, in practical terms, worth it. Clearly, when one is interested in determining both second virial coefficients and activity coefficients of the highest accuracy, one must. But for many practical engineering applications (e.g., see Brandani [151] and Tassios [152]), the use of only a fugacity correction is probably well justified.

TABLE 3

Comparison of γ_{13}^{∞} Values by Gas Chromatography and Static Isotherm (^{S}) Techniques

Stationary phase	Solute	Temp. (°C)	γ_{13}^{∞}	Ref.
Squalane	n-Pentane	25	0.655	31
		30	0.641^{S}	68
		30	0.640^{S}	41
		30	0.620^{S}	84
		30	0.658	81
		60	0.704	83
		78	0.53	73
		80	0.701	83
	2,2-dimethyl-butane	25	0.697	31
		25	0.692^{S}	31
		30	0.690	31
		30	0.690^{S}	41
		30	0.675	68

TABLE 3 (Continued)

Stationary phase	Solute	Temp. (° C)	γ_{13}^{∞}	Ref.
Squalane (Continued)	Cyclohexane	30	0.536	31
		30	0.538^{s}	41
		30	0.510^{s}	84
		30	0.511	68
		30	0.511^{s}	68
	n-Hexane	25	0.684	31
		25	0.650^{s}	31
		30	0.687	31
		30	0.649^{s}	31
		30	0.646	68
		30	0.681	81
		30	0.640	75
		30	0.640^{s}	84
		30	0.658^{s}	69
		50	0.689	31
		50	0.649^{s}	31
		60	0.689	31
		60	0.711^{s}	31
		60	0.776	32
		60	0.772	83
		78	0.58	73
		80	0.72	83
	n-Heptane	30	0.681	68
		30	0.669^{s}	84
		60	0.736	83
		78	0.61	73

TABLE 3 (Continued)

Stationary phase	Solute	Temp.	γ_{13}^{∞}	Ref.
Squalane (Continued)	n-Heptane (Continued)	80	0.649	78
		80	0.736	83
	Benzene	30	0.681	68
		30	0.733	81
		30	0.698^{s}	84
		30	0.697^{s}	41
		50	0.652	72
		60	0.62	32
		70	0.638	84
		70	0.629	71
		70	0.640	71
		78	0.58	73
		80	0.592	78
	Toluene	70	0.659	82
		70	0.657	71
		70	0.662	71
		78	0.60	73
		80	0.622	78
	n-Octane	30	0.700^{s}	84
		60	0.756	83
		65	0.752	72
		80	0.753	83
		80	0.687	78
	o-Xylene	70	0.698	82
		70	0.690	71

TABLE 3 (Continued)

Stationary phase	Solute	Temp.	γ_{13}^{∞}	Ref.
Squalane (Continued)	o-Xylene (Continued)	70	0.696	71
		80	0.670	78
	m-Xylene	70	0.692	82
		70	0.683	71
		70	0.696	71
	p-Xylene	70	0.679	82
		70	0.667	71
		70	0.686	71
Hexadecane	Hexane	20	0.904	31
		20	0.89	91
		25	0.908	31
		30	0.904	31
		30	0.900^{s}	31
		30	0.904	89
		30	0.88	91
		40	0.910	81
		40	0.87	91
		50	0.902	81
		50	0.891	32
		50	0.892	92
		50	0.905	92
		50	0.86	91
		60	0.889	31
		60	0.895^{s}	31
		60	0.910	81
		60	0.86	91

TABLE 3 (Continued)

Stationary phase	Solute	Temp.	γ_{13}^{∞}	Ref.
Hexadecane (Continued)	Benzene	20	1.110	88
		25	1.109	88
		30	1.106	88
		30	1.007	81
		40	0.989	81
		50	0.995	88
Heptadecane	Hexane	50	0.887	93
		50	0.89	94
		50	0.94	91
Octadecane	Benzene	32	1.000	88
		32	1.101	51
		35	0.986	95
		35	0.986	88
		40	0.966	88
		40	0.966	51
		50	0.927	88
		50	0.918	72
		50	0.927	51
	Octane	35	0.914	95
		40	0.891	72
$C_{22}H_{46}$	Benzene	50	0.816	88
		60	0.796	88
		67.5	0.778	88
$C_{24}H_{50}$	Pentane	55	0.764	100
		60	0.761	102

TABLE 3 (Continued)

Stationary phase	Solute	Temp.	γ_{13}^{∞}	Ref.
$C_{24}H_{50}$ (Continued)	Pentane (Continued)	60	0.760	100
		65	0.762	100
	Hexane	55	0.783	100
		60	0.795	102
		60	0.795	101 (in 102)
		60	0.776	100
		65	0.776	100
	Benzene	50	0.782	88
		55	0.769	100
		55	0.769	88
		60	0.755	100
		60	0.757	88
		60	0.774	102
		60	0.757	101 (in 102)
		65	0.745	100
		65	0.745	88
	Heptane	20	0.813	100
		60	0.815	101 (in 102)
		60	0.824	102

TABLE 3 (Continued)

Stationary phase	Solute	Temp.	γ_{13}^{∞}	Ref.
$C_{24}H_{50}$ (Continued)	Octane	60	0.845	102
		60	0.851	101 (in 102)
		60	0.832	100
	Cyclohexane	60	0.641	102
		60	0.641	101 (in 102)
		60	0.640	100
	Heptene-1	60	0.824	102
		60	0.818	101 (in 102)
		60	0.809	100
	Cyclohexane	60	0.641	102
		60	0.641	101 (in 102)
		60	0.640	100
	Heptene-1	60	0.824	102
		60	0.818	101 (in 102)
		60	0.809	100
	1-Chloropentane	76	0.880	102
		76	0.904	101a (in 102)

TABLE 3 (Continued)

Stationary phase	Solute	Temp.	γ_{13}^{∞}	Ref.
Dinonyl-phthalate	Hexane	25	1.122	31
		25	1.120[s]	31
		30	1.201[s]	84
		37.4	1.119	31
		37.4	1.118[s]	31
		50	1.116	31
		50	1.115[s]	31
		60	1.131	125
		65.4	1.114	31
		65.4	1.112[s]	31
	Benzene	20	0.566	31
		20	0.559[s]	31
		30	0.549[s]	84
		30	0.564	31
		30	0.559[s]	31
		40	0.564	31
		40	0.559[s]	31
		50	0.555	31
		50	0.556[s]	31
		50	0.552	72
		60	0.588	119
		60	0.552	125
	Cyclohexane	30	0.931[s]	84
		30	0.942	31

TABLE 3 (Continued)

Stationary phase	Solute	Temp.	γ_{13}^{∞}	Ref.
Dinonyl-phthalate (Continued)	Heptane	30	1.311	75
		30	1.313[s]	84
		60	1.224	125
	Toluene	60	0.588	125
		60	0.634	119
		60	0.773	121
	Ethyl benzene	60	0.654	125
		60	0.709	119
		60	0.865	121
	o-Xylene	60	0.624	125
		60	0.676	119
		60	0.824	121
	p-Xylene	60	0.638	125
		60	0.692	119
		60	0.844	121
	m-Xylene	60	0.640	125
		60	0.695	119
		60	0.848	121
Di-n-octyl ketone	Hexane	50	1.09	94
		50	1.066	93
		50	1.19	91

TABLE 3 (Continued)

Stationary phase	Solute	Temp.	γ_{13}^{∞}	Ref.
Di-n-octyl ether	Hexane	50	0.96	94
		50	0.906	93
		50	1.06	91
Diethylene glycol	Benzene	50	7.17	112
		50	6.35	108
		70	6.56	112
		70	6.24	108
Thiodipropio-nitrile	Benzene	25	3.43	111
		30	3.38^{s}	69
		30	3.45	120
		60	3.390	119
	Toluene	60	5.122	119
		60	6.247	121
	Ethyl benzene	60	7.809	119
		60	9.518	121
	p-Xylene	60	7.767	119
		60	9.473	121
	m-Xylene	60	7.514	119
		60	9.164	121
	Cyclohexane	30	42.7	120
		30	40.5^{s}	69

TABLE 4

Comparison of γ_{13}^{∞} Determined by Peak Maximum (PM), GC Isotherm (GCI), and Static Isotherm (SI)[a]

Solute	30° C			40° C				50° C			
	PM[b]	GCI[b]	SI[c]	PM[b]	GCI[d]	GCI[b]	SI[c]	PM[b]	GCI[d]	GCI[b]	SI[c]
Squalane solvent											
1,1-Dichloroethane	0.889	0.929	0.914	0.820	0.836	—	0.867	0.830	0.815	0.867	0.851
1,2-Dichloroethane	1.23	1.26	1.30	1.14	1.09	—	1.19	1.06	1.06	1.07	1.11
1,1,1-Trichloroethane	0.612	0.634	0.692	0.586	—	0.598	0.676	0.565	—	0.585	0.652
cis-1,2-Dichloroethylene	0.759	0.813	0.798	0.738	0.771	0.782	0.780	0.703	0.753	0.764	0.767
trans-1,2-Dichloroethylene	0.593	0.644	0.621	0.597	0.594	0.573	0.551	0.564	0.582	0.520	0.541
Carbon tetrachloride	0.527	0.544	0.541 0.552[e]	0.517	0.552	—	—	0.515	0.552	—	—
Dichloromethane	0.904	0.931	1.077	0.830	0.826	—	1.052	0.789	0.796	—	0.955
Chloroform	0.647	0.681	0.653	0.626	0.658	0.631	0.640	0.608	0.650	0.625	—
Dinonylphthalate solvent											
1,1-Dichloroethane	0.436	0.408	0.433	0.453	—	0.425	0.453	0.447	—	0.438	0.446
1,2,-Dichloroethane	0.450	0.460	0.466	0.460	—	0.425	0.453	0.455	—	0.434	0.478
1,1,1-Trichloroethane	0.485	0.513	0.529	0.490	—	0.481	0.535	0.473	—	0.493	0.535
cis-1,2-Dichloroethylene	0.289	0.287	0.299	0.301	—	0.287	0.322	0.312	—	—	0.341

TABLE 4 (Continued)

Solute	30° C			40° C				50° C			
	PM[b]	GCI[b]	SI[c]	PM[b]	GCI[d]	GCI[b]	SI[c]	PM[b]	GCI[d]	GCI[b]	SI[c]
				Dinonylphthalate solvent (Continued)							
trans-1,2-Dichloroethylene	0.437	0.494	0.408	0.456	—	—	0.421	0.443	—	—	0.436
Carbon tetrachloride	0.585	0.557	0.600	0.581	—	0.550	0.594	—	—	—	—
Dichloromethane	0.329	—	0.379	0.333	—	—	0.380	0.328	—	—	—
Chloroform	0.257	0.255	0.251	0.269	—	0.272	0.266	—	—	—	—

[a]This table is taken from Sewell and Stock [77].
[b]Sewell and Stock [77].
[c]Sewell and Stock [79].
[d]Martire and Pollara [67].
[e]Freeguard and Stock [42].

2. Solubilities Related to Chromatographic Retention

From Eq. (3) it is clear that one can derive distribution coefficients from solute retention volumes. This was demonstrated early in the development of GLC [5, 153, 154], where K^0 values in agreement with static measurements were obtained for a variety of systems. Generally, however, the values desired are those free of experimental conditions, i.e., those corrected to zero pressure. Thus, for precise, fully corrected K^0 values the retention volume has to be corrected as above. Again, for many practical applications lesser accuracies are acceptable and a simpler form of the retention volume equation can be used. This is certainly true for poorly defined solvents, such as Solar Oil, in which the partition equilibria of light hydrocarbons were studied [155].

An indirect method for determining liquid-liquid distribution coefficients was devised by Sheehan and Langer [78]. The specific retention volumes of several hydrocarbons were measured chromatographically in squlane and four polar solvents immiscible with squalane. The weight fraction distribution coefficients between a polar solvent and squalane were then estimated from the ratio of the solute specific retention volumes in the two solvents. Again, GLC provides a far quicker method than conventional techniques. Liquid-liquid chromatography, of course, provides such parameters directly, although for practical reasons LLC would be inapplicable to the particular systems studied [78].

Similarly, Henry's constants (H_{13}) can be derived directly from the retention volume as was shown for light hydrocarbons in heavier n-alkanes [90, 96] and for light hydrocarbons in polar solvents [142]. Henry's constants are directly proportional to the solute limiting activity coefficient from

$$H_{13} = f_1^0(p_1^0 = 0)\, \gamma_{13}^{\infty}(P = 0, T) \tag{19}$$

which enables calculation of activity coefficients from the data indicated in Table 2.

Another chemical engineering parameter, which is of great importance to extractive distillation calculations, is the selectivity of a polar solvent toward altering the partial vapor pressure ratio of an alkane-alkene pair. This selectivity is generally expressed as the ratio of the pair's infinite dilution activity coefficients in the polar solvent. Röck [156] was apparently the first to use GLC data to determine which entrainer (polar solvent) to use for the separation of hydrocarbons. Soon after Warren et al. [157] made similar application. Döring [158] showed good agreement between statically determined [159] and GLC-determined selectivities for a variety of polar solvents, as did Vernier et al. [141]. More recently, Tassios

[152] estimated activity coefficient ratios from relative retention times and also [160] by using another, unconventional method proposed earlier [161]; the precise details are not clear from the description.

Another approach to selectivities was taken by Harris and Prausnitz [74], who used a form of inverse chromatography to establish a scale of Lewis acidities. The limiting activity coefficients of 22 polar organic solvents were measured on columns containing squalane, squalene, benzyl diphenyl, and octadecane. For a given solute, the relative Lewis acidity K_a was calculated from

$$K_a = \frac{\gamma_1^{\infty}\text{ (squalane or octadecane)}}{\gamma_1^{\infty}\text{ (squalene or benzyl diphenyl)}} - 1 \tag{20}$$

Practicing chemical engineers often feel more at home characterizing phase equilibria by means of a K value defined as $K = y/x$, where y and x are the equilibrium mole fractions of solute in the vapor and liquid phases, respectively. Preston [162] showed that this parameter is related to the chromatographic K^0 [Eq. (3)] by

$$K = \frac{1}{K^0}\left[\frac{zRT}{PM_3}\right] \tag{21}$$

where z is the compressibility of the vapor phase, P is the average system pressure, and M_3 is the stationary phase molecular weight. Lopez-Mellado and Kobayashi [114] first determined K values over a wide range of pressures and temperatures for the n-butane-dodecane system, and the five C_4 hydrocarbons-furfuraldehyde systems. Their results were in good agreement with tabulated statically determined values [163].

Cruickshank et al. [52] called attention to the effect of carrier gas solubility on the interpretation of retention data [Eq. (8)], which for their purposes was regarded as much a nuisance as anything. However, Stalkup and Kobayashi [103] set out to study high-pressure phase equilibria with multicomponent liquid phases, using as a carrier fluid a single-component or multicomponent mixture appreciably soluble in the stationary phase. The carrier fluid is at equilibrium with the stationary phase, thus producing a multicomponent stationary liquid. Vapor-liquid equilibrium K values for solutes not initially present in the mobile phase have greater engineering applicability than solute K values between a liquid and a fixed or rare gas. For such systems as these, the K values are calculated from experimental net retention volumes from the expression [103]

$$K_i = \frac{y_i}{x_i} = \frac{zRTn}{P(V_R - V_2)(1 - y_3/K_3 - y_4/K_4 - \cdots - y_m/K_m)} \tag{22}$$

where $n/1 - y_3/K_3 - \cdots)$ is the number of moles of nonvolatile liquid component in the stationary phase in the column and $x_3 = y_3/K_3, \cdots$, are the mole fractions of the soluble components dissolved in the stationary liquid $(i \neq m)$. The use of this equation assumes that the partition coefficient for component i is constant throughout the range of concentrations encountered, i.e., linear isotherms or small sample sizes, so that K is effectively an infinite dilution value. In addition the pressure drop along the column is constant; i.e., the pressure drop is negligible. The latter assumption may be approximated experimentally by using short columns packed with coarse supports and low flow rates. Presumably a more sophisticated mathematical treatment would allow precise accounting for the pressure drop. The gas phase compressibility factor z is used in calculating the flow rate under the column conditions from the flow meter conditions, as well as in calculating the K value from Eq. (22). The increase in stationary phase volume and corresponding decrease in V_2 resulting from dissolution of the soluble components at high pressures must be taken into account.

For a one-component, soluble carrier gas (component 2), into which is injected a very small sample of solute (component 1), Eq. (22) becomes

$$K_1 = \frac{zRTw}{p(V_{R_1} - V_2)(1 - 1/K_2)} \tag{23}$$

K_1 values were determined for propane and ethane in the system CH_4-n-decane from -30° to 20° C, and pressures from 15 to 2000 psia [103]. The required K_2 values for CH_4 were obtained from the static data of Sage and Lacey [164]. There was good agreement between their K_1 values for butane at infinite dilution in the CH_4-decane system and those reported by Sage and Lacey [165], the NGAA [163], and in later experimental determinations [166]. Limiting activity coefficients at atmospheric pressure of solute in the CH_4-decane system were calculated from

$$\gamma_{13}^{\infty} = \frac{Pf_1(P, T)K_1}{p_1^0 f_1^0 (p_1^0, T)} \tag{24}$$

where $f_1(P, T)$ is the fugacity coefficient of solute vapor (component 1) in the gas phase at column temperature T and pressure P, and $f_1^0(p_1^0, T)$ is the fugacity coefficient of the saturated component 1 in the gas phase at T and pressure equal to the saturation vapor pressure p_1^0. These values were in good agreement with calculated values for all systems studied. Similar measurements were made for ethane, propane, and butane in the CH_4- decane system [167].

Kobayashi et al. have extended this work to finite concentrations of solute, as will be discussed below, and to gas-solid systems such as silica gel-methane, with C_2-C_4 n-paraffins as adsorbates [168].

3. Derived Thermodynamic Quantities

Since

$$\ln \gamma_{13}^{\infty} = \frac{\overline{h}_{13}^{\infty}}{RT} - \frac{\overline{s}_{13}^{\infty}}{R} = \frac{\overline{\mu}_{13}^{\infty}}{RT} \tag{25}$$

a plot of $\ln {}_{13}^{\infty}$ versus 1/T will yield the infinite dilution partial molar enthalpy of mixing $\overline{h}_{13}^{\infty}$, and thus the corresponding entropy $\overline{s}_{13}^{\infty}$. These values are of great use in interpreting activity coefficients and thus in testing solution theories. In addition, as with γ_{13}^{∞}, many practical engineering calculations require knowledge of these quantities.

Blu et al. [169] present calculations indicating that for small temperature increments, if one can measure retention time with a precision of $\pm 0.01\%$, one can obtain $\overline{h}_{13}^{\infty}$ to better than 1%. Indeed, Wicarova et al. [81] and Pacakova and Ullmannova [99] claim uncertainties in $\overline{h}_{13}^{\infty}$ values of 1 to 5%. Notwithstanding, since a differentiation of the data is required, the uncertainty in the derived quantity can be expected to be roughly 10 times the error in the original data [170]. Estimates by several other groups bear this out. For benzene in n-hexadecane at 25° C, Gainey and Young [88] estimate an uncertainty of the order of 10% in $\overline{h}_{13}^{\infty}$ versus about $\pm 1\%$ in γ_{13}^{∞}. Clark and Schmidt [106] give similar estimations. These precisions are probably typical of those to be expected even when quite precise, fully corrected measurements are made. It is to be borne in mind that $\overline{h}_{13}^{\infty}$ relates to the temperature variation of γ_{13}^{∞}, not of V_N.

Waksmundzki and Suprynowicz [171] devised a method for estimating small excess free energies of mixing of miscible liquids. Solutes are chromatographed on columns containing various mole fractions of the two liquids. If the pair forms a nonideal solution, the mole fraction-based distribution coefficient in the mixed solvent, ${}^{x}k$, should vary with mole fraction according to [172]

$$\log {}^{x}k = x_1 \, {}^{x}k_1 + x_2 \, {}^{x}k_2 + \frac{g^E}{4.6T} \tag{26}$$

where ${}^{x}k_1$ and ${}^{x}k_2$ are the distribution coefficients in pure solvents 1 and 2, and g^E is the excess free energy of mixing of the two liquids at T. The latter is obtained from deviations from linearity in $\log {}^{x}k$ with x.

GLC data for members of homologous series can generally be represented by linear plots of $\log V_N$ versus carbon number or quantities directly related to the carbon number. This was early recognized by Pierotti et al. [5, 173] who developed functional group contributions to $\ln \gamma_{13}^{\infty}$. GLC has proved to be a rapid and simple means of providing data to establish these very useful correlations. More recently Scheller et al. [174] have compiled extensive data using GLC.

B. Gas-Liquid Chromatography at Finite Concentrations

Thermodynamic properties of solutions at infinite dilution are of much theoretical and practical interest. In particular, from an engineering point of view it is of considerable importance that the mixing functions can be represented by various empirical or semiempirical expressions, such as the van Laar, Margules, Redlich-Kister, or Wilson equations [170]. Given the end values of the property-composition diagram (i.e., the infinite dilution or limiting values), properties such as the excess partial molar free energy of mixing can be calculated over the entire concentration range. Similarly, theoretical expressions for activity coefficients simplify considerably in the limit of infinite dilution, so it is again these values that are of greatest importance in testing theory. In addition, because conventional gravimetric methods for studying solutions are best suited for finite concentrations near $x_{solute} = 0.5$, GLC is the technique of choice for infinite dilution measurements. Nonetheless, GLC can also be operated at finite concentrations, often to advantage in terms of speed. Glueckauf [2, 175], again, first showed how to determine adsorption isotherms from liquid-solid chromatograms. James and Phillips [176] applied this technique of frontal analysis to determine gas-solid isotherms, using what has turned out to be a somewhat simplified approach.

For infinite dilution operation of GLC, a small, discrete sample is introduced in the form of a very narrow plug; the retention time at the top of the peak (or at some other location) is the datum of interest. Solution thermodynamic properties at finite concentrations, on the other hand, are studied by introducing a much larger sample, either all at once or suddenly replacing part or all of the carrier gas with the solute under study. What is measured here is the alteration of the shape of the solute concentration profile between the time it enters the column and the time it elutes. The sample is retarded in the column because the soluble components partition themselves between the two phases. The shape of the boundary of moving solute vapor is modified mainly because of the nonlinearity of the distribution isotherm, but also because of nonequilibrium spreading mechanisms, mobile phase velocity changes attending sorption and desorption of the moving solute band, and local temperature variations accompanying sorption and desorption.

Thus, the quantitative treatment of finite concentration GLC is rather more complicated since it must take explicitly into account (a) the nonlinear distribution isotherm, (b) flow rate variations caused by flux of solute molecules between phases (the sorption effect), (c) nonideality of the gas phase, (d) gas phase compressibility, (e) chromatographic mass transfer nonequilibrium, (f) temperature and mobile phase viscosity changes accompanying the passage of a solute front, (g) swelling of the stationary phase as solute dissolves into it, and (h) the possibility in some cases of interfacial adsorption on the liquid surface or on the solid support. Good reviews of these problems have been given by Chen and Parcher [32] and Conder [18].

Conder and Purnell [177, 178] have taken explicitly into account the most important of these complicating factors, i.e., the first four, in developing a generalized retention theory for GLC. Using basically a mass balance approach, they show that for a point of constant concentration c on a front, the retention volume equation is

$$V_R^0 = V_R j = V_2 + V_3(1 - ajy_0)\left(\frac{\partial q}{\partial c}\right)_P \tag{27}$$

For GSC, V_3 is replaced by the weight of adsorbent or some related parameter. In this equation, $(\partial q/\partial c)_P$ is the slope of the sorption isotherm at mean system pressure P. The term $1 - ajy_0$ arises from the sorption effect [179] and the nonideality of the gas phase. y_0 is the gas phase solute mole fraction at the column outlet corresponding to solute gas phase concentration c:

$$j = J_3^2\left[1 + \frac{y_0^2 p_o B_{22}}{RT}(J_3^2 - 1)\right] \tag{28}$$

$$a = \frac{b_2^1}{b_3^2}\left[1 + \frac{2y_0 p_o B_{22}}{RT}(1 - y_0)J_2^1\right] \tag{29}$$

$$b_n^m = 1 + k(1 - y_0 J_n^m) \tag{30}$$

and k is the mass distribution coefficient, the equilibrium ratio of the number of moles of solute in the stationary phase to that in the mobile phase, at concentration c.

For gases deviating not too far from ideality, $a = 1$, $j = J_3^2$, and $P = p_0 J_3^4$. In general, the equation is valid for gases of no more than moderate imperfections, if high solute concentrations ($y < 0.6$-0.95) are used, at pressures up to a few atmospheres, and with small (<1 atm) column pressure drop. In practice these are not severe limitations on the measurement of the thermodynamic properties of adsorption or solution.

Again, it is also assumed in developing this equation that viscosity and temperature changes are negligible, that interfacial solute adsorption is absent, and that solute equilibration in both phases and between phases is achieved more rapidly than the rate of downstream migration (i.e., no mass transfer resistance). How well these assumptions hold will clearly depend on the nature of the system. Generally, fewer complications will arise in strictly nonpolar or only slightly polar systems.

Qualitatively the most important conclusion is that for nonlinear partition isotherms, $V_R = f(c)$. In consequence, solute boundaries are produced which are either diffuse or self-sharpening [177, 178], depending on the curvature of the isotherm. For a typical adsorption isotherm of the Langmuir type, the frontal boundary will be self-sharpening since as the solute concentration increases in the passing front, $\partial q/\partial c$ decreases and thus V_R decreases, producing a nearly vertical front profile. Conversely, the rear boundary will be diffuse because as the concentration drops the solvent tries increasingly to hold onto the solute, which is reflected in increased V_R and tailed boundaries. For a typical anti-Langmuir solution partition isotherm, the opposite is true: Fronts are diffuse and tails are self-sharpening. Linear isotherms produce frontal and rear boundaries of identical form since $K \neq f(c)$.

In order to use the generalized retention theory Eq. (27) must be integrated to produce the sorption isotherm, i.e., the relationship between q and c. From the isotherm can be deduced the properties of interest such as solute activity coefficients as functions of concentration. As shown by Conder and Purnell [177], the zero-pressure activity coefficient $\gamma_{13}(0)$ is calculated from information obtained from the isotherm from

$$\ln \gamma_{13}(0) = \ln \frac{RT(qV_3 + n_3)}{p_1^0 K V_3} + \frac{p_o J_3^4}{RT}\left[2y_0 J_4^3 B_{11} + 2(1 - y_0 J_4^3)B_{13} - \bar{v}_1\right] - \frac{p_1^0}{RT}(B_{11} - v_1^0) \tag{31}$$

which reduces to Eq. (17) when $y_0 \longrightarrow 0$ and $q \longrightarrow C$. From γ as an f(T), other thermodynamic parameters of interest can be generated.

Conder and Purnell [180] consider several different experimental methods for determining isotherms. Conder has described and discussed these in detail in his review article [18] and elsewhere [180a]. The latter compares and contrasts earlier work with the more recent developments, and considers in detail some possible simplifications.

The first technique is called frontal analysis by characteristic point (FACP), and derives from Glueckauf's LC method [2]. It gives the complete isotherm in a single chromatographic experiment. The diffuse front or tail of a frontal chromatogram is used, as sketched in Fig. 1 for a diffuse tail. The input profile MLK is determined from the signal produced by a detector at the column inlet. R is a typical characteristic point. The boundary height y_{MN} must be sufficient to encompass the concentration range under study. The detector response must be calibrated. Various R are selected on the chromatogram and lines LR are drawn parallel to the baseline. The product of LR in units of time and flow rate F(y) corrected for the sorption effect is the retention volume at that concentration c corresponding to R. Equation (27) can be integrated for several values of R to find q:

$$qV_3 = \int_0^c \frac{V_R^0 - V_2}{1 - ajy_0}\, dc \qquad (32)$$

The integration is performed graphically by plotting the integrand versus c (which is calculated from y_0 using a virial expansion for an imperfect gas). Alternatively, the equation can be integrated mathematically if certain assumptions are made, and with the loss of some accuracy. It turns out that q can be calculated in this case directly from the area KLRQ.

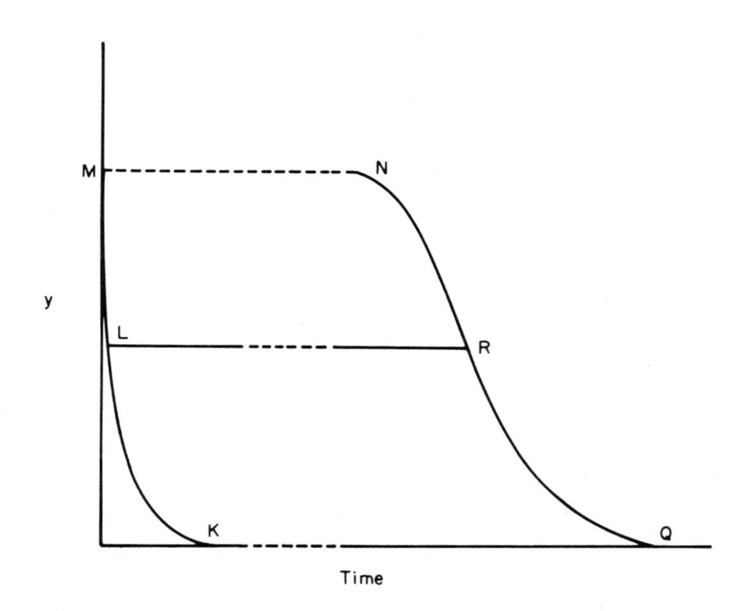

FIG. 1. Analysis of frontal analysis chromatogram.

The second method is frontal analysis. Here a number of chromatographic experiments are required equal to the number of R taken in FACP, since instead of the characteristic point R, the plateau concentration N is used. Areas KLMNRQ are measured to obtain q. Either the diffuse or self-sharpening boundary can be used. This method has also been used by Chen and Parcher [32], who devised their own integration procedure.

Elution on a plateau is another method considered by Conder and Purnell [178], which derives from a suggestion of Reilley et al. [181]. Here a frontal plateau is generated and maintained. Small samples of the solute forming the plateau are injected to give small positive peaks on the top of the plateau; alternatively, small carrier gas samples can be injected to produce small negative peaks of the same retention time. This procedure is repeated over a series of plateau concentrations as in frontal analysis. For reasons that are not intuitively obvious, the retention volume measured in this case is not related to $K^0 = q/c$, but to the gradient of the partition isotherm $\partial q/\partial c$ at the plateau concentration c [182]. Thus, V_R at several concentrations c must be measured to determine $\partial q/\partial c$ as an f(c), so that q can be obtained as above by graphical integration. Stalkup and Kobayashi [103] (see below) earlier devised an iteration procedure to carry out the integration.

Other finite concentration GC techniques are discussed by Conder [18], who evaluated each in detail. Conder and Purnell [75] experimentally tested these three methods, and obtained results which agreed with each other and with static determinations. For the systems n-hexane/squalane and n-heptane/dinonylphthalate, liquid interfacial adsorption and any sorption-desorption temperature effects were found to be negligible. Neither chromatographic nonideality nor stationary phase volume increase significantly affected the accuracy of the results. GC accuracy was found to be independent of the concentration whereas that of the gravimetric method deteriorated rapidly with decreasing solute concentration. Static methods are also slower because of the reduced rate of equilibration.

Parcher et al. [32, 92, 94] have continued such measurements using frontal analysis and also get generally good agreement with static measurements (cf. Table 3). They conclude that GC will not supplant the McBain balance but is certainly quicker. Parcher and Hussey [94] showed how frontal analysis can be used to generate activity coefficient data to test a simplified form of Wilson's equation [183]. This is one of the more powerful semiempirical representations of γ-x data. Their simplified version requires only one adjustable parameter and one measured variable, γ_{13}, as opposed to two of each in the original equation. For nonpolar systems, at least, the simplified equation represents the experimental data from frontal analysis almost as well as the original Wilson equation. One hopes that the people who use equations such as this are able to locate this report since Analytical Chemistry is not on the usual reading list of chemical engineers.

In addition to studying high-pressure phase equilibria at infinite dilution with soluble carrier gases, Kobayashi et al. have pioneered in devising methods for studying finite concentrations. Equations (22) and (23) pertain to a solute not present in the moving phase which is in equilibrium with the stationary, nonvolatile liquid. They have also shown [103], similar to theory proposed by Stalkup and Deans [184], that when a negligibly small volume of one of the soluble moving fluid components is injected into the mobile phase, it produces a perturbation which moves through the column with a retention volume given by

$$V_{R_1} - V_2 = \frac{zRTwK_1K_2}{P}\left\{(K_1 - 1)y_1\left[1 - y_2\frac{d\ln K_2}{dy_2}\right] + (K_2 - 1) \times y_2\left[1 - y_2\frac{d\ln K_2}{dy_2}\right]\right\}/[K_2K_1 - K_1 + y_1(K_1 - K_2)]^2 \qquad (33)$$

This equation is evidently equivalent to Eq. (27), except that it incorporates the solubility of carrier gas in the stationary phase and the effect of this on V_R, a complication unnecessary for the systems studied by Conder and Purnell [177, 178]. If in Eq. (33) an additional greatly simplifying assumption is made that the conditions justify neglecting the terms in $y_i(d\ln K_i/dy_i)$, i.e., that the perturbation velocity is independent of the vapor phase composition and thus is not affected by the dissolved carrier gas component(s), then for two components Eq. (33) reduces to

$$V_{R_1} - V_2 = \frac{zRTwK_1K_2}{P}\left\{\frac{K_2 - 1 + y_1(K_1 - K_2)}{[K_1K_2 - K_1 + y_1(K_1 - K_2)]^2}\right\} \qquad (34)$$

To use Eq. (34) to determine K_1, again one must independently measure K_2. Where $y_2(d\ln K_2/dy_2)$ cannot be neglected, an iterative procedure has been developed [103, 159, 166] using both Eqs. (33) and (34), to determine the K_1 versus y_1 curve. It was pointed out [103] that in some cases a finite value of $y_2(d\ln K_2/dy_2)$ does not appreciably affect the final value of K_1 obtained, thus enabling the use of the simpler Eq. (34).

Equation (34) was applied to the determination of $K_{propane}$ in the system methane-propane-decane [103]. The use of this equation is valid in this case since it was found that $K_{propane}$ did not vary much at all with changing concentration of propane in the stationary liquid phase [166], as might be expected for a purely hydrocarbon system. This may not of course hold true in more complex systems. In this work the concentration of propane in the moving phase was varied from 7 to 26 mole %, but no independent K data were available with which to compare the results.

It is evident that Kobayashi's method differs from the elution on a plateau technique of Conder and Purnell [180] in that there is no insoluble carrier gas component. Chueh and Ziegler [108] synthesized the two procedures by using a binary carrier gas with helium as one component, and analyzed the data with extensions of Stalkup's equations [103]. For the system benzene-diethylene glycol at 50°, 70°, and 90° C, and n-hexane-1,2,4-trichlorobenzene at 30° C, they obtained good agreement (±5%) with static equilibrium measurements.

Kobayashi and associates [103, 184] proposed in addition that if the system (binary soluble moving phase in equilibrium with the nonvolatile liquid) is perturbed with a small but distinguishable sample of one of the components in the mobile phase, e.g., a radioactive tracer, its retention volume will generally be different from that of a finite concentration pulse. Thus, the tracer pulse is at infinite dilution but its retention volume is characteristic of its physical behavior at the finite concentration of that component in the carrier fluid. This phenomenon has been treated in elegant detail by Helfferich et al. [182, 185-187] and is of course one of the finite concentration methods considered by Conder [18, 180]. If labeled compounds and flow counting equipment are available, this method for determining finite concentration equilibria has the distinct advantage that a retention volume measurement for the labeled component at a particular concentration of the same but unlabeled compound can be made, and a point q/c on the isotherm is directly calculable. This is in contrast to the method [180, 181] of elution on a plateau of a concentration pulse (an unlabeled component or small sample of carrier gas), whose retention volume is related to the slope of the partition isotherm at concentration c, $(dq/dc)_c$ [Eq. (27)]. In the latter case, again, a series of these values must be determined over a range of concentrations and then integrated to obtain the isotherm. The real advantage of the tracer pulse method is thus in the greater ease of calculation in determining isotherms or K values.

In terms of Kobayashi's equations, the concentration pulse retention volume is related to Eq. (33), whereas that of the distinguishable tracer pulse is given by Eq. (23). The tracer pulse method was used [166] to determine the interstitial volume V_2 of the methane-decane system, by injecting small volumes of tritiated methane. For this one-component carrier gas,

$$V_2 = V_{R_1} - \frac{zRTw}{P(K_1 - 1)} \tag{35}$$

where K_1 refers to the known (independently determined) K value of methane in decane.

In addition, the tracer pulse method was used to determine K values for propane at infinite dilution in the systems methane-decane and methane-propane-decane [166] over a range of temperature from -30° to +20° C and pressures from 20 to 1000 psia. Good agreement with the infinite dilution values determined by the concentration pulse method [103] was obtained, as well as with the NGAA values [163] and other earlier work cited by Koonce et al. [166]. More measurements on these systems and on the methane-propane-heptane ternary [188]; on the methane-ethane-heptane and methane-propane-toluene systems from -73° to -18° C and pressures from 100 to 1600 psia [189]; and on the CO_2-methane-octane and H_2S-methane-octane systems from -40° to +20° C and 20 to 1500 psia [190] were reported.

More recently, the tracer pulse method was compared directly with the approximate [Eq. (34)] perturbation method by Khoury and Robinson [191] for the CO_2-methane-octane system from -40° to +20° C and for pressures up to 1000 psia. The tracer results were in good agreement with those of Asano et al. [190] and with static values [192] for both $K_{methane}$ and K_{CO_2}. The $K_{methane}$ values determined by both chromatographic methods agreed well, but the K_{CO_2} values derived from the approximate perturbation method deviated by 6 to 15% from the true (tracer pulse) results. Thus, the solubility of CO_2 in octane with dissolved methane present does vary with the exact composition. Although in this work only a single mobile phase composition was used, dK_i/dy_i could be determined if measurements were made at several different carrier gas compositions. That $K_{methane}$ did not seem to vary with dissolved CO_2 indicates that for this system the term derived by Cruickshank et al. [59], $(\partial \ln \gamma_1^{\infty}/\partial x_2)_0$, must be negligible, whereas this term for CO_2 must be appreciable. This could presumably be demonstrated explicitly using Eqs. (24) and (33) initially.

Korol [19] reviewed the work of Zhukhovitskii (all in Russian) on the determination of solubilities of substances by vacancy chromatography. This is related to the elution on a plateau method, except there is no inert carrier gas component. Perturbations are created by injecting air or a sample of one or more components in the carrier fluid. An air sample produces a series of negative peaks, whereas a small sample of one of the components yields a small positive peak. The retention volume can be related, as in the methods discussed above, to the slope of the partition isotherm at the plateau concentrations.

C. Liquid Chromatography at Infinite Dilution

Although LC has been around in one or another form far longer than GC, it has never become as powerful a tool for analysis or for physical measurements. However, the analytical aspect is improving rapidly (e.g., see Snyder and Kirkland [35]). Some initial work has been done on the latter [34] which can only ultimately benefit from the recently available improved instrumentation. LLC should have certain advantages over GLC for physical measurements because low-molecular-weight solvents are used; the mobile phase is incompressible at ordinary pressures; all mobile phase interactions are lumped together into the solute activity coefficient in that solvent; and there is the possibility of directly combining GLC and LLC data. However, the simplest and most readily studied form of liquid chromatography, LLC with immiscible liquids, is a rather restricted area in terms of the number of systems accessible to study. Not only must the two phases be immiscible in this case, but both must have sufficient solvency for the solutes to produce retention volumes neither too close to the interstitial volume nor too large. Interfacial adsorption and mass transfer problems may exist. Finite solute concentrations are often out of the question because of solubilization of the two phases by the solute. There has as yet been no published theoretical treatment of partially miscible systems amenable to LLC application, and given the state of solution theory it is unlikely there will be any but a largely empirical description available. The finer theoretical details of LLC are not worked out although a start has been made on some aspects [193, 194]. Finally, there is again the very real experimental problem of the lack of a sensitive, universal detector for liquid chromatography.

Indeed, the only reported physical measurements by LLC are my own and more recently those of Alessi. For two immiscible liquid phases, it has been shown [195] that to a good approximation, the net retention volumes of a solute at infinite dilution is given by

$$V_N = \frac{\gamma_{13}^{m,\infty} n_3}{\gamma_{13}^{s,\infty} n_2} V_2 \tag{36}$$

where m and s refer to mobile and stationary phases, respectively, and V_2 is the volume of mobile phase in the column. Knowing solute activity coefficients in one phase, one can combine these with experimental retention volumes to calculate the corresponding values in the second phase, at the same temperature.

This was done for the C_1-C_4 alcohols in glycerol [196] and a wide variety of hydrocarbons in acetonitrile [197]. In the former case, the eluent was n-heptane, for which the alcohol activity coefficient data were available from static measurements, while for the acetonitrile eluent, squalane served as the stationary phase. Many activity coefficients for hydrocarbons in squalane are available from GLC measurements (cf. Table 2). More recently, Alessi and Kikic [57, 198] studied the system aniline-squalane and aniline or acetonitrile-Apiezon L [199].

Comparison with independent, static activity coefficients was possible in the case of the acetonitrile data [197]. For $\gamma_{13}^{m,\infty}$ values less than about

15, good agreement was generally obtained. Although such agreement was by no means as precise as that obtained for, say, n-hexane in squalane by fully corrected GLC, given the relatively ill-defined nature of LLC systems, the need for independent data for one phase, the effect, unaccounted for, of partial miscibility, and the possible existence of complicating mass transfer factors, the agreement can be called good. This then indicates the promise of LLC for this application, if not the ultimate delivery of the goods.

Enthalpies of transfer of solute from phase to phase, and the partial molar excess quantities were also calculated from the distribution coefficient and activity coefficient data, respectively, obtained over a range of temperatures [196, 197]. In the case of the partial molar enthalpies of mixing at infinite dilution, reasonable agreement was obtained between the LLC values and static determinations.

It was clear in this work [196, 197] that for solutes with small solubilities in one or both phases, the apparent activity coefficients calculated from the retention data were too large compared with static values, the discrepancy increasing with decreasing solubility. The simple model [Eq. (36)] did not accurately describe the system. In both cases, I attributed this to solute adsorption at the liquid-liquid interface, justifying such attribution on the basis of similar observations in GLC for solutes with large (> 5) activity coefficients [200] (Sec. IX). All the requisite conditions for interfacial adsorption are present: the temperature is usually near ambient, one of the phases is always polar, and the solute solubilities may be small in at least one of the phases. Again, the discrepancies between the LLC and static measurements were most pronounced for higher values of $\gamma_{13}^{m,\infty}$, i.e., lower solubilities. If the solute is "squeezed out" of solution, it has no place to go but to the interface. Eon et al. [194] have found a direct correlation between V_N and the interfacial tension of the two liquid phases, but precisely how this relates to the problem of interfacial adsorption in the present context is not immediately clear, except to support its potential existence.

Alessi and Kikic [198, 199], on the other hand, ascribe the problem to the small mutual solubilities of the two phases; the activity coefficients determined by LLC refer not to the pure solvent but to that solvent saturated with the second phase. In this work, solute activity coefficient ratios in the two phases were calculated from LLC retention volumes, and also from GLC-derived data, for the aniline-squalane system. The mutual solubility here, as shown in Table 5, is smaller than for the acetonitrile-squalane system [197]. The GLC-determined activity coefficient ratios should correspond to the pure solvents, whereas the LLC values pertain to the mutually saturated phases. For 10 hydrocarbons, these ratios differed by up to 10%. However, if the LLC activity coefficients were corrected by using a phase-average molecular weight and density, and compared with the GLC

TABLE 5

Mutual Phase Solubilities[a]

Stationary phase	Mobile phase	Mole fraction solubility (25° C)	
		Stationary in mobile	Mobile in stationary
Squalane	Acetonitrile	0.0005	0.115
Squalane	Aniline	0.0001	0.0807

[a]Data from Alessi and Kikic [198].

values determined using squalane saturated with aniline and aniline saturated with squalane as the stationary phases, the two sets of activity coefficient ratios for all solutes agreed to better than 1%. Additional support for this view was obtained by lowering the mutual phase solubility. First, at 2° C, the mutual solubilities are much smaller, and indeed the pure solvent ratios (by GLC) agreed with the LLC values when measured at this lower temperature. Second, using Apiezon L stationary phase for LLC and GLC, and acetonitrile or aniline as eluent in LLC, activity coefficients at infinite dilution in the eluent solvents were derived that were in good agreement with static values [199].

This does not account for the previous observation [197] that discrepancies between static and LLC activity coefficients in acetonitrile did not seem to appear except at large values. All of this illustrates that LLC is not as finely worked out as GLC for these purposes; additional work is clearly needed, and one anticipates far wider application of LLC to physical measurements in all areas.

An important new type of LC stationary phase is comprised of high-performance siliceous particles, the surfaces of which are covered with organic groups chemically bonded onto the surface Si-OH groups. These modified adsorbents have a large number of advantages over liquid-liquid systems and liquid-high surface area adsorbents for analytical applications [201]. Recently I considered [202] these chemically bonded phases from the point of view of Everett's thermodynamic theory of adsorption [203, 204]. It transpires that for bonded nonpolar hydrocarbons such as n-C_{18} or a naphthyl group on pellicular or microparticle supports, the relative retention is to a good approximation given by

$$\alpha = \frac{V_{N_2}}{V_{N_1}} = \frac{\gamma_2^{m,\infty}}{\gamma_1^{m,\infty}} \tag{37}$$

where the $\gamma_i^{m,\infty}$ refer to the limiting solute activity coefficients in the mobile phase for solutes 1 and 2. Since for small solubilities,

$$\gamma_i^{m,\infty} = \frac{1}{x_i^m} \tag{38}$$

where x_i^m is the solute mole fraction solubility in the mobile phase, this suggests a simple method for estimating small solubilities, if the solubility of one of a series of compounds is independently known. From the relative retentions one determines the relative solubilities (relative activity coefficients) and thus the mole fraction solubilities by proportion. This was done for a series of aromatic hydrocarbons [202]. Chromatographic data were obtained from several sources, and the independently determined molar water solubilities of benzene and fluoranthene were used to establish the slope of a plot of log(relative retention time) versus -log(solubility). The solubility results are given in Table 6, and given the extreme smallness of the values, the agreement is again adequate.

Both GLC and LLC seem to offer a relatively versatile, rapid, simple, and reasonably accurate means for determining thermodynamic properties

TABLE 6

Determination of Molar Water Solubility (20° C) Using Bonded Phase Liquid Chromatography[a]

Compound	From LC (mol/liter)	From the literature (mol/liter)
Benzene		2.3×10^{-2}
Naphthalene	4.7×10^{-3}	8.0×10^{-4}
Phenanthrene	1.6×10^{-5}	9.0×10^{-6}
Anthracene	5.9×10^{-6}	4.2×10^{-6}
Fluoranthene	—	1.2×10^{-6}
Pyrene	5.1×10^{-7}	8.2×10^{-7}
Chrysene	6.0×10^{-9}	6.6×10^{-9}
Benz-e-pyrene	4.4×10^{-10}	
Benz-a-pyrene	1.9×10^{-10}	1.6×10^{-9}

[a]From Locke [202].

of the solutions involved. By far GLC is the more sophisticated technique, capable of study of a variety of systems over a range of solute concentrations, temperatures, and pressures. LLC shows promise in this regard but has not yet been as well developed. In any case one should beware the apparent simplicity of chromatography. There are a number of potential complicating factors for which corrections are required if reliable data are to be obtained. As with any technique one must use restraint and intelligence in its application, and always recognize that although anyone can calculate a retention volume of some compound injected into a chromatograph, it is far more hazardous to relate unequivocally and unambiguously this datum to some desired property of the system.

VI. STUDY OF VAPOR PHASE INTERACTIONS BY GAS-LIQUID CHROMATOGRAPHY

Originally in GC, the carrier gas was regarded as an inert, noninteracting fluid whose only function was to push the solute molecules through the column. In 1961, Klesper et al. [205] showed that porphyrin mixtures would move down columns packed with polyethylene glycol using as carrier gases dichlorofluoromethane or chlorodifluoromethane at temperatures above their critical temperatures (112° and 96° C, respectively), and at pressures up to 2300 psia. Clearly, the dense carrier fluid was solubilizing the nonvolatile porphyrins and causing their downstream migration. This was not news to power plant engineers, who have long been well aware of the ability of supercritical steam to lift silica, quartz, and inorganic salts. The analytical advantages of high-pressure operation were realized by Giddings et al. [206-210] who suggested that the separating potential of GLC should merge with that of LLC at pressures of about 10^3 atm. They also showed [209] how to determine solubility parameters of higher-molecular-weight materials such as Carbowaxes 1000 and 400, in good agreement with calculated values. Sie and Rijnders [211] used supercritical CO_2, pentane, and 2-propanol to increase greatly the speed of separation of high-molecular weight polynuclear aromatic hydrocarbons and dialkylphthalates. Gouw and Jentoft have made several interesting and valuable contributions using supercritical fluids, and have summarized this new field in a recent review article [33]. Although there are serious instrumental and operational drawbacks to supercritical fluid chromatography at high pressures, there are clear advantages in speed, selectivity, and range of application over both ordinary pressure GC and LC.

Another aspect of the influence of ordinary carrier gases on chromatographic retention was noted by Desty et al. [212] at ordinary column pressures. This small alteration of retention time with the nature and absolute pressure of the carrier gas was related theoretically to the interaction second virial coefficient B_{12} of the solute vapor-carrier gas mixture [58]. Subsequently, for insoluble carrier gases, Cruickshank et al. [52] made a more exact evaluation, resulting in Eq. (17). Later work treated the situation with a soluble carrier gas [59]. A required correction thus became the source of additional, valuable information. It was in any case first evident from the work of Desty et al. [58] that for insoluble carrier gases, measurements of log V_N over a range of mean column pressures should allow evaluation of B_{12}. Reduced-pressure GC [213] was also used with the same equation to estimate these parameters. The most extensive application of GLC to this measurement has been made by the group at the University of Bristol [52, 59, 72, 88, 89, 95, 51, 100] for a variety of hydrocarbons using as carrier gases He, H_2, N_2, O_2, Ar, CO, CH_4, CO_2, and N_2O. They have clearly demonstrated that it is possible to measure B_{12} values independent of the stationary phase and operating conditions.

The types of systems studied are listed in Table 7, and a comparison of the values from several laboratories and with the few published static values is given in Table 8. The uncertainties are given in Table 7, and it is seen that the agreement is for the most part quite good. Static measurements of B_{12} are far more tedious than static measurements of activity coefficients, so this GLC application represents a truly useful advance. Values of B_{12} calculated from the theory of corresponding states are given in several of these papers [31, 51, 58, 59, 72, 88, 89, 95, 100, 104, 213].

Equation (14), which is used to determine B_{12} values from the pressure dependence of the retention volume, is an approximate equation. In particular, it requires the use of small pressure drops and, more important, carrier gases that are either insoluble in the stationary liquid phase or whose presence in the stationary phase does not affect the solubility of other vapors, as was apparently the case in some of the systems studied by Kobayashi et al. as noted above. However, the K values determined by Kobayashi may be far less sensitive to the presence of dissolved carrier gas than B_{12} values derived from retention data, as is suggested by the work of Pecsok and Windsor [216]. However, using the methods of Kobayashi, or an equivalent description, with mixed insoluble-soluble carrier gases, one should be able to evaluate $(\partial \ln \gamma_1^{\infty}/\partial x_2)$ [Eq. (13)].

Up to now, this term has not been measured or calculated, and Cruickshank et al. regard ignorance of it as a limiting factor in the application of GLC to the determination of B_{12}. Thus, when they were seeking to measure B_{12} for the benzene-CO_2 mixture [59], they would have preferred to use squalane as the stationary phase. In squalane, benzene has a retention volume of optimum magnitude, and mass transfer effects are negligible. However, CO_2 is appreciably soluble in squalane, and presumably affects the solubility of benzene. Ignoring this contribution to the slope of the experimental curve would produce an estimated error equivalent to about ± 75 cc/mole in B_{12}, or about 40% for this particular case [59], in

TABLE 7

Systems in Which B_{12} Values Have Been Determined by GLC

Gases	Solutes	Precision (cc/mole)	Temperature (° C) range	Ref.
H_2, N_2, O_2, CO_2	20 hydrocarbons	±30	25	58
He, Ar, CO_2	Five hydrocarbons	±30	50	213
N_2, H_2, Ar	Six hydrocarbons	±6	25-65	31
CO_2	Nine organics	±20	40	214
N_2, Ar, H_2	17 hydrocarbons	±10	35	95
N_2, Ar, H_2	Benzene	±10	20-75	88
N_2	10 hydrocarbons	±10	55-75	100
He, H_2, N_2, O_2, Ar, CO, CH_4	Benzene	±8	50	72
H_2, CO	n-Octane	±8	40-65	72
H_2	Isooctane	±8	40-65	72
N_2	Four hydrocarbons	±10	30-60	89
Ar	Five hydrocarbons	±20	25	215
CH_4, C_2H_6	Five hydrocarbons	±40	25, 50	216
N_2	Benzene, nine fluorobenzenes	±8	32-50	96
N_2, CO_2, H_2	Benzene	±6	50	59
N_2	21 hydrocarbons	±6	40	104
N_2, Ar, CO_2	60 organics	±10-30	25-80	217
N_2	Benzene, cyclohexane, n-hexane	±30	30-70	218
N_2	Seven hydrocarbons	±10, 15	25	219
N_2	Six hydrocarbons	±10	30	220

TABLE 8

Comparison of B_{12} Values (cc/mole) by GLC and Static Measurements[a]

Compound	58 (25)	31	95 (35)	Nitrogen carrier gas		104 (40)	217 (80)	218	219 (25)	220 (30)
				100	89					
Butane			84	77 (60)	78 (30) 69 (50)					
Isopentane	105		91	64 (55) 76 (60) 81 (65)		92				
Pentane	105	103 (25)	85	66 (55) 72 (60) 78 (65)	85 (40) 80 (60)	86	60		90	85
Hexane	128	112 (25) 103 (30) 79 (50)	113	90 (55) 98 (60) 104 (65)	108 (30) 107 (40)	110	69	131 (30) 102 (40) 58 (50)	118	107
Heptane	154		132	101 (60) 87 (75)	111 (40)	110	81		142	111
Octane			143	119 (60) 92 (75)		134	98		146	143
2-Methyl pentane	127					106				93
2,2-Dimethyl butane	105	94 (25)	60			83				

2,4-Dimethyl pentane	130	109		115				109
Cyclohexane	120	122	80 (60)	116	63	141 (30) 98 (40) 110 (50) 100 (60)	130	

Nitrogen carrier gas

Compound	58 (25)	31	88	100	72 (50)	96	59 (50)	104 (40)	217 (80)	218	219 (25)
Benzene	117	122 (30) 108 (40)	109 (25) 107 (30) 104 (35) 94 (50) 91 (60) 85 (65) 74 (70)		87	103 (32) 96 (40) 90 (50)	98	100	74	129 (30) 112 (40) 91 (50) 82 (60) 77 (70)	126
Hexafluoro-benzene				117 (55) 110 (65) 118 (75)		150 (32) 130 (40) 111 (50)		126			

Argon carrier gas

Compound	31	95 (35)	88	72 (50)	215 (25)	Static 215 (25)	217
Benzene	88 (50)		135 (32) 126 (40) 90 (50)	79, 85			112 (25) 117 (50) 105 (80)
Hexane	130 (20) 127 (25)				124 107 (50)	80-180 108-148 (50)	106 (50) 81 (80)

TABLE 8 (Continued)

Compound		95 (35)	88	72 (50)	215(25)	Static 215 (25)	217
				Argon carrier gas (Continued)			
Hexane (Continued)	104 (37) 107 (50) 100 (65)						
Pentane	98 (25)				98	125-136	80 (50) 68 (80)
2,2-Dimethyl butane	115 (25)	98			115	156	108 (25)
2-Methyl pentane	125 (25)				125	147	115 (25)
2-Methyl butane	94 (25)	78			94	178	

Hydrogen carrier gas

Compound	58 (25)	88 (40)	72	59 (50)	Static 221	Static 222
Benzene	13	9	7.5 (50)	6.0	7.5 (50) -1.6 (100)	
Octane			-9 (40) -22 (65)		-18.1 (100)	
Isooctane			-4 (40) -6 (65)		-18.7 (75)	-37.6 (50) -55.7 (75)

Carbon dioxide carrier gas

Compound	58 (25)	216 (40)	59 (50)	215 (80)
Benzene	251 ± 30	288 ± 20	250 ± 15	216 ± 20

Methane carrier gas

Compound	216 (25)	216 (50)	Static 223 (25)	Static 223 (50)
Pentane	204	138	222	190
Hexane	292	280	261	225

Ethane carrier gas

Compound	216 (25)	Static 224 (25)
Pentane	414 ± 171	448 ± 16

[a] The numbers in parentheses are temperatures (° C); the column heading is the reference for the data.

contradiction to the results of Sie et al. [214]. The latter found experimentally substantial CO_2 solubility in squalane which increased linearly with pressure up to at least 80 atm. Negligible solubility of CO_2 in glycerol was found. Nonetheless, they obtained identical slopes of their log K^0 versus mean column pressure plots for squalane and glycerol with the CO_2 carrier gas, for a variety of organic compounds, and thus concluded that carrier gas solubility plays only a minor role. Guided by their theory, Cruickshank et al. [59] selected glycerol as the stationary phase because the solubility of CO_2 in glycerol is 30 times less than that in squalane. However, because of its slight solubility in glycerol, benzene elutes close to the air peak, which decreases the precision of the retention volume measurement; corrections were required for interfacial adsorption of benzene (Sec. IX); substantial corrections were necessary for the large resistance to mass transfer of benzene in glycerol because of the solvent's self-association by hydrogen bonding; and there were difficulties in estimating the infinite dilution partial molar volume of the solute. Nonetheless, a value of B_{12} was derived from the data albeit with far greater effort and with less accuracy than for cleaner systems.

In the CH_4-squalane system studied by Pecsok and Windsor [216], comparison of derived B_{12} values for hydrocarbons with static values showed considerable discrepancy in the absence of a carrier gas solubility correction (which was of the same order of magnitude as predicted by Cruickshank et al. [59]); taking into account the methane solubility, the GC results were in excellent agreement with the static values. Precisely how to correct was again not clear since the term $\partial \ln \gamma_1^\infty / \partial x_2$ is an unknown quantity. Pecsok and Windsor thus assumed that $1 - \partial \ln \gamma_1^\infty / \partial x_2$ lies between 0 and 1, so they estimated a contribution of 0.5 ± 0.5. The carrier gas solubility λ was calculated assuming that CH_4 displays ideal behavior. The main drawback of this procedure is that the uncertainty in derived B_{12} values is approximately doubled at lower temperatures, from about ± 20 to roughly ± 40 cc/mole. Thus, the solubility problem is the ultimate limitation on the application of GLC to the measurement of B_{12}. This does not represent as great a problem in determining γ_{13}^∞ values since these are obtained by extrapolation to zero pressure; in any case the magnitude of the correction is only about 1% in γ_{13}^∞ which is generally about the same as experimental error.

For soluble carrier gases, Vigdergauz and Semkin [217] pointed out that if for a series of similar solutes, the B_{12} value is independently known for one of the solute vapors in the carrier gas, the apparent GC interaction virial coefficient value B_{eff}, obtained by ignoring the carrier gas solubility, will differ from the true value B_{12} by

$$B_{eff} - B_{12} = 0.5\,\lambda \left(1 - \frac{\partial \ln \gamma_1^\infty}{\partial x_2}\right) RT \tag{39}$$

If the right-hand side can be assumed constant for a series of similar compounds, appropriate corrections can be made with the aid of the one known B_{12} value. Whether this is more or less accurate than the procedure of Pecsok and Windsor [216] remains to be seen.

Vigdergauz and Semkin [217] also showed that if B_{12} values for a series of normal paraffins in an insoluble carrier gas are known, one can use the pressure dependence of the retention index to calculate B_{12} values for isomeric hydrocarbons eluting among the normals.

Gas-solid chromatography, in contrast to GLC, is probably not a suitable method for the determination of B_{12}, because of uncertainty in the exact nature of the carrier gas adsorption isotherm [225].

It has been proposed [226] that one might use the tracer pulse method [103, 182] using pure carrier gas to determine values of B_{22} for that gas. No measurements have been reported.

VII. POLYMER STUDIES USING CHROMATOGRAPHY

A. Gas-Liquid Chromatography

Polymeric stationary phases such as the silicone oils, polyesters, and polyglycols are among the most common stationary phases for analytical GLC. Polymers can also be identified by pyrolysis followed by GC analysis of the volatile breakdown products [227]; different chemical types of polymers produce pyrolysis-GC patterns as unique and characteristic as infrared spectra if the pyrolytic and chromatographic conditions are constant and reproducible.

Only recently has GLC been used to study the physical characteristics of polymers, using volatile solutes as molecular probes [228]. The polymer is generally coated onto a solid support such as Chromosorb in the usual fashion, and solute molecules which interact with the polymer (e.g., dissolve in it) are injected as in analytical GC. The solute retention characteristics are related to physical properties of the polymer solution. This subject has been thoroughly reviewed recently by Guillet [229], himself a prime original contributor to these studies, and only papers published since this review will be cited here.

An important area accessible to GLC study is solution thermodynamics, i.e., weight-fraction activity coefficients at infinite dilution and at finite concentrations (mole fraction-based activity coefficients cannot be determined because there is no unique molecular weight for a polydisperse polymer), Henry's constants, Flory-Huggins interaction parameters, partial molar hearts of solution, and Hildebrand solubility parameters.

Covitz and King [230] studied the absorption of 13 organic solvents by molten polystyrene, determining activity coefficients, interaction parameters, and enthalpies of solution. Newman and Prausnitz [231] extended their earlier solution thermodynamic work to 24 binary systems of hydrocarbon solutes at infinite dilution in polyethylene, polyisobutylene, and ethylene-vinyl acetate and ethylene-propylene copolymers.

Very careful, fully corrected measurements of Flory-Huggins interaction parameters were made by Leung and Eichinger [219] on polyisobutylene-hydrocarbon solutions at 25° C; their values agree well with other chromatographically determined as well as statically determined values, as shown in Table 9. Interaction second virial coefficients were also determined from the retention data. Polymers are advantageous for this latter application in that their low volatility allows a wider variation in column temperature than do simpler stationary phases such as squalane and hexadecane.

Brockmeier et al. [235] used the Conder-Purnell elution on a plateau method to determine solubility isotherms and thermodynamic properties of solutions of hexane and isooctane in high-density polyethylene, at pressures up to 70 atm and temperatures to 250° C. Semicrystalline, stereoregular

TABLE 9

Infinite Dilution Flory-Huggins Interaction Parameters at 25° C in Polyisobutylene-Hydrocarbon Solutions

Solute	GC[a]	Static[b]	GC[c]	Static[d]
n-Pentane	0.90 ± 0.02	0.85	0.87	0.88 ± 0.03
n-Hexane	0.79	0.72		
n-Heptane	0.69	0.69		
n-Octane	0.63	0.67		0.65[e]
n-Nonane	0.59	0.64		
Benzene	0.99	1.00	1.03	1.13
Cyclohexane	0.67	0.55	0.55	0.52

[a]From Leung and Eichinger [219].
[b]From Hammers and de Ligny [232].
[c]From Newman and Prausnitz [233].
[d]From Eichinger and Flory [234].
[e]Static, cited by Leung and Eichinger [219].

polypropylene solutions with n-hexane, cyclohexane, and the 80° C n-hexane-2-propanol azeotropic mixture were subsequently studied [236].

A more applied study of solution measurements was made by Varsano and Gilbert [237, 238] who studied the interaction of various typical organic solvents and several bacteriostatic agents such as methyl and propyl paraben, with packaging-grade polyethylene and polyvinyl chloride. Useful information concerning the migration of such chemicals into packaging materials results from such measurements. Similar properties of commercially important plastics such as Acryloid A11, Lucite 2044, and Eponol 55 were studied using 21 typical solvents by Newman and Prausnitz [239].

Other examples of the application of GLC to the study of polymer solutions are reviewed by Guillet [229]. For example, from discontinuities in plots of log V_N versus 1/T for compounds which are not solvents for a polymer used as the stationary phase, one can determine glass transition temperatures of amorphous polymers. At this transition temperature, changes in the physical structure allow the noninteracting solute molecules to penetrate the polymer, and thus increase the retention volume. The determination of crystallinity, the study of surface properties, and the determination of diffusion coefficients of solutes in polymers are other areas covered by Guillet [229].

B. Liquid Chromatography

One of the more powerful tools for polymer characterization is gel permeation chromatography (GPC). Since separation here is based on molecular size, through appropriate calibration GPC is used to determine molecular weights and molecular weight distributions for many different kinds of synthetic and natural polymers. The rapidity of the technique and the ease of comparison of chromatograms for different samples enable the analytical application of GPC to fundamental polymer studies in synthesis, solution interactions, reactor design, polymerization kinetics, degradation by irradiation, and chain lengths in single crystals, in many of the various applied polymer areas. Such applications are discussed in several recent books and review articles [240-248].

The retention volume equation for GPC has the same form as the equation for other chromatographic techniques:

$$V_R = V_2 + KV_p = V_e \tag{40}$$

where V_e is the elution volume, the usual GPC designation for the retention volume, V_2 is the retention volume of a molecule totally excluded from the pores of the gel, V_p is the internal pore volume of the solvent-swollen gel, and K is a sort of distribution coefficient, defined in various ways equivalent to

$$K = \frac{V_p(\text{accessible})}{V_p} \tag{41}$$

where V_p(accessible) is the gel pore volume accessible to a nonexcluded molecule. In the absence of adsorptive or other interactions between gel and solutes, $0 \leq K \leq 1$. This equation assumes that diffusional equilibrium is reached, since this is the only mechanism by which solutes can enter and leave the pores of the gel.

A relationship between V_e and molecular weight is generated by chromatographing calibration standards. The columns, conditions, temperature, and solvent must be the same. Cazes [244] has listed suppliers for various well-characterized polymer standards, which are samples of the polymer of interest having a narrow molecular weight distribution. A plot of log(average molecular weight) versus V_e is made. The use of this calibration plot for the determination of average molecular weights of unknown samples has two drawbacks: (a) the calibration standards are not in fact monodisperse, and (b) the elution volume at the top of the peak V_e does not correspond exactly to the weight average M_w, the number average M_n, or any other of the usual molecular weight averages, but instead is in between M_w and $(M_w M_n)^{1/2}$ [248]. Assuming V_e corresponds to M_w leads to errors of 5% or more for polydisperse samples. In addition, very few calibration standards exist.

Because of the paucity of standards, a universal GPC calibration is clearly desirable. A large number of such correlations have been developed between V_e and molecular size-related parameters [242, 245, 248]. The most universal of these in terms of independence from chemical type and molecular shape, seems to be the log(hydrodynamic volume) = $\log(M_w[\eta])$, where $[\eta]$ is the polymer's intrinsic viscosity [249]. Thus, a primary universal calibration curve can be prepared using the best characterized reference samples, a series of polystyrenes available from the National Bureau of Standards. Weight-average molecular weights of unknown polymers can be obtained from the calibration plot in conjunction with viscosity measurements on eluted fractions. Alternatively, the Mark-Houwink equation can be applied to relate the calibration plot made using polystyrene standards to that for another polymer [245].

The conventional molecular weight averages are calculated from the molecular weight distribution, as shown for example by Cazes [240, 244]. The molecular weight distribution in turn is generated from the calibration plot and the experimental GPC chromatogram, again all conditions constant.

Computer programs are available for the calculation [240, 248], which print out the various molecular weight averages, degree of polymerization, the distribution of molecular weight, and so forth. These calculations ignore distortion of the peak arising from longitudinal diffusion, mixing in extra column volumes in the detector or injector, column overload for excessively large samples, channeling effects, nonequilibrium, and so on. Some corrections are possible [250], although they are not wholly satisfactory [245].

It has been observed [251, 252] that for small molecules, GPC measures their effective volume in solution. Thus, molecules which can self-associate or associate with the solvent through hydrogen bonding, complexation, ion-pair formation, etc., have elution volumes that are different from those that occur in the absence of association. This suggests that quantitative molecular size measurements in these cases are possible, along with information derivable from the effective size in solution.

VIII. STUDY OF COMPLEX EQUILIBRIA

A. Gas-Liquid Chromatography*

Selectivity, the ability of a chromatographic system to pull apart two solutes, can be made very large if the stationary phase contains a substance which interacts in a specific manner with a certain functional group in one of the solutes. Thus, columns prepared with $AgNO_3$ dissolved in a polar solvent selectively retard olefins [253] relative to saturates. This is explained in terms of the equilibrium reaction between

$$\text{olefin} + Ag^+ \rightleftharpoons Ag^+\text{-olefin complex} \tag{42}$$

Similarly, Phillips et al. [133] found that amines were selectively absorbed by columns containing molten transition metal stearates; the nature of the metal ion had a large effect on selectivity, Ni^{2+} being the most and Cu^{2+} the least interactive. Gil-Av and Herzberg-Minzly [254] found that N_2 and O_2 could be separated at 30° to 40° C on 2-m columns packed with deoxygenated blood from sheep, cows, or humans. The O_2 interacts specifically with the hemoglobin component of the blood. The precise selectivity (as indicated by the relative retention) depended, among other factors, on the recent medical history of the donor. Gil-Av and Herling [135] also first pointed out the possibility of using GLC for quantitative measurement of complexation. Many other interesting examples involving charge transfer complexation, shape factors, interlamellar (inclusion) complexes, hydrogen bonding, etc., are to be found in the analytical literature of GLC.

Purnell [255] was the first to develop in a systematic manner the full extent of the possibilities inherent in using GLC for the quantitative study of

---

*See Addendum of this volume for further information.

a variety of complex equilibria. The theory and application of GLC to complexing reactions have recently been reviewed thoroughly by Wellington [256], so that only the very recent additions to the literature are covered here.

The most straightforward type of reaction accessible to GLC is one in which a 1:1 complex is formed between a volatile solute molecule and a nonvolatile ligand. Broadly, two distinct methods for studying such equilibria have evolved, one due to Cadogan and Purnell [71] and the other to Martire and Riedl [93]. The two methods have recently been compared in detail by Martire et al. [257].

1. Cadogan-Purnell Method

The ligand B is dissolved in a nonvolatile, nonreactive solvent I. Reactive solute A is injected and partitioned between gas (g) and liquid (l) phases,

$$A(g) \underset{}{\overset{K_R^0}{\rightleftharpoons}} A(l) \tag{43}$$

and when dissolved in the stationary phase reacts to form a 1:1 complex with B:

$$A(l) + B(l) \overset{K_1}{\rightleftharpoons} AB(l) \tag{44}$$

K_R^0 is the distribution coefficient of uncomplexed A between liquid and vapor phases, and K_1 is the thermodynamic formation constant,

$$K_1 = \frac{a_{AB}}{a_A a_B} = \frac{c_{AB}}{c_A c_B} \frac{\gamma_{AB}}{\gamma_A \gamma_B} \tag{45}$$

The GC distribution coefficient on a column containing pure I would K_R^0, whereas on a column containing dissolved ligand B it is

$$K_R = \frac{V_N}{V_3} \tag{46}$$

where V_N is (ideally) the net retention volume fully corrected for gas phase and chromatographic nonideality, liquid surface effects, etc., and V_3 is the total volume of stationary phase (I + B). Combining these equations.

$$K_R = K_R^0(1 + K_1 a_B) \tag{47}$$

where B and I form an ideal solution, $\gamma_B = 1$, and

$$K_R = K_R^0(1 + K_1 c_B) \tag{48}$$

Martire et al. [257] have suggested GC methods for determining nonideal solution behavior. Where this equation holds, to evaluate K_1 one must prepare several five or six columns of varying c_B, and evaluate K_1 and K_R^0 from the slope and intercept of the linear experimental K_R versus c_B plot. The K_R^0 value should agree with the value determined for pure I, and can be used as above for the determination of γ_{AI}^{∞} values.

The Cadogan-Purnell method has been applied recently to hydrogen-bonded complexes between aliphatic alcohols and didecyl sebacate [258], bicyclic alcohols, and tris(p-tert-butylphenyl) phosphate [259]; to charge transfer complexes between aromatic hydrocarbon electron donors and di-n-propyl tetrachlorophthalate [71], tetra-n-butyl pyromellitate [82], 2,4,7-trinitro-9-fluoroenone [130], 1,3,5-trinitrobenzene [125], and dienes and aromatic hydrocarbons with 2,4,7-trinitrofluorenone and 2,3-dichloro-5,6-dicyanobenzoquinone [260]; and to π complexes between the nitroxide radical (from dissociation in solution of tert-butyl mesityl nitroxide) in squalane solvent and aromatic hydrocarbons and the monomers styrene, methyl methacrylate, methyl acrylate, and methacrylonitrile [261].

2. Martire-Riedl Method*

Rather than a solution of ligand in an inert solvent, pure ligand (B) is used as the stationary phase. Retention volumes of complexing (A) and noncomplexing (N) solutes are compared on the ligand column with those on columns containing a noninteracting reference solvent (R) of the same molecular shape, size, and polarizability as the ligand B. Thus, to study interactions between di-n-octyl ether and various haloalkanes, the corresponding alkanes were used as the inert solutes and octadecane was chosen as the reference solvent [98]. For this situation, Martire and Riedl show

$$\frac{K_R^{A,B}}{K_R^{A,R}} \frac{K_R^{N,R}}{K_R^{N,B}} = 1 + K_1 a_B \tag{49}$$

*See Martire and Riedl [93].

in which $a_B = 1/v_B^0$ for $\gamma_B = 1$ (based on the infinite dilution convention for γ_B, i.e., independent of the complex formed in solution).

Thus, the Martire-Riedl approach is experimentally simpler since retention data are required on only two columns, but it requires additional assumptions regarding the similarity of the reference and interacting species. It is also restricted to 1:1 complexes, whereas the Cadogan-Purnell method can be extended to higher orders. The Martire-Riedl method was applied to hydrogen bonding between alcohols and di-n-octyl ether and di-n-octyl ketone [93]; and to molecular association of haloalkanes with di-n-octyl ether and di-n-octyl thioether [98], with di-n-octylmethylamine and tri-n-hexylamine [116], and with di-n-octylmethylamine [257].

As noted above, these two approaches have been compared by Martire et al. [116, 257] for three di-n-octylamine-haloalkane systems at 40° C, and as shown in Table 10, the two methods produce results in experimental agreement; they should, since Martire et al. [257] pointed out the fundamental relationship between the two methods.

De Ligny et al. [262, 263] were interested in studying Ag^+-alkene complexation, which requires the use of a solvent for the Ag^+ salt. A theoretical description incorporating elements of both these methods was derived independently [262]. This is required because experimentally one must vary the concentration of Ag^+ in solution, but because K_R^0 would change with changing ionic strength, reference columns with a noninteracting salt in the same solvent are used. Thus, for the Ag^+-alkene studies, de Ligny et al. [263] used 0.25, 1, 2, and 3 M $AgNO_3$ in ethylene glycol. The reference solutes were the alkanes corresponding to the alkenes under study, while the reference solvents were solutions of $LiNO_3$ in ethylene glycol of the same molarity as the $AgNO_3$ columns. Heavily loaded columns (40 wt %) were used to minimize interfacial adsorption effects. Their results agree well with earlier chromatographic measurements of Cvetanovic et al. [264], but not with those of Muhs and Weiss [265]; the discrepancy was attributed to solute adsorption on the solid support in the case of the more lightly loaded columns of Muhs and Weiss.

Sievers et al. [266, 267] have made relative retention measurements on the interesting systems of organic nucleophiles with the lanthanide NMR shift reagents such as tris(1,1,1,2,2,3,3-heptafluoro-7,7-dimethyl-4,6,-octandianato)europium (III). Unfortunately no quantitative interpretation of the data is possible. An approach such as one of those discussed above was not (inexplicably) used, but instead only retention times in squalane and in squalane plus shift reagent were measured.

How do GLC results agree with independently determined values? Purnell and Srivastava [130] have made the only direct comparison. The π complexes between 2,4,7-trinitro-9-fluorenone (dissolved in various di-n-butyl esters) and aromatic hydrocarbons were measured by GLC, UV

TABLE 10

Thermodynamic Association Constants with Di-n-octylamine at 40.0° C

Solute	Method of Cadogan and Purnell	Martire-Riedl method	
		Ref. 257	Ref. 116
$CHCl_3$	0.405 ± 0.019	0.403	0.394
CH_2Cl_2	0.179 ± 0.014	0.187	0.183
CH_2Br_2	0.222 ± 0.004	0.219	0.224

spectroscopy, and NMR spectroscopy. For these particular systems, the spectroscopic methods were shown to produce data inconsistent and unreliable, both among themselves and between the NMR and UV data. The GLC values, on the other hand, while not proven accurate, were reasonable, self-consistent, and amenable to rational interpretation. Clearly, further comparative studies in additional systems are warranted.

B. Liquid Chromatography

There are many examples here, too, of the analytical use of specific molecular interactions for increasing selectivity in LC. As in GC, one of the more familiar examples is the use of Ag^+-π electron interactions for resolving unsaturated compounds. This technique, sometimes called argentation chromatography [268], uses an Ag^+ salt dissolved in glycol or adsorbed onto silica gel. Mikes et al. [269] observed a proportionality between retention times of butenes and the stability constants with Ag^+ ions. Similarly, aromatic compounds are often separated in the liquid phase with the aid of charge transfer complexation. Thus, Godlewicz [270] was the first to impregnate silica gel with compounds such as trinitrobenzene and picric acid, to obtain selective retention of aromatic hydrocarbons. Tye and Bell [271] used solutions of 1,3,5-trinitrobenzene in glycols for the LLC separation of polynuclear aromatic hydrocarbons; the additive was shown to be essential to the separation. Ayres and Mann [272] synthesized a poly-[3-(2,4-dinitrobenzyl)dinitrostyrene] resin for the separation of similar compounds. More recently, Orr separated organic sulfides by LLC using stationary phases containing mercuric acetate dissolved in 5 to 70% aqueous acetic acid solutions [273], or 67% aqueous $ZnCl_2$ [274], with hexane eluents. Many other examples are at hand, and in each case the analytical separation depends on the formation of a molecular complex, an Ag^+-olefin "salt," a hydrogen-bonded "complex," or a charge transfer complex. In each case no separation was gained in the absence of reactant to retain certain solutes selectively or to extract them selectively from the stationary phase. That in the $AgNO_3$-butene analysis [269], chromatographic K values agreed with independently measured values, indicates the possibility of determining complex association constants from LLC retention volumes. Again, by studying the fundamental basis for selectivity, methods are developed for studying the molecular association phenomena, the equilibrium constants, heats of reaction, and related quantities. Until recently, quantitative LC measurements were restricted to inorganic aqueous systems [275]; however liquid-liquid partition (solvent extraction) has long been a standard technique for investigating molecular association [276-278].

At the 1969 Eastern Analytical Symposium in New York City, I presented a discussion of how one could apply Purnell's treatment [255] of GC complexing to LC. As might be expected, results similar in form were obtained for the several different cases considered; for example, an equation of the exact form as Eq. (48) resulted for the analogous case of solute-ligand association in the stationary phase using a noninteracting eluent insoluble in the stationary phase. No data were found in the literature that could be directly compared to the results obtained using these equations. However, the equations were shown to be able to represent correctly the form of data for various systems, and reasonable values for equilibrium constants were derived. Thus, for the symmetrical trinitrobenzene-polynuclear aromatic hydrocarbon system [271], reasonable values of K_1 for fluoranthene and pyrene, which showed the expected variation with solvent, were deduced. For the acetonitrile-aromatic hydrocarbon solute systems [197], using essentially what has become the Martire-Riedl approach, values of K_1 for the association of light aromatics with acetonitrile of the order of 0.45 liter/mole were calculated, which are reasonable for weak charge transfer complexes [204, 278]. This work has not yet been published.

More recently, Freeman and co-workers [279-282] have devised a new and systematic basis for chemical separations using interactive gel networks with liquid eluents. Interactive gels are solvent-swollen cross-linked polymers containing interactive or reactive groups. The separation is governed by a combination of steric exclusion and chemical interactions among the gel, the solute molecules, and the solvent. A comprehensive theory considering solute-gel, solvent-gel, solvent-solute, and solute-solute interactions in relation to interactive gel chromatography was derived [280]. GPC (in which the gel in noninteractive), ion exchange chromatography (in which the gel is reactive), and interactive gel chromatography (in which some sort of acceptor-donor complex is formed) are unified in this approach [281]. Thus, in GPC, the distribution coefficient K can be defined as the equilibrium ratio of solute (A) inside and outside the pores of the gel; using Purnell's notation,

$$K_R^0 = \frac{c_A^{gel}}{c_A^{liq}} \tag{50}$$

If an interactive site (G_{gel}) inside the pores is accessible to A in the gel, the acceptor-donor equilibrium can be represented as

$$A\ (gel) + G_{gel} \overset{K_1}{\rightleftharpoons} AG_{gel} \tag{51}$$

if the solvent does not participate. Thus, as in GLC, if K_R represents the effective chromatographic partition coefficient, in this case

$$K_R = K_R^0 (1 + K_1 c_{G_{gel}}) \tag{52}$$

where $c_{G_{gel}}$ represents the molar concentration of the functional interacting moiety in the gel. With a polystyrene-divinylbenzene gel, a K_1 value of 0.39 liter/mole was obtained [279] for the I_2-aryl complex, which is a reasonable value in comparison with the value 0.20 liter/mole for the I_2-xylene complex. Similarly, reasonable values were deduced for alcohol and carboxylic acid solutes on a basic poly-[2-methyl-5-vinylpyridine] gel cross-linked with divinyl benzene. Solvent effects were noted and accounted for quantitatively.

IX. LIQUID INTERFACIAL EFFECTS IN CHROMATOGRAPHY

Martin [283] was the first to suggest that solute adsorption at the gas-liquid interface can contribute to retention in certain systems. Although initially greeted with skepticism, if not outright incredulity, his hypothesis has since been borne out and elaborated upon by a number of groups. Clearly, if one is interested in studying bulk liquid phenomena, any additional retention mechanisms will either obviate the study or require corrections. Martin [283] proposed a method to isolate interfacial adsorptive contributions: he defined a corrected (for column pressure drop) retention volume per gram of column packing (stationary liquid plus solid support), $V_{R_g}^0$, and set at infinite dilution

$$V_{R_g}^0 = K_L V_L' + K_I A_L' \tag{53}$$

where K_L is the molar bulk liquid partition coefficient; K_I is a surface partition coefficient, the ratio of the equilibrium number of moles of solute per unit area of surface in excess of that in the bulk, to the number of moles of solute per unit volume of gas phase; V_L' and A_L' are the volume of stationary liquid phase per gram of packing and available surface area of liquid per gram of packing, respectively. This equation suggested a test of the hypothesis: A plot of $V_{R_g}^0/A_L$ versus V_L'/A_L gives K_I as the intercept and K_L from the slope. Indeed, for two polar stationary phases, β, β'-thiodipropionitrile and 1-chloronaphthalene, 13 nonpolar hydrocarbon solutes gave straight-line plots and allowed evaluation of K_L and K_I values, which were independent of the solid support used. Interfacial adsorption accounted for 8 to 98% of the retention volume.

Martire [47] critically reviewed the subsequent chromatographic and static work which confirmed and amplified Martin's hypothesis. To summarize briefly, careful static measurements using a unique apparatus were made by Martire and associates [69, 120], of the bulk and surface partition coefficients. The former agreed well with the chromatographic K_L values while the K_I values showed substantial discrepancy, possibly because of the use of erroneous surface areas in the GLC calculations. Martire [111] studied C_1-C_4 alcohols on glycerol and β, β'-oxydipropionitrile; only methanol appeared to adsorb on the liquid surface. Consideration of his own and other data led Martire to conclude that interfacial liquid adsorption can be expected in systems in which either (a) solutes, especially nonpolar ones, are slightly soluble in a polar liquid phase or (b) both solute and solvent are of comparable high polarity, solubility is high (low γ_{13}^{∞}), and adsorption results from surface hydrogen bonding or strong dipolar attraction. Pecsok and Gump [43], using apparatus similar to that of Martire et al. [120], calculated from their static data that for polar solutes-nonpolar solvents there should be a large K_I contribution in GLC, in contradiction to the experimental measurements of Martin [283] and Urone and Parcher [284] on such systems. Pecsok and Gump could not from static measurements confirm the occurrence of liquid interfacial adsorption in the polar solute-squalane system, although there were indications. However, they did verify the profound effect of the solid support on solute uptake, especially at low ($< 2\%$) loadings, in support of the work of Urone and Parcher [284, 285]. Conder et al. [286] have shown that the apparent discrepancy between static and GLC results can in principle be reconciled. But to continue, Martire and Riedl [93] noted peak asymmetry and pronounced retention variation with sample size for alcohol solutes on n-$C_{17}H_{36}$ stationary phase on a deactivated support. They adopted a somewhat empirical procedure for extrapolation to zero sample size. Because V_g^0 values were the same on 10 and 15% (w/w) columns, they concluded that liquid interfacial adsorption was negligible (it has since been shown [286], however, that this use of only two columns is not a sufficient criterion), and attributed the sample size effects to residual support adsorption. Subsequently, Liao and Martire [287] did find a V_g^0 dependence on liquid loading for 2-propanol and 2-methyl-1-propanol on n-$C_{18}H_{38}$ solvent. They thus concluded that through some small systematic error, their previous work [93] had overlooked a small liquid interfacial effect. K_I values in the same region as those of Pecsok and Gump [43] were estimated from their data.

Parcher, who has all the while maintained that any adsorptive effects for polar solute-nonpolar solvent are attributable to the solid support alone, recently obtained further support from both elution and frontal analysis measurements. Parcher and Hussey [94] found no dependence of V_g^0 on V_L. They calculated γ_{13}^{∞} values for n-hexane and 2-propanol in n-heptadecane in good agreement with those of Martire and Riedl [93] and through frontal analysis determined that for a deactivated support coated with varying amounts of n-$C_{17}H_{36}$, $q = Kc$. They thus concluded that for 2-propanol on n-heptadecane at 50°C, there is no liquid surface adsorption.

Although it is clear that mixed retention mechanisms occur in a variety of systems (although which ones are not always predictable), the problem is how to deal with this circumstance if one is interested in K_L. In addition, one might be interested in how to study interfacial phenomena chromatographically. Several groups have suggested approaches to these problems [59, 76, 93, 286, 288-293]. The method discussed here is one of the more comprehensive considerations [76, 286, 288].

Equation (53) accounts for bulk liquid and liquid interfacial adsorption only. It also assumes that conditions are at effectively infinite dilution for all equilibria. However, common solid supports are weak siliceous adsorbents and can interact with polar solute molecules, unless totally deactivated by chemical treatment. The total number of moles of solute sorbed n in equilibrium with a concentration c in the gas phase is

$$n = q_L V_3 + q_I A_I + q_S A_S = \sum_i q_i \phi_i \tag{54}$$

where the q terms are the equilibrium solute concentrations in the bulk liquid (q_L, mole/cc), in the gas-liquid interface (q_I, mole/cm^2), and on the exposed support or accessible liquid-support interface (q_S, mole/cm^2); V_3 is the total volume of stationary liquid phase in the column; A_I and A_S are the solute-accessible surface areas of the liquid phase and solid support, respectively; and ϕ_i represents V_3, A_I, or A_S. The contribution of each retention mechanism to V_N of a solute zone whose (ideal) gas phase concentration is c, is [18]

$$V_{N,i} = (1 - jy_0)K_i' \phi_i \tag{55}$$

where $K_i' = dq_i/dc_i$, the slope of the appropriate distribution isotherm, and $jy_0 \ll 1$ for elution chromatography with very small samples. Only in the limit of true effective infinite dilution does $K_i' = K_i^0$. Thus, the observed net retention volume can be represented by

$$V_N = \sum_i V_{N,i} = \frac{dn}{dc} = K_{obs} V_3 \tag{56}$$

and

$$\frac{V_N}{V_3} = K_{obs} = K'_L + \frac{K'_I A_I}{V_3} + \frac{K'_S A_S}{V_3} \tag{57}$$

and at infinite dilution,

$$\frac{V_N}{V_3} = K_{obs} = K_L + \frac{K_I A_I}{V_3} + \frac{K_S A_S}{V_3} \tag{58}$$

which is applicable when symmetrical peaks are observed (barring chromatographic nonequilibrium or extra-column effects). The curvature of K'_I and K'_S may be far greater than that of K'_L, so that infinite dilution overall may correspond to a much smaller concentration than that in the bulk liquid.

These equations suggest that to determine K_L, which leads to γ_{13}^{∞} and other bulk solution fundamental information, one can plot V_N/V_3 versus $1/V_3$. This procedure avoids the considerable problem of measuring A_I; it works even when all three mechanisms exist simultaneously; and in cases where the partition isotherm becomes linear at sample sizes larger than those for the other two isotherms, it yields K_L values for sample sizes at which K'_I and K'_S have not reached their infinite dilution values. This approach also suggests that experiments in which the solvent/support ratio is varied (more than two different loadings are generally required) can best be studied at constant concentration c using the procedure of Conder [288]. Finally, this treatment [286] suggested several qualitative diagnostic procedures to indicate which mechanisms might be participating.

A suitable test of these considerations was made [76, 104] using the worst of all systems, polar solutes (alcohols) in a nonpolar stationary phase (squalane). Multiple sorption produces peak asymmetry persistent even at the smallest sample sizes. It was found that peak maximum retention volumes coincide with a point at the same height on the diffuse edge of a peak of larger size. One could thus construct this edge of a large, asymmetrical peak from a plot of retention volume versus peak height. Conder [288] suggests that a series of columns identical in all aspects except V_3 be prepared, and arbitrarily numbered $j = 0, 1, 2, 3, \cdots$, where column 0 is the most heavily loaded. Small (0.1 μl) samples are injected into each column to produce a large asymmetrical peak. All detector conditions and recorder settings must remain the same. Conder shows that a set of net retention volumes $V_{N,i}$ corresponding to elution at the same vapor phase concentration c is generated by locating corresponding points h_j on the diffuse side of each peak at which the ratio $h_j/V_{N,i}$ is constant. In practice, one can pick a point on the peak from column number 0, of height h_0, corresponding to a retention volumn $V_{N,0}$ at that point. For the other columns, one must find sets of h_j such that

$$\frac{h_0}{V_{N,0}} = \frac{h_j}{V_{N,j}} \tag{59}$$

Given the four or so chromatograms from columns numbered 0, 1, 2, 3, ··· and a slide rule, the procedure is simple in practice. These values of $V_{N,0}$, $V_{N,1}$, ···, $V_{N,j}$ for the j columns of different V_3 are then corrected for the column pressure drop and plotted versus $1/V_3$. This plot, which is nonlinear, is then extrapolated to $1/V_3 = 0$, i.e., to an infinite solvent volume which by definition must correspond to infinite dilution conditions for the solute. As an aid to and a check on the extrapolation, since the values obtained at any h_0 correspond to a finite gas phase concentration c, this procedure should be repeated at several, say four, values of h_0 to generate a family of curves (e.g., see Cadogan et al. [76, Fig. 1], or Liao and Martire [287, Figs. 3 and 4]). The intercept at $1/V_3 = 0$ is K_L^0, the true infinite dilution value. Clearly, the procedure is tedious and introduces experimental error and some subjective judgment into the results, but is is relatively simple given the true complexity of these systems, and allows bringing the considerable advantages of GLC to otherwise inaccessible systems. In addition, if A_I and/or A_S are known, K_L^0 can be evaluated as above and

$$V_N - K_L^0 V_3 = K_I A_I + K_S A_S \tag{60}$$

plotted versus A_I or A_S will produce a curve which will yield values of K_I or K_S. Conder has also suggested another procedure [288].

Other routes to K_L^0 have been put forth. Martire and Riedl [93] found that in a similar system, as sample size was decreased, the initial retention volume remained constant at first and then increased, while over the same range the peak maximum retention volume decreased linearly before increasing. These decreasing peak maximum values at constant initial retention volumes were thus plotted versus sample size and extrapolated to zero sample size. The actual sample sizes were in the 3- to 0.5-μl range. Here, only one column is required, which represents a considerable saving in time and effort. A comparison of the two methods [287] showed that the Martire-Riedl method gave K_L^0 values consistently a few percent larger than those generated by the Conder procedure, but within experimental error. In the latter work [287] an alternative procedure based on Conder's method was developed. Several columns are required, but a set of h_j values on the diffuse side of the peak is determined by computer iteration which both satisfies Eq. (59) and gives the best fit to the bulk solution equation $V_N = K_L{}^0 V_3$. This produces a linear extrapolation with possibly less opportunity for subjective judgment than the original Conder method. The resulting $K_L{}^0$ values were in agreement with both of the other methods, within experimental error.

As part of a continuing study of the surface properties of water by GC [38], Karger and Liao [294] studied the adsorption and partition phenomena of nonelectrolytes with water and aqueous tetraalkylammonium bromide solutions. Values of K_L and K_I were determined using the method of Martin [283]. For surface area A_I, the value calculated by Martire et al. [120] for thiodipropionitrile was assumed to apply approximately to a column packed with 20% by weight of H_2O stationary phase. Since the water solubility of octane is trivial, all retention of octane on the water column was attributable to surface adsorption on the liquid, enabling calculation of K_A. Although these values are small (e.g., 1% of that of cyclohexane on thiodipropionitrile), it is probably because $K_L \cong 0$ [see Eq. (61)]. K_A value for columns of smaller water loadings, or with columns with different solid supports, or with columns containing aqueous salt solutions as stationary phase, was assumed constant, so that the ratio of octane's retention volume on different columns was used to estimate their A_I value. The solutes CCl_4, CH_2Cl_2, $CHCl_3$, benzene, and toluene were found to have appreciable K_L values. Both K_A and K_L were found to increase from water to tetramethylammonium bromide solutions to tetraethylammonium bromide solutions. Salting-in constants were also calculated. For the same system, Weiner et al. [293] used the powerful statistical technique called factor analysis [295] to determine how many retention mechanisms contribute to retention. As presumed previously, they found that for octane and other saturated alkane solutes, retention data could be represented by a single factor, whereas for the polar and unsaturated solutes two terms were required to represent the data within experimental error.

Cruickshank et al. [51, 59] used a variation of the V_N/V_3 versus $1/V_3$ plot method. Here, a zero flow rate molar retention volume, V_N/n_3, is plotted against the weight of solid support per mole of solvent, and extrapolated in the same fashion to achieve infinite dilution. They also noted the effect of interfacial adsorption of carrier gas, CO_2, onto glycerol.

GLC measurements can be used as well to study interfacial phenomena. James et al. [296] suggested a method for estimating interfacial mass transfer resistance coefficients (also called transmission or accommodation coefficients) from peak broadening on columns dominated by this effect.

Surface adsorption constants can be determined in principle. Although K_L values are readily accessible, to derive K_I requires knowledge of A_I.

This is not at all straightforward. The usual BET method uses an inert gas adsorbate, which because of the nature of the porous structure of chromatographic supports, probably sees a greater surface area than would a larger molecule [297]. The BET method also requires an initial high-vacuum, high-temperature outgassing, which is clearly impossible with coated GLC supports, but omission of this step introduces errors into the results. Adsorption is generally carried out at low temperatures to condense the adsorbate. With GLC liquids, freezing and subsequent cracking could produce spurious area values. There is, in short, no generally acceptable direct method for the accurate determination of liquid surface areas of GLC packings. Probably the best method available [69] is the independent static determination of K_I combined with the GLC $K_I A_I$ value.

If K_I values are known, the limiting slope of the surface tension-mole fraction diagram is directly obtained, since K_I can be shown [69] to be related to this by

$$K_I = -\frac{K_L^0 M}{RT\rho}\left(\frac{\partial \Gamma}{\partial x}\right)^{\infty} \tag{61}$$

where M and ρ are the molecular weight and density of the solvent, respectively, and $(\partial \Gamma/\partial x)^{\infty}$ is the limiting value for the variation of surface tension with solution composition.

Eon and Guiochon [291] showed how to calculate infinite dilution interfacial "phase" activity coefficients from K_I values, and presented these activity coefficients, bulk limiting activity coefficients, and values of $(\partial \Gamma/\partial x)^{\infty}$ for a large number of solutes in or on thiodipropionitrile, and three halocarbons on water. Interfacial adsorption isotherms and parameters calculable from these were deduced by Suprynowicz et al. [292].

In liquid chromatography, as discussed (Sec. V.C.), the problem of interfacial adsorption may well exist, although no direct measurements have been made. Techniques presumably similar to those used in GLC will probably be required. The considerations of Conder et al. [286] are applicable to all forms of liquid partition chromatography, GLC and LLC, which suggests similar methods will handle interfacial effects in both.

X. SURFACE PROPERTIES OF SOLIDS AND CATALYSTS; REACTION KINETIC STUDIES BY GAS CHROMATOGRAPHY

Gas chromatography has been widely used to study physical and chemical properties of solid surfaces, especially those used as catalysts. Surface areas, adsorption isotherms, kinetics and energetics of adsorption, pore structure and pore-size distribution, and diffusivities within catalysts are physical parameters accessible to GC determination. GC columns packed with reactive or catalytic materials can be used in a variety of ways not only to study the catalyst, but to determine the kinetics of reactions occurring on its surface.

Since there have recently been published several detailed and comprehensive reviews of the subject, only these will be cited here. A recent book [298] covers the larger area of GSC. Physical and chemical properties of catalysts and the GC study of catalysis are the subjects of two recent comprehensive reviews [299, 300]. Steingaszner reviewed microreaction GC techniques [301]. Van Swaay [302] covered the study of reaction kinetics through distortion of chromatographic elution peaks. Phillips, who has made many valuable contributions to this area of chromatography, summarized his and related work on what he calls "the chromatography of reactions" [303]. Most recently, Langer and Patton [304] covered in a most informative, lengthy article the use of the GC column as a chemical reactor to study reactions occurring during the passage of compounds through it. Articles on adsorption chromatography and on the study of reaction kinetics were published in the preceding volume of this series [305, 306].

XI. MEASUREMENT OF DIFFUSION COEFFICIENTS USING CHROMATOGRAPHIC APPARATUS

Among the factors contributing to chromatographic peak spreading are mobile phase longitudinal diffusion and stationary phase mass transfer resistance. Both processes are diffusion controlled. Quantitative treatment of chromatographic nonideality [49] relates the observed spreading of peaks in GLC or LLC columns explicitly to mobile and stationary phase solute diffusion coefficients. In general, however, to calculate these parameters from peak broadening requires too many approximations for accurate determination. The mathematical analysis of spreading of a solute zone in an empty tube leads to a far simpler expression, and opens the way to accurate, rapid determination of diffusion coefficients. Most of this work has been done using gases as the mobile phase, although recently Grushka and Ouano [332-336] have used liquid carriers to study liquid phase diffusion. The subject has been recently reviewed by Grushka [307] in Volume 12 of this series, and therefore will not be covered here.

XII. MISCELLANEOUS PHYSICAL PROPERTIES BY CHROMATOGRAPHY

In addition to its use as an analytical tool for determining physical properties of compounds, retention behavior can be related, usually empirically and approximately, to various properties. Thus, although not as accurate as conventional methods, chromatography is fast and convenient, and since it is a separation techinque, often one does not have to use very high purity compounds. Only trivial amounts are required, a definite advantage for research materials.

A. Latent Heat of Vaporization*

Where Eq. (7) represents V_g^0 with sufficient accuracy, it is readily shown that the specific retention volume is related to the infinite dilution partial molar heat of evaporation of solute from solution, $\Delta\bar{h}_s$, by

$$\ln V_g^0 = \frac{\Delta\bar{h}_s}{RT} + c \tag{62}$$

Since

$$\Delta\bar{h}_s = \Delta h_v - \bar{h}_{13}^{\infty} \tag{63}$$

in solutions with partial molar enthalpies of mixing, $\bar{h}_{13}^{\infty}$, much smaller than the solute heat of vaporization Δh_v, the slope of a linear plot of $\ln V_g^0$ versus $1/T$ gives Δh_v. This method is generally applicable to nonpolar stationary phases. One practical example of this application was given by Sie et al. [309] who determined the heats of vaporization of 18 inorganic chlorides and oxychlorides using Kel-F 40, a fluorinated polymer, as the stationary phase. The GC values were in substantial agreement with literature values determined from vapor pressures. Heats of solution correlate empirically with induction energies, and GLC-derived values have been used to calculate these quantities [310].

B. Boiling Point

Trouton's rule can be applied directly to the Δh_v data, since

$$\frac{\Delta h_v}{T_b} = 23 \text{ cal/mole} \cdot \text{deg} \tag{64}$$

The use of Trouton's rule is better in conjunction with Eq. (62). Since

*See Purnell [308].

$$T_b = a\,\Delta h_v \tag{65}$$

$$\ln V_g^0 = bT_b + d \tag{66}$$

where a, b, and d are constants for a given column, conditions, and homologous series of compounds. Thus, the slope of a plot of $\ln V_g^0$ versus T_b can be established for two or more members of the series, and the boiling points of the others can be interpolated. The linearity of these plots gives rise to the designation "boiling point column" for nonpolar stationary phases used to separate compounds of the same functional group. On such columns, there is a tendency for either $\bar{h}_{13}^{\infty} \ll \Delta h_v$ or $\bar{h}_{13}^{\infty}$ to remain constant for the group of compounds studied, so the approximations involved are good. This is also the basis for the technique of GC distillation [311]. Recently, Sojak et al. [312] calculated the boiling points of 11 straight-chain dodecenes from retention data. Earlier, Baumann et al. [313] measured the boiling points of 35 C_{10}-C_{15} straight-chain alkyl benzene isomers from GLC data. If empirical corrections for $\bar{h}_{13}^{\infty}$ were made, the GC boiling points were estimated to be accurate to $\pm 1.0°$ C.

C. Vapor Pressure

For systems with small deviations from thermodynamic solution equilibrium ideality, it can be shown [308] that at constant temperature,

$$\ln V_g^0 = -e \ln p_1^0 + f \tag{67}$$

where e and f are again constants. As before, the straight-line $\ln V_g^0/\ln p_1^0$ plots for a given family of compounds can be applied to predict p_1^0. An example is given by Rose and Schrodt [314] who applied the GC method to the determination of vapor pressures of various series of fatty esters, fatty acids, alcohols, hydrocarbons, and chloroalkanes. These values also allowed Poizat molecular structure factors to be determined. The latter were used to predict boiling points in good agreement with independent measurements.

Duty and Mayberry [315] used Eq. (7) in conjunction with Dalton's law to determine the total vapor pressure above solutions of hydrocarbons, and the boiling points of the mixtures. Good agreement between the latter and static values was obtained, although there was substantial, unaccounted for discrepancy in the case of the total pressure.

D. Freezing Points and Transition Temperatures

Since a discontinuity in solution properties and thus GLC retention occurs at the freezing point, study of retention as a function of temperature in the vicinity of this temperature provides a means for its estimation. The solute serves, again, as a probe, as in the case of the polymer studies referred to (Sec. VII). Other sorbent transitions can be studied in a similar way, as for example phase changes in liquid crystals. This subject has been comprehensively reviewed by McCrea [316].

E. Solvent Molecular Weight

Martire and Purnell [317] applied the Flory-Huggins theory to γ_{13}^{∞} in conjunction with Eq. (7), to show that if two solutes x and y are chromatographed on two chemically similar stationary phases A and B, the molecular weight of one of which is known, then approximately

$$\frac{1}{M_B} = \frac{\rho A}{\rho_B M_A} - \frac{\ln[(V_x/V_y)_A/(V_x/V_y)_B]}{\rho_B[(v_1^0)_y - (v_1^0)_x]} \tag{68}$$

where the V_i terms are net retention volumes. Thus, for x = n-hexane, y = n-heptane, A = eicosane, and B = squalane, M_B is determined by GLC to be 422 g/mole versus the known value of 422.8 g/mole. For x = n-heptane, y = n-octane, A = polypropylene glycol 400, and B = polypropylene glycol 1200, the measured M_B is 1220 versus 1260 g/mole by a cryoscopic method. The accuracy of this GC method deteriorates with molecular weight increasing above about 1500, but below this could be applied with advantage to liquid silicones, polyglycols, paraffin oils, and similar compounds of low polarity.

F. Molecular Structure Determination

In much the same fashion as the correlations above, for homologous series of compounds $C_nH_{2n+1}X$,

$$\ln V_g^0 = gn + h \tag{69}$$

for GLC and LLC, and for both homologous series of solutes on a given stationary phase and series of solvents with a given solute [128]. Such plots have long been used for identification of unknown compounds. Thus, among many examples James [318] determined the degree of unsaturation of long-chain fatty acids from log-log plots of relative retention of Apiezon versus that on polyethylene glycol adipate. A similar grid pattern was found by Phillips and Timms [319] for silanes, germanes, and silicogermanes. The considerable advantages of GLC are quite apparent in this very interesting work, since these highly reactive compounds were generated in small quantities, and handled and studied in a closed system with an inert atmosphere.

Structural assignment can also be based on the regularity of GLC retention behavior. Thus, for example, Hively [320] was able to determine the cis/trans configurations of monoolefin pairs, since he found experimentally that the ratio of retention times of the cis/trans isomers was greater on polar columns than on nonpolar columns. This led to a correction of the American Petroleum Institute (API) published assignments of 3-methyl-2-pentene and 3,4-dimethyl-2-pentene.

G. Miscellaneous

For much the same reasons as those that form the bases for these methods, linear relationships have been noted between log V_N and molar volume [321], polarizability, etc., for various homologous series.

Laub and Pecsok [260] found an empirical inverse correlation between charge transfer formation constants as determined by GLC as described above, and the solute vertical ionization potential, for solvents in which no steric effects interfere. Values accurate to ± 0.1 eV were calculated. The method is regarded as applicable also to the determination of electron affinities, but probably more useful is its application in sorting out steric effects on complex formation equilibria.

XIII. USE OF ANCILLARY EQUIPMENT

This section is concerned with a few examples of physical measurements made using components normally associated with gas chromatographs, but whose operation is independent of the chromatographic process; here the column serves as a sample purification and inlet system.

A. Thermal Conductivity Detector for Estimation of Critical Volumes

In a series of papers, Rosie et al. [322-325] related the relative molar response of the thermal conductivity cell to the molecular diameter and then to the solute's critical volume. The resulting equation was

$$V_{c_i} = \left(\frac{1.05RMR}{A} - 3.86\right)^3 \tag{70}$$

where RMR is the molar response of the detector toward compound i relative to that toward benzene, and

$$A = \frac{M_i - M_{He}}{M_{Bz} - M_{He}} \tag{71}$$

where M is molecular weight and Bz stands for benzene. For a wide variety of substances the agreement between values calculated from the thermal conductivity detector response and independently measured values was quite good, as might be anticipated for transport properties such as thermal conductivity. Group contributions to V_c were also determined which again proved quite accurate predictors.

B. Molecular Weights by Gas Density Detector

The gas density balance detector invented by Martin and James [326] responds to vapors in a manner directly related to vapor density and thus to molecular weight. When a weight w of vapor of molecular weight M is passed through the gas density detector in a constant flowing stream of carrier gas of molecular weight m, the integral response (such as total peak area A) is

$$w = kA \frac{M}{M - m} \tag{72}$$

where k is a constant for a given detector. This can be used by comparing peak areas for equal weights of two compounds of known and unknown molecular weight, in two different carrier gases, as was done by Liberti et al. [327]. They claimed molecular weights accurate to within 4% of their true values. Phillips and Timms [328] improved considerably upon this determination by combining the gas density detector response with a pressure-volume measurement. Molecular weights to better than 1% were obtained. This was subsequently used by them to identify positively the silane and germane isomers eluted from GLC columns [319].

C. Electron Capture Detector and Electron Affinity

The response of the electron capture detector of Lovelock [329] is based on the attachment of thermal electrons by molecules with large electron affinities. Wentworth and Becker [330] related this response to the absolute electron affinity, i.e., the energy associated with the process in which a gaseous molecule captures an electron to form a gaseous ion. Values obtained for some aromatic hydrocarbons correlated well with polarographically determined values, and agreed with theoretical calculations.

D. Spectroscopic Techniques

The use of combined GC or LC with mass spectroscopy, NMR, IR, UV, photoelectron spectroscopy, and so on, is the most powerful method for analysis of unknown mixtures. The GC serves as an inlet to the spectrometer (or the spectrometer serves as an expensive detector, depending on your point of view), presenting it with highly purified compounds, mixed of course with carrier gas. All the structural and other information from the spectrometer is available. McFadden [331] has written and edited a recent thorough discussion of this subject, which is beyond the scope of this chapter.

XIV. SUMMARY AND CONCLUSIONS

A great deal of diverse physicochemical information is available from chromatographic measurements. There are many advantages to this application. Although GLC may often appear to be "idiot proof" as an analytical tool, for physical measurements the technique must be used with considerable care and discretion, to avoid obtaining results that are "idiotic." There are definite limitations; contrary to popular opinion chromatography is not magic, and skill and imagination are required for its proper application.

The future would seem to lie in wider application of GLC and in developing the application of LC. It is encouraging for the former that such eminent chemical engineers as J. M. Prausnitz have started using chromatography as more than an analytical tool. Although only preliminary results have been obtained from LC, there are growing indications of wider application, and I for one am expanding upon earlier work in the area of solution thermodynamics.

ACKNOWLEDGMENTS

The support of the National Science Foundation under Grant 17551 is gratefully acknowledged, as well as the existence of the excellent facilities at the New York Public Library's Division of Science and Technology, the Mid-Manhattan Library, and the Chemist's Club Library.

GLOSSARY

Symbol	Definition or Equation Where First Used
a	0.175b; Eq. (29); thermodynamic activity.
b	B_{22}/RT
b_n^m	Eq. (30).
c	$(C_{222} - B_{22})^2/(RT)^2$; mobile phase solute concentration; Eq. (62).
c'	Chart speed, Eq. (2).
d_r	Distance on recorder chart from point of solute injection to top of the eluted peak.
f_1^0	Solute fugacity.
g^E	Free energy of mixing; Eq. (26).
$\overline{h}_{13}^{\infty}$	Partial molar enthalpy of mixing of solute with stationary phase at infinite dilution.
h_j	Height on peak; Eq. (59).
$\Delta\overline{h}_s$	Solute partial molar heat of solution.
Δh_v	Heat of vaporization.
j	Eq. (28).
x_k	Mole fraction partition ratio; Eq. (26).
m	Mobile phase.
n_3	Number of moles of stationary liquid phase.
n	Carbon number; Eq. (69).
p_{H_2O}	Vapor pressure of water.
p_i, p_o	Inlet and outlet GC column pressures.
p_1^0	Solute vapor pressure.
q	Solute concentration in stationary phase.
s	Stationary phase.
$\overline{s}_{13}^{\infty}$	Partial molar excess entropy of mixing of solute with stationary phase at infinite dilution.
t_r	Retention time.
t_m	Retention time of nonsorbed component: "dead time."

Symbol	Definition or Equation Where First Used
v_1^0	Solute molar volume.
v_{13}^{∞}	Solute partial molar volume at infinite dilution in stationary phase.
w_3	Weight of stationary phase in column.
w	Eq. (23).
x	Liquid phase mole fraction.
y	Vapor phase mole fraction.
y_0	Solute mole fraction in vapor phase at column outlet.
z	Gas compressibility.
A	Eq. (71); peak area.
A_L'	Surface area of stationary liquid phase per gram of column packing.
A_L	Surface area of stationary liquid phase.
A_S	Surface area of solid support accessible to solutes.
B	Second virial coefficient.
B_{eff}	Eq. (39).
C	Third virial coefficient.
F_c	Flow rate corrected to column outlet conditions.
F_a	Flow rate at measurement conditions.
H_{13}	Henry's constant.
J_m^n	Eq. (10).
K	Vapor-liquid equilibrium constant (= x/y).
K^0	Equilibrium zero column pressure distribution coefficient.
K_a	Eq. (20).
K_R	Eq. (43).
K_1	Eq. (44).
K_R	Eq. (46).
K_I, K_L	Eq. (53).
K_S	Eq. (58).

Symbol	Definition or Equation Where First Used
M_3	Stationary phase molecular weight.
M_w, M_n	Weight-average and number-average molecular weights.
P	Total system pressure.
P_a	Flow meter pressure.
R	Gas constant.
T	Temperature.
T_b	Boiling point temperature.
V_R^0	Peak maximum retention volume corrected for column pressure drop.
V_N	Net retention volume ($= V_R^0 - V_2$).
V_2	Column interstitial volume = retention volume corrected for pressure drop of a nonsorbed component = "dead volume."
V_3	Volume of stationary phase in column.
V_g^0	Specific retention volume.
V_p, V_e	Polymer pore volume, elution volume in Eq. (40).
V_L'	Volume of stationary phase in column per gram of packing; Eq. (53).
V_c	Critical volume; Eq. (70).
1	Solute.
2	Mobile phase.
3	Stationary phase.
γ_{13}^{∞}	Infinite dilution (limiting) solute activity coefficient.
γ_1^{∞}	Eq. (13).
β'	Eq. (13).
ζ', κ'	Eq. (14).
λ, ϕ, ψ	Eq. (15).
$\overline{\mu}_{13}^{\infty}$	Partial molar excess free energy of mixing; Eq. (25).
α	Eq. (37).

Symbol	Definition or Equation Where First Used
$[\eta]$	Intrinsic viscosity.
Γ	Eq. (61).
ρ	Density.
∞	Infinite dilution.

REFERENCES

1. A. J. P. Martin and R. L. M. Synge, Biochem. J., 35, 1358 (1941).
2. E. Glueckauf, Nature, 156, 748 (1945).
3. M. R. Hoare and J. H. Purnell, Research (London), 8, 541 (1955).
4. A. B. Littlewood, C. S. G. Phillips, and D. T. Price, J. Chem. Soc., 1955, 1480.
5. P. E. Porter, C. H. Deal, and F. H. Stross, J. Am. Chem. Soc., 78, 2999 (1956).
6. S. P. Cram and R. S. Juvet, Anal. Chem., 46 (5), 101R (1974); 44 (5), 213R (1972); 42 (5), 1R (1970).
7. J. H. Purnell, Annu. Rev. Phys. Chem., 18, 81 (1967).
8. T. Gaumann, Annu. Rev. Phys. Chem., 15, 125 (1965).
9. I. Brown, Annu. Rev. Phys. Chem., 15, 147 (1965).
10. H. W. Habgood, Annu. Rev. Phys. Chem., 13, 259 (1962).
11. R. A. Keller, G. H. Stewart, and J. C. Giddings, Annu. Rev. Phys. Chem., 11, 347 (1960).
12. C. J. Hardy and F. H. Pollard, J. Chromatogr., 2, 1 (1959).
13. J. H. Purnell, Endeavour, 23 (90), 142 (1964).
14. R. Kobayashi, P. S. Chappelear, and H. A. Deans, Ind. Eng. Chem., 59 (10), 63 (1967).
15. J. C. Giddings and K. L. Mallik, Ind. Eng. Chem., 59 (4), 19 (1967).
16. E. Cremer, Z. Anal. Chem., 170, 219 (1959).
17. C. L. Young, Chromatogr. Rev., 10, 129 (1968).
18. J. R. Conder, in Progress in Gas Chromatography (J. H. Purnell, ed.), Wiley (Interscience), New York, 1968, p. 209.

19. A. N. Korol, Russ. Chem. Rev., 41, 174 (1972).

20. M. S. Vigdergauz and R. I. Izmailov, Application of Gas Chromatography to the Determination of Physico-Chemical Properties of Substances, Izd. Nauka, Moscow, 1970, 159 pp., in Russian.

21. M. Goedert and G. Guiochon, Anal. Chem., 42, 962 (1970).

22. M. Goedert and G. Guiochon, Anal. Chem., 45, 1180 (1973).

23. M. Goedert and G. Guiochon, Anal. Chem., 45, 1188 (1973).

24. B. E. Bowen, S. P. Cram, J. E. Leitner, and R. L. Wade, Anal. Chem., 45, 2185 (1973).

25. H. Barth, E. Dallmeier, G. Courtois, H. E. Keller, and B. L. Karger, J. Chromatogr., 83, 289 (1973).

26. J. E. Oberholtzer and L. B. Rogers, Anal. Chem., 41, 1234 (1969).

27. R. Kaiser, Chromatographia, 2, 215 (1969).

28. B. Versino, Chromatographia, 3, 231 (1970).

29. C. A. Cramers, J. A. Luyten, and J. A. Rijks, Chromatographia, 3, 441 (1970).

30. O. Wicarova, J. Novak, and J. Janak, J. Chromatogr., 51, 3 (1970).

31. A. J. B. Cruickshank, M. L. Windsor, and C. L. Young, Proc. R. Soc. A 295, 271 (1966).

32. C. J. Chen and J. F. Parcher, Anal. Chem., 43, 1738 (1971).

33. T. H. Gouw and R. E. Jentoft, J. Chromatogr., 68, 303 (1972).

34. D. C. Locke, Adv. Chromatogr., 8, 47 (1969).

35. L. R. Snyder and J. J. Kirkland, Modern Liquid Chromatography, Wiley, New York, 1974.

36. A. Kwantes and G. W. A. Rijnders, in Gas Chromatography 1958 (D. H. Desty, ed.), Butterworth, London, 1958, p. 125.

37. P. E. Barker and A. K. Hilmi, J. Gas Chromatogr., 5, 119 (1967).

38. A. Hartkopf and B. L. Karger, Acc. Chem. Res., 6, 209 (1973).

39. J. R. Conder, Anal. Chem., 43, 367 (1971).

40. A. J. B. Cruickshank and D. H. Everett, J. Chromatogr., 11, 289 (1963).

41. A. J. Ashworth and D. H. Everett, Trans. Faraday Soc., 56, 1609 (1960).

42. G. F. Freeguard and R. Stock, in Gas Chromatography 1962 (M. van Swaay, ed.), Butterworth, London, 1962, p. 102.

43. R. L. Pecsok and B. H. Gump, J. Phys. Chem., 71, 2202 (1967).

44. J. Serpinet, J. Chromatogr., 68, 9 (1972).

45. J. Serpinet, J. Chromatogr., 77, 289 (1973).

46. J. Serpinet, Chromatographia, 8, 18 (1975).

47. D. E. Martire, in Progress in Gas Chromatography (J. H. Purnell, ed.), Wiley (Interscience), New York, 1968, p. 93.

48. R. P. W. Scott, Anal. Chem., 35, 481 (1963).

49. J. C. Giddings, Dynamics of Chromatography, Marcel Dekker, New York, 1965.

50. A. B. Littlewood, Gas Chromatography, 2nd ed., Academic Press, New York, 1970, p. 169 ff.

51. A. J. B. Cruickshank, B. W. Gainey, and C. L. Young, in Gas Chromatography 1968 (C. L. A. Harbourn, ed.), Inst. of Petroleum, London, 1969, p. 76.

52. A. J. B. Cruickshank, B. W. Gainey, C. P. Hicks, T. M. Letcher, R. W. Moody, and C. L. Young, Trans. Faraday Soc., 65, 1014 (1969).

53. D. H. Everett and C. T. H. Stoddart, Trans. Faraday Soc., 57, 746 (1961).

54. E. Kucera, J. Chromatogr., 19, 237 (1965).

55. S. N. Chesler and S. P. Cram, Anal. Chem., 45, 1354 (1973).

56. C. P. Hicks, Ph.D. Thesis, University of Bristol, 1970.

57. P. Alessi and I. Kikic, Gazz. Chim. Ital., 104, 739 (1974).

58. D. H. Desty, A. Goldup, G. R. Luckhurst, and W. T. Swanton, in Gas Chromatography 1962 (M. van Swaay, ed.), Butterworth, London, 1962, p. 67.

59. A. J. B. Cruickshank, M. L. Windsor, and C. L. Young, Proc. R. Soc. A, 295, 259 (1966).

60. J. R. Conder and S. H. Langer, Anal. Chem., 39, 1461 (1967).

61. S. Evered and F. H. Pollard, J. Chromatogr., 4, 451 (1960).

62. D. H. Desty and A. Goldup, in Gas Chromatography 1960 (R. P. W. Scott, ed.), Butterworth, London, 1960, p. 162.

63. R. A. Keller and H. Freiser, in Gas Chromatography 1960 (R. P. W. Scott, ed.), Butterworth, London, 1960, p. 301.

64. D. H. Desty and W. T. Swanton, J. Phys. Chem., 65, 766 (1961).

65. A. B. Littlewood, J. Gas Chromatogr., 1 (5), 6 (1963).

66. G. F. Freeguard and R. Stock, Trans. Faraday Soc., 59, 1655 (1963).

67. D. E. Martire and L. Z. Pollara, J. Chem. Eng. Data, 10, 40 (1965).

68. A. J. B. Cruickshank, D. H. Everett, and M. T. Westaway, Trans. Faraday Soc., 61, 235 (1965).

69. D. E. Martire, R. L. Pecsok, and J. H. Purnell, Trans. Faraday Soc., 61, 2496 (1965).

70. E. C. Pease and S. Thorburn, J. Chromatogr., 30, 344 (1967).

71. D. F. Cadogan and J. H. Purnell, J. Chem. Soc., 1968, 2133.

72. D. H. Everett, B. W. Gainey, and C. L. Young, Trans. Faraday Soc., 64, 2667 (1968).

73. C. Bighi, A. Betti, G. Saglietto, and F. Dondi, J. Chromatogr., 35, 309 (1968).

74. H. G. Harris and J. M. Prausnitz, J. Chromatogr. Sci., 7, 685 (1969).

75. J. R. Conder and J. H. Purnell, Trans. Faraday Soc., 65, 839 (1969).

76. D. F. Cadogan, J. R. Conder, D. C. Locke, and J. H. Purnell, J. Phys. Chem., 73, 708 (1969).

77. P. A. Sewell and R. Stock, J. Chromatogr., 50, 10 (1970).

78. R. J. Sheehan and S. H. Langer, IEC Process Des. Dev., 10, 44 (1971).

79. P. A. Sewell and R. Stock, Trans. Faraday Soc., 67, 1617 (1971).

80. A. N. Korol, J. Chromatogr., 67, 213 (1972).

81. O. Wicarova, J. Novak, and J. Janak, J. Chromatogr., 65, 241 (1972).

82. J. P. Sheridan, M. A. Capless, and D. E. Martire, J. Am. Chem. Soc., 94, 3298 (1972).

83. H.-H. Guermouche and J.-M. Vergnaud, J. Chromatogr., 81, 19 (1973).

84. A. J. Ashworth, J. Chem. Soc. Faraday I, 1973, 459.

85. T. M. Reed, Anal. Chem., 30, 221 (1958).

86. A. B. Littlewood, Anal. Chem., 36, 1441 (1964).

87. P. S. Snyder and J. I. Thomas, J. Chem. Eng. Data., 13, 527 (1968).

88. B. W. Gainey and C. L. Young, Trans. Faraday Soc., 64, 349 (1968).

89. C. P. Hicks and C. L. Young, Trans. Faraday Soc., 64, 2675 (1968).

90. P. M. Cukor and J. M. Prausnitz, J. Phys. Chem., 76, 598 (1972).

91. C. L. Hussey and J. F. Parcher, Anal. Chem., 45, 926 (1973).

92. C. L. Hussey and J. F. Parcher, J. Chromatogr., 92, 47 (1974).

93. D. E. Martire and P. Riedl, J. Phys. Chem., 72, 3478 (1968).

94. J. F. Parcher and C. L. Hussey, Anal. Chem., 45, 188 (1973).

95. A. J. B. Cruickshank, B. W. Gainey, and C. L. Young, Trans. Faraday Soc., 64, 337 (1968).

96. S. Ng, H. G. Harris, and J. M. Prausnitz, J. Chem. Eng. Data, 14, 482 (1969).

97. A. J. B. Cruickshank, B. W. Gainey, C. P. Hicks, T. M. Letcher, and C. L. Young, Trans. Faraday Soc., 65, 2356 (1969).

98. J. P. Sheridan, D. E. Martire, and Y. B. Tewari, J. Am. Chem. Soc., 94, 3294 (1972).

99. V. Pacakova and H. Ullmannova, Chromatographia, 7, 75 (1974).

100. C. L. Young, Trans. Faraday Soc., 64, 1537 (1968).

101. Y. B. Tewari, D. E. Martire, and J. P. Sheridan, J. Phys. Chem., 74, 2345 (1970).

101a. Y. B. Tewari, J. P. Sheridan, and D. E. Martire, J. Phys. Chem., 74, 3263 (1970).

102. E. F. Meyer, K. S. Stec, and R. D. Hotz, J. Phys. Chem., 77, 2140 (1973).

103. F. I. Stalkup and R. Kobayashi, Am. Inst. Chem. Eng. J., 9, 121 (1963).

104. B. W. Gainey and R. L. Pecsok, J. Phys. Chem., 74, 2548 (1970).

105. S. H. Langer and J. H. Purnell, J. Phys. Chem., 67, 263 (1963).

106. R. K. Clark and H. H. Schmidt, J. Phys. Chem., 69, 3682 (1965).

107. W. E. Hammers and C. L. de Ligny, Rec. Trav. Chim., 88, 961 (1969).

108. C. F. Chueh and W. T. Ziegler, Am. Inst. Chem. Eng. J., 11, 508 (1965).

109. R. E. Pecsar and J. J. Martin, Anal. Chem., 38, 1661 (1966).

110. A. B. Littlewood and F. W. Willmot, Anal. Chem., 38, 1076 (1966).

111. D. E. Martire, Anal. Chem., 38, 244 (1966).

112. P. Alessi, I. Kikic, and G. Tlustos, Chim. Ind., 53, 925 (1971).

113. R. K. Kuchhal and K. L. Mallik, J. Chem. Eng. Data, 17, 49 (1972).

114. G. Lopez-Mellado and R. Kobayashi, Pet. Refiner, 39 (2), 125 (1960).

115. C. J. Hardy, J. Chromatogr., 2, 490 (1959).

116. J. P. Sheridan, D. E. Martire, and F. P. Banda, J. Am. Chem. Soc., 95, 4788 (1973).

117. D. E. Martire, Anal. Chem., 33, 1143 (1961).

118. H. H. Smiley, J. Chem. Eng. Data, 15, 413 (1970).

119. D. L. Meen, F. Morris, J. H. Purnell, and O. P. Srivastava, J. Chem. Soc. Faraday I, 69, 2080 (1973).

120. D. E. Martire, R. L. Pecsok, and J. H. Purnell, Nature, 203, 1279 (1964).

121. D. L. Meen, F. Morris, and J. H. Purnell, J. Chromatogr. Sci., 9, 281 (1971).

122. E. R. Adlard, M. A. Khan, and B. T. Whitham, in Gas Chromatography 1960 (R. P. W. Scott, ed.), Butterworth, London, 1960, p. 251.

123. E. R. Adlard, M. A. Khan, and B. T. Whitham, in Gas Chromatography 1962 (M. van Swaay, ed.), Butterworth, London, 1962, p. 84.

124. G. Burrows and F. H. Preece, J. Appl. Chem., 13, 430 (1963).

125. R. C. Castells, Chromatographia, 6, 57 (1973).

125a. D. E. Martire, in Gas Chromatography 1963 (L. Fowler, ed.), Academic Press, New York, 1963, p. 33.

126. S. H. Langer and H. Purnell, J. Phys. Chem., 70, 904 (1966).

127. N. Petsev and Chr. Dimitrov, J. Chromatogr., 23, 382 (1966).

128. N. Petsev and Chr. Dimitrov, J. Chromatogr., 20, 15 (1965).

129. S. H. Langer, B. M. Johnson, and J. R. Conder, J. Phys. Chem., 72, 4020 (1968).

130. J. H. Purnell and O. P. Srivastava, Anal. Chem., 45, 1111 (1973).

131. D. Carter and G. L. Esterson, J. Chem. Eng. Data, 18, 167 (1973).

132. P. Urone, J. E. Smith, and R. J. Katnik, Anal. Chem., 34, 476 (1962).

133. D. W. Barker, C. S. G. Phillips, G. F. Tusa, and A. Verdin, J. Chem. Soc., 1959, 18.

134. J. W. King and P. R. Quinney, J. Chromatogr., 49, 161 (1970).

135. E. Gil-Av and J. Herling, J. Phys. Chem., 66, 1208 (1962).

136. P. Vernier, C. Raimbault, and H. Renon, J. Chim. Phys., 66, 690 (1969).

137. I. Brown, I. L. Chapman, and G. J. Nicholson, Aust. J. Chem., 21, 1125 (1968).

138. A. Apelblat and A. Hornik, Trans. Faraday Soc., 63, 185 (1967).

139. A. Apelblat, J. Inorg. Nucl. Chem., 31, 483 (1969).

140. A. Apelblat, J. Inorg. Nucl. Chem., 32, 3647 (1970).

141. P. Vernier, C. Raimbault, and H. Renon, J. Chim. Phys., 66, 429 (1969).

142. J.-Y. Lenoir, P. Renault, and H. Renon, J. Chem. Eng. Data, 16, 340 (1971).

143. D. L. Shaffer and T. E. Daubert, Anal. Chem., 41, 1585 (1969).

144. D. E. Martire, P. A. Blasco, P. F. Carone, L. C. Chow, and H. Vicini, J. Phys. Chem., 72, 3489 (1968).

145. L. C. Chow and D. E. Martire, J. Phys. Chem., 73, 1127 (1969).

146. W. L. Zielinski, D. H. Freeman, D. E. Martire, and L. C. Chow, Anal. Chem., 42, 176 (1970).

147. L. C. Chow and D. E. Martire, J. Phys. Chem., 75, 2005 (1971).

148. D. G. Willey and G. H. Brown, J. Phys. Chem., 76, 99 (1972).

149. L. C. Chow and D. E. Martire, Mol. Cryst. Liquid Crystl., 14, 293 (1971).

150. H. T. Peterson, D. E. Martire, and W. Lindner, J. Phys. Chem., 76, 596 (1972).

151. V. Brandani, IEC Fundamentals, 13, 154 (1974).

152. D. P. Tassios, IEC Process Des. Dev., 11, 43 (1972).

153. J. R. Anderson, J. Am. Chem. Soc., 78, 5692 (1956).

154. J. R. Anderson and K. H. Napier, Aust. J. Chem., 10, 250 (1957).

155. J. Simon, Chromatographia, 4, 98 (1971).

156. H. Röck, Chem.-Ing.-Tech., 28, 489 (1956).

157. G. W. Warren, R. R. Warren, and V. A. Yarborough, Ind. Eng. Chem., 51 (12), 1475 (1959).

158. C. Döring, Z. Chem., 11, 347 (1961).

159. J. A. Gerster, J. A. Gorton, and R. B. Eklund, J. Chem. Eng. Data, 5, 423 (1960).

160. D. Tassios, Hydrocarbon Process., 49 (7), 114 (1970).

161. M. R. Sheets and J. M. Marcello, Hydrocarbon Process. Pet. Refiner, 42 (12), 99 (1963).

162. S. T. Preston, Proceedings, Natural Gasoline Association of America, Dallas, Texas Meeting, April 22-24, 1959, p. 33.

163. Natural Gasoline Association of America, Equilibrium Ratio Data Book, NGAA, Tulsa, Oklahoma, 1957.

164. B. H. Sage and W. N. Lacey, Monograph on American Petroleum Institute Research Project 37, API, New York, 1950.

165. B. H. Sage and W. N. Lacey, Monograph on API Research Project 37, API, New York, 1955.

166. K. R. Koonce, H. A. Deans, and R. Kobayashi, Am. Inst. Chem. Eng. J., 11, 259 (1965).

167. F. I. Stalkup and R. Kobayashi, J. Chem. Eng. Data, 8, 564 (1963).

168. H. B. Gilmer and R. Kobayashi, Am. Inst. Chem. Eng. J., 10, 797 (1964).

169. G. Blu, L. Jacob, and G. Guiochon, J. Chromatogr., 50, 1 (1970).

170. H. C. Van Ness, Classical Thermodynamics of Non-Electrolyte Solutions, Pergamon, London, 1964, Chapter 6.

171. A. Waksmundzki and F. Suprynowicz, J. Chromatogr., 18, 232 (1965).

172. W. Kemula and H. Buchowski, Bull. Acad. Pol. Sci., Ser. Sci. Chem., 9, 601 (1961).

173. G. J. Pierotti, C. H. Deal, E. L. Derr, and P. E. Porter, J. Am. Chem. Soc., 78, 2989 (1956).

174. W. A. Scheller, J. L. Petricek, and G. C. Young, IEC Fundamentals, 11, 53 (1972).

175. E. Glueckauf, Nature, 160, 301 (1947).

176. D. H. James and C. S. G. Phillips, J. Chem. Soc., 1954, 1066.

177. J. R. Conder and J. H. Purnell, Trans. Faraday Soc., 64, 1505 (1968).

178. J. R. Conder and J. H. Purnell, Trans. Faraday Soc., 64, 3100 (1968).

179. G. J. Krige and V. Pretorious, Anal. Chem., 37, 1186, 1191, 1195, 1202 (1965).

180. J. R. Conder and J. H. Purnell, Trans. Faraday Soc., 65, 824 (1969).

180a. J. R. Conder, Chromatographia, 1, 387 (1974).

181. C. N. Reilley, G. P. Hildebrand, and J. W. Ashley, Anal. Chem., 34, 1198 (1968).

182. D. L. Peterson and F. Helfferich, J. Phys. Chem., 69, 1283 (1965).

183. G. M. Wilson, J. Am. Chem. Soc., 86, 127 (1964).

184. F. I. Stalkup and H. A. Deans, Am. Inst. Chem. Eng. J., 9, 106 (1963).

185. F. Helfferich and D. L. Peterson, Science, 142, 661 (1963).

186. D. L. Peterson, F. Helfferich, and R. J. Carr, Am. Inst. Chem. Eng. J., 12, 903 (1966).

187. F. Helfferich, J. Chem. Educ., 41, 410 (1964).

188. T. R. Koonce and R. Kobayashi, J. Chem. Eng. Data, 9, 494 (1964).

189. L. D. Van Horn and R. Kobayashi, J. Chem. Eng. Data, 12, 294 (1967).

190. K. Asano, T. Nakahara, and R. Kobayashi, J. Chem. Eng. Data, 16, 16 (1971).

191. F. Khoury and D. B. Robinson, J. Chromatogr. Sci., 10, 683 (1972).

192. Natural Gas Processers and Suppliers Association, Engineering Data Book, 8th ed., Tulsa, Oklahoma, 1967.

193. M. Martin, G. Blue, and G. Guiochon, J. Chromatogr. Sci., 11, 641 (1973).

194. C. Eon, B. Novosel, and G. Guiochon, J. Chromatogr., 83, 77 (1973).

195. D. C. Locke and D. E. Martire, Anal. Chem., 39, 921 (1967).

196. D. C. Locke, J. Gas Chromatogr., 5, 202 (1967).

197. D. C. Locke, J. Chromatogr., 35, 24 (1968).

198. P. Alessi and I. Kikic, J. Chromatographia, 7, 299 (1974).

199. P. Alessi and I. Kikic, J. Chromatographia, 97, 15 (1974).

200. D. E. Martire and L. Z. Pollara, Adv. Chromatogr., 1, 335 (1965).

201. D. C. Locke, J. Chromatogr. Sci., 11, 120 (1973).

202. D. C. Locke, J. Chromatogr. Sci., 12, 433 (1974).

203. D. H. Everett, Trans. Faraday Soc., 61, 2478 (1965).

204. D. H. Everett, Trans. Faraday Soc., 60, 1803 (1964).

205. E. Klesper, A. H. Corwin, and D. A. Turner, J. Org. Chem., 27, 700 (1962).

206. J. C. Giddings, J. Gas Chromatogr., 1, 73 (1966).

207. J. C. Giddings, M. N. Myers, L. McLaren, and R. A. Keller, Science, 162, 67 (1968).

208. M. N. Myers and J. C. Giddings, in Progress in Separation and Purification, Vol. 3 (E. S. Perry and C. J. van Oss, eds.), Wiley (Interscience), New York, 1970, p. 133.

209. J. C. Giddings, M. N. Myers, and J. W. King, J. Chromatogr., 7, 276 (1969).

210. J. J. Czubryt, M. N. Myers, and J. C. Giddings, J. Phys. Chem., 74, 4260 (1970).

211. S. T. Sie and G. W. A. Rijnders, Anal. Chim. Acta, 38, 31 (1967).

212. D. H. Desty, A. Goldup, and W. T. Swanton, in Gas Chromatography 1961 (N. Brenner, J. E. Callen, and M. D. Weiss, eds.), Academic Press, New York, 1962, p. 105.

213. D. C. Locke and W. W. Brandt, in Gas Chromatography, Fourth International Symposium (L. Fowler, ed.), Academic Press, New York, 1963, p. 55.

214. S. T. Sie, W. van Beersum, and G. W. A. Rijnders, Sep. Sci., 1, 459 (1966).

215. E. M. Dantzler, C. M. Knobler, and M. L. Windsor, J. Chromatogr., 32, 433 (1968).

216. R. L. Pecsok and M. L. Windsor, Anal. Chem., 40, 1238 (1968).

217. M. Vigdergauz and V. Semkin, J. Chromatogr., 58, 95 (1971).

218. B. K. Kaul, A. P. Kudchadker, and D. Devaprabhakara, J. Chem. Soc. Faraday I, 1973, 1821.

219. Y.-K. Leung and B. E. Eichinger, J. Phys. Chem., 78, 60 (1974).

220. C. L. Young, Thesis, University of Bristol, 1967. Cited by Gainey and Pecsok [104].

221. J. F. Connolly, Phys. Fluids, 4, 1494 (1961).

222. J. M. Prausnitz and P. R. Benson, Am. Inst. Chem. Eng. J., 5, 161 (1959).

223. E. M. Dantzler, C. M. Knobler, and M. L. Windsor, J. Phys. Chem., 72, 676 (1968).

224. Sh. D. Zaalishvili, Zh. Fiz. Khim., 30, 1891 (1956).

225. J. J. Czubryt, H. D. Gesser, and E. Bock, J. Chromatogr., 53, 439 (1970).

226. R. B. Spertell and G. T. Chang, J. Chromatogr. Sci., 10, 60 (1972).

227. V. G. Berezkin, Analytical Reaction Gas Chromatography, Plenum, New York, 1968.

228. J. E. Guillet, J. Macromol. Sci. Chem., A-4, 1669 (1970).

229. J. E. Guillet, in New Developments in Gas Chromatography (J. H. Purnell, ed.), Wiley (Interscience), New York, 1973, p. 187.

230. F. H. Covitz and J. W. King, J. Polymer Sci. A1, 10, 689 (1972).

231. R. D. Newman and J. M. Prausnitz, Am. Inst. Chem. Eng. J., 19, 704 (1973).

232. W. E. Hammers and C. L. de Ligny, Rec. Trav. Chim., 90, 912 (1971).

233. R. D. Newman and J. M. Prausnitz, J. Phys. Chem., 76, 1492 (1972).

234. B. E. Eichinger and P. J. Flory, Trans. Faraday Soc., 64, 2035 (1968).

235. N. F. Brockmeier, R. E. Carlson, and R. W. McCoy, Am. Inst. Chem. Eng. J., 19, 1133 (1973).

236. N. F. Brockmeier, R. W. McCoy, and J. A. Meyer, Macromolecules, 6, 176 (1973).

237. J. L. Varsano and S. G. Gilbert, J. Pharm. Sci., 62, 87 (1973).

238. J. L. Varsano and S. G. Gilbert, J. Pharm. Sci., 62, 92 (1973).

239. R. D. Newman and J. M. Prausnitz, J. Paint Technol., 45 (585), 33 (1973).

240. J. Cazes, J. Chem. Educ., 43, A567, A625 (1966).

241. K. H. Altgelt and J. C. Moore, Polymer Fractionation, Academic Press, New York, 1967, p. 123.

242. H. Determann, Gel Chromatography, Springer-Verlag, Berlin and New York, 1968.

243. J. F. Johnson and R. S. Porter, eds., Polymer Symposium, J. Polymer Sci., Part C, No. 21, 1968.

244. J. Cazes, J. Chem. Educ., 47, A461, A505 (1970).

245. D. D. Bly, Science, 168, 527 (1970).

246. K. H. Altgelt and L. Segal, eds., Gel Permeation Chromatography, Marcel Dekker, New York, 1971.

247. N. M. Bikales, Characterization of Polymers, Wiley, New York, 1971.

248. A. R. Cooper, J. F. Johnson, and R. S. Porter, Am. Lab., May 1973, 12.

249. Z. Grubisic, P. Rempp, and H. Benoit, J. Polymer Sci. B, 5, 753 (1967).

250. R. N. Kelley and F. W. Billmeyer, Sep. Sci., 5, 291 (1970).

251. J. Cazes and D. R. Gaskill, Sep. Sci., 2, 421 (1967).

252. J. G. Hendrickson, Anal. Chem., 40, 49 (1968).

253. B. W. Bradford, D. Harvey, and D. E. Chalkley, J. Inst. Pet., 41, 80 (1955).

254. E. Gil-Av and Y. Herzberg-Minzly, J. Am. Chem. Soc., 81, 4749 (1959).

255. J. H. Purnell, in Gas Chromatography 1966 (A. B. Littlewood, ed.), Inst. of Petroleum, London, 1967, p. 3.

256. C. A. Wellington, in New Developments in Gas Chromatography (J. H. Purnell, ed.), Wiley (Interscience), New York, 1973, p. 237.

257. H.-L. Liao, D. E. Martire, and J. P. Sheridan, Anal. Chem., 45, 2087 (1973).

258. D. F. Cadogan and J. H. Purnell, J. Phys. Chem., 73, 3849 (1969).

259. R. Vivilecchia and B. L. Karger, J. Am. Chem. Soc., 93, 6598 (1971).

260. R. J. Laub and R. L. Pecsok, Anal. Chem., 46, 1214 (1974).

261. L. Batt, G. M. Burnett, G. C. Cameron, and J. Cameron, Chem. Commun., 1971, 29.

262. C. L. de Ligny, J. Chromatogr., 69, 243 (1972).

263. C. L. de Ligny, T. van't Verlaat, and F. Karthaus, J. Chromatogr., 76, 115 (1973).

264. R. Cvetanovic, F. J. Duncan, W. E. Falconer, and R. S. Irwin, J. Am. Chem. Soc., 87, 1827 (1965).

265. M. A. Muhs and F. T. Weiss, J. Am. Chem. Soc., 84, 4697 (1962).

266. B. Feibush, M. F. Richardson, R. E. Sievers, and C. S. Springer, J. Am. Chem. Soc., 94, 6717 (1972).

267. J. J. Brooks and R. E. Sievers, J. Chromatogr. Sci., 11, 303 (1973).

268. L. J. Morris, J. Lipid Res., 7, 717 (1966).

269. F. Mikes, V. Schurig, and E. Gil-Av, J. Chromatogr., 83, 91 (1973).

270. M. Godlewicz, Nature, 164, 1132 (1949).

271. R. Tye and Z. Bell, Anal. Chem., 36, 1612 (1964).

272. J. T. Ayres and C. K. Mann, Anal. Chem., 38, 861 (1966).

273. W. L. Orr, Anal. Chem., 38, 1558 (1966).

274. W. L. Orr, Anal. Chem., 39, 1163 (1967).

275. V. Carunchio and G. G. Strazza, Chromatogr. Rev., 8, 260 (1966).

276. T. S. Moore, F. Shepherd, and E. Goodall, J. Chem. Soc., 1931, 1447.

277. H. D. Anderson and D. L. Hammick, J. Chem. Soc., 1950, 1089.

278. L. J. Andrews and R. M. Keefer, Molecular Complexes in Organic Chemistry, Holden-Day, San Francisco, 1964.

279. D. H. Freeman and D. P. Enagonio, Nature, 230, 135 (1971).

280. D. H. Freeman, Anal. Chem., 44, 117 (1972).

281. D. H. Freeman, J. Chromatogr. Sci., 11, 175 (1973).

282. D. H. Freeman, R. M. Angeles, D. P. Enagonio, and W. May, Anal. Chem., 45, 768 (1973).

283. R. L. Martin, Anal. Chem., 33, 347 (1961).

284. P. Urone and J. F. Parcher, Anal. Chem., 38, 270 (1966).

285. P. Urone and J. F. Parcher, Adv. Chromatogr., 6, 299 (1968).

286. J. R. Conder, D. C. Locke, and J. H. Purnell, J. Phys. Chem., 73, 700 (1969).

287. H.-L. Liao and D. E. Martire, Anal. Chem., 44, 498 (1972).

288. J. R. Conder, J. Chromatogr., 39, 273 (1969).

289. V. G. Berezkin, J. Chromatogr., 65, 227 (1972), and references cited therein.

290. Z. Suprynowicz, A. Waksmundzki, and W. Rudzinski, J. Chromatogr., 67, 21 (1972).

291. C. Eon and G. Guiochon, J. Colloid Interface Sci., 45, 521 (1973).

292. Z. Suprynowicz, A. Waksmundzki, W. Rudzinski, and J. Rayss, J. Chromatogr., 91, 67 (1974).

293. P. H. Weiner, H.-L. Liao, and B. L. Karger, Anal. Chem., 46, 2182 (1974).

294. B. L. Karger and H.-L. Liao, Chromatographia, 7, 288 (1974).

295. P. H. Weiner, E. R. Malinowski, and A. R. Levinstone, J. Phys. Chem., 74, 4537 (1970).

296. M. R. James, J. C. Giddings, and H. Eyring, J. Phys. Chem., 69, 2351 (1965).

297. R. H. Perrett and J. H. Purnell, J. Chromatogr., 7, 455 (1962).

298. A. V. Kiselev and Ya. I. Yashin, Gas Adsorption Chromatography, Plenum, New York, 1969.

299. N. C. Saha and D. S. Mathur, J. Chromatogr., 81, 207 (1973).

300. V. R. Choudhary and L. K. Doraiswamy, IEC Product Res. Dev., 10, 218 (1971).

301. P. Steingaszner, in Ancillary Techniques of Gas Chromatography (L. S. Ettre and W. H. McFadden, eds.), Wiley, New York, 1969, p. 13.

302. M. van Swaay, Adv. Chromatogr., 8, 363 (1969).

303. C. S. G. Phillips, in Gas Chromatography 1970 (R. Stock, ed.), Inst. of Petroleum, London, 1971, p. 1.

304. S. H. Langer and J. E. Patton, in New Developments in Gas Chromatography (J. H. Purnell, ed.), Wiley (Interscience), New York, 1973.

305. C. Vidal-Madjar, M.-F. Gonnard, and G. Guiochon, Adv. Chromatogr., 13, 177 (1975).

306. M. Suzuki and J. M. Smith, Adv. Chromatogr., 13, 213 (1975).

307. V. R. Maynard and E. Grushka, Adv. Chromatogr., 12, 99 (1975).

308. J. H. Purnell, Gas Chromatography, Wiley, New York, 1962.

309. S. T. Sie, J. P. A. Bleumer, and G. W. A. Rijnders, Sep. Sci., 1, 41 (1966).

310. E. F. Meyer and R. A. Ross, J. Phys. Chem., 75, 831 (1971).

311. L. E. Green, L. J. Schmauch, and J. C. Worman, Anal. Chem., 36, 1513 (1964).

312. L. Sojak, J. Hrivnak, A. Simkoricova, and J. Janak, J. Chromatogr., 71, 243 (1972).

313. F. Baumann, A. E. Strauss, and J. F. Johnson, J. Chromatogr., 20, 1 (1965).

314. A. Rose and V. N. Schrodt, J. Chem. Eng. Data, 8, 9 (1963).

315. R. C. Duty and W. R. Mayberry, J. Gas Chromatogr., 1, 115 (1966).

316. P. F. McCrea, in New Developments in Gas Chromatography (J. M. Purnell, ed.), Wiley (Interscience), New York, 1973, p. 87.

317. D. E. Martire and J. H. Purnell, Trans. Faraday Soc., 62, 710 (1966).

318. A. T. James, J. Chromatogr., 2, 552 (1959).

319. C. S. G. Phillips and P. L. Timms, Anal. Chem., 35, 505 (1963).

320. R. A. Hively, Anal. Chem., 35, 1921 (1963).

321. M. Wurst and J. Churasek, J. Chromatogr., 70, 1 (1972).

322. E. F. Barry and D. M. Rosie, J. Chromatogr., 59, 269 (1971).

323. E. F. Barry and D. M. Rosie, J. Chromatogr., 63, 203 (1971).

324. E. F. Barry, R. S. Fischer, and D. M. Rosie, Anal. Chem., 44, 1559 (1972).

325. E. F. Barry, R. Trakimas, and D. M. Rosie, J. Chromatogr., 73, 226 (1972).

326. A. J. P. Martin and A. T. James, Biochem. J., 63, 138 (1956).

327. A. Liberti, L. Conti, and V. Crescenzi, Nature, 178, 1067 (1956).

328. C. S. G. Phillips and P. L. Timms, J. Chromatogr., 5, 131 (1961).

329. J. E. Lovelock, Nature, 189, 729 (1961).

330. W. E. Wentworth and R. S. Becker, J. Am. Chem. Soc., 84, 4263 (1962).

331. W. H. McFadden, ed., Techniques of Combined Gas Chromatography and Mass Spectroscopy, Wiley (Interscience), New York, 1973.

332. E. Grushka and E. J. Kitka, J. Phys. Chem., 78, 2297 (1974).

333. E. Grushka and E. J. Kitka, J. Am. Chem. Soc., 98, 643 (1976).

334. A. C. Ouano, IEC Fundamentals, 11, 268 (1972).

335. A. C. Ouano and J. A. Carothers, J. Phys. Chem., 79, 1314 (1975).

Chapter 5

GAS-LIQUID CHROMATOGRAPHY IN DRUG ANALYSIS

W. J. A. VandenHeuvel

Merck Sharp & Dohme Research Laboratories
Rahway, New Jersey

A. G. Zacchei
Merck Sharp & Dohme Research Laboratories
West Point, Pennsylvania

I. INTRODUCTION

The role of gas-liquid chromatography (GLC) in the analysis of drugs and their metabolites in a variety of matrices has grown to the extent that this and related methods have literally been applied from A (amitriptyline [1] to Z (zearalenones [2]). It is interesting to note that although the inventors of GLC were initially concerned with compounds of biological interest [3, 4], it was in the field of petroleum that the resolving power of GLC columns was first exploited extensively. Later, lipid biochemists and analysts in the fats and oils industry recognized the great advances GLC methods for qualitative and quantitative fatty acid analyses represented over the approaches that were then standard. The foundation for the ultimate plethora of papers concerned with the GLC of drugs and related compounds which began in the mid and late 1960s was actually laid more than five years earlier with the introduction of thin-film column packings prepared from deactivated support and thermostable liquid phases [5, 6], which made it possible to chromatograph compounds of considerable molecular weight (e.g., 300-400) and polarity at column temperatures <250° C. The concept of derivatization (functional group changes on the microgram scale) developed in the early 1960s should not be underestimated either, for without conversion to less polar derivatives many drugs would not be amenable to GLC analysis. Separation patterns can be markedly improved by derivative formation. Derivatization also allows the introduction of halogen-containing moieties to permit sensitive and selective detection by electron capture of compounds that are otherwise nondetectable by this detector. Indeed, it is the variety of highly useful detectors (especially the hydrogen flame ionization and electron capture detectors, which also date from the beginning of the last decade) which complement the separation aspects of GLC and make it admirably suited for drug analysis.

GLC methods are often both more sensitive and more specific than colorimetric, spectrometric, and spectrofluoimetric assays, and can be employed for assay purposes per se or to validate the simpler methods. In turn, normal GLC methods can be tested and often improved upon by use of a mass spectrometer as a sensitive and selective detector. GLC should not be considered as the only avenue to follow when a method is required for the identification or estimation of a drug or related substance. Other chromatographic approaches, especially "high pressure" or "high performance" liquid chromatographic systems, are now available, as are techniques involving the quantitation by polarography of drugs containing reducible functional groups. The speed, resolution, and sensitivity of GLC separation and detection methods, especially with compounds possessing electroncapture properties per se and requiring no or straightforward derivatization, mkae these highly attractive for drug analysis problems ranging from bioavailability and tissue residue studies to assay of pharmaceutical preparations. Even when the necessary derivatization involves relatively sophisticated microscale organic chemistry, a GLC approach is often the only way to obtain the desired information. Thus, authors of review articles, such as this one, will not lack for subject matter for a number of years to come. Riedmann [7] has discussed some of these techniques in his paper on GLC analysis of drugs and drug metabolites.

In general, the analysis of drugs and their metabolites by gas-chromatographic techniques requires the following steps: (a) the drug (and/or metabolites) of interest must be isolated from the biological specimen; (b) for those compounds that are not readily adaptable for GLC analysis, a suitable volatile derivative must be prepared; (c) a proper choice of gas-chromatographic operating conditions (i.e., column temperature, isothermal or programmed stationary phase, and detector) must be determined; (d) identification and quantitation of the compounds; and (e) data acquisition.

Discussions of the methods of sample isolation, derivative formation, choice of detectors, quantitation, pitfalls of GLC methodology, and the applications of GLC to the analysis of a number of selected problems of biological interest are presented in this chapter. The choice of sample inlet systems, columns, and supports will not be discussed.

Although it is assumed that all readers of this chapter possess at least a rudimentary knowledge of GLC, a brief description of the GLC process is presented prior to any discussion of the analysis of drugs by this methodology.

Introduction of a sample (usually several microliters of a solution of appropriate concentration) into the injector or vaporizer zone of the chromatograph results in rapid vaporization of the sample molecules, which are transported through the column by an inert carrier gas. During their passage the molecules undergo a partitioning process between the carrier gas and the stationary phase. As the components of the sample are eluted individually (one hopes) from the column as a function of time they are detected by one of a variety of detection systems. The resulting signal is either displayed on a strip chart to produce an analog report (gas chromatogram) and/or is fed into an integrator or other data system providing numerical data reduction to give a digital report. Tentative identification of a component can be made on the basis of its retention time. Since the areas under the eluting peaks are dependent on the amount of emerging sample component, those peak areas (following proper calibration) can be related quantitatively to the amount or concentration of the individual components present.

II. SAMPLE PREPARATION

A. Isolation

The determination of drugs or metabolites cannot usually be achieved by GLC alone. It is essential in most instances to employ one or more procedures to isolate a crude fraction of the material from the biological specimen. Solvent extraction, liquid column chromatography (adsorbents and resins), and thin-layer chromatography(TLC) techniques have been utilized as initial "cleanup" procedures. The extent to which preliminary separation is required depends on the concentration of the compound to be analyzed and the degree to which endogenous substances lead to possible interference. When the concentration of drug or metabolite(s) is high, only the most rudimentary sample preparation is necessary. However, when one is trying to analyze nanogram quantities of compounds that are closely related to endogenous constituents, extensive purification is required. Extensive cleanup is often required when the electron capture detector is employed to measure low levels of drug (thereby making the method very laborious).

The most commonly used drug isolation procedures are based on extraction techniques using different solvents at selected pH values, thereby permitting separation of the drugs or metabolites, or both, into neutral, basic, acidic, and amphoteric fractions. The highly ionized components are water soluble, whereas the nonionized drug can usually be extracted into benzene, ether, chloroform, or ethyl acetate. When the drugs are extensively bound to tissue protein, denaturation procedures may need to be employed prior to extraction to free the bound drug. Injection of the organic extract into GLC columns without further purification has succeeded in a number of cases [8-17]; however, in some instances undesirable biological components are concomitantly injected, thereby causing a high baseline and at times resulting in interfering peaks from previous injections. If possible, the organic extract should be further purified by back-extraction and subsequent extraction into the same or different organic solvent. The organic phase can then be concentrated prior to injection. A number of investigators [1, 18-22] have used this approach.

When the levels of drug and metabolite at the therapeutic dose are low (below 1 μg/ml), care must be exercised to prevent volatilization or adsorption of the compound at the final concentration step. To prevent such volatilization during this step the sample may be converted to a salt form by the addition of a trace amount of base or acid. The deleterious effects of such losses are precluded by the addition of an internal standard (analog of parent drug is preferred) at the initial step in the purification.

Many acidic metabolites or urinary acids [23] have been shown to be extracted with solvents such as ethyl acetate; however, all acids (e.g., glucuronides and amphoteric compounds) were not extracted under these conditions. Therefore, in these situations the result of the GLC analysis (in a metabolic profile study) is not necessarily a reflection of metabolic processes but rather that of a poor extraction process. Horning and Horning [23, 24] used an anion exchange resin to collect the acids; their procedure required considerably more time than simple solvent extraction since the eluate had to be lyophilized prior to derivative formation. In most instances the neutral drugs are quantitatively extracted; however, the oxidative metabolites which may form are readily water soluble and thereby may require ion exchange resins. The use of salt-solvent pairs for the isolation of drugs and metabolites from biological fluids has been proposed by Horning et al. [25]. Nonionic resins (i.e., XAD-2, XAD-4) have provided extremely clean samples [26, 27]. These resins bind the glucuronide and sulfate conjugates and readily separate them from the aqueous phase. These conjugates can then be hydrolyzed with acid or enzyme prior to subsequent solvent extraction. Acid hydrolysis is rapid but may lead to partial or total destruction of the compound; this method should thus be confined to acid-stable compounds. Enzyme hydrolysis, although slower, is generally preferred.

Sohn et al. [27] discuss the methods currently employed in the cleanup procedure in drug screening. The most commonly employed methods are solvent extraction followed by TLC or GLC, or both. The authors also state that resin-loaded papers provide clean samples particularly if multiple extractions at various pH levels are performed. Dole et al. [28] recommend extractions at pH values of 2.2, 9.3, and 11. However, Mule [29] states that only 2.4% of labeled pentobarbital and 2.1% of labeled amphetamine were recovered using the Dole ion exchange paper technique. The combination of solvent extraction, TLC, and then GLC has been employed routinely by many investigators [30-32]. The major emphasis now appears to be the employment of a nonionic resin column (XAD-2) for the absorption and concentration of drugs followed by TLC and then GLC. The aforementioned discussion on drug isolation can be illustrated in several specific examples from a simple method to an extremely laborious procedure.

Ramsey and Campbell [16] describe an ultrarapid method for the extraction of lipid-soluble drugs from biological fluids. The method was applied to the detection and measurement in urine of four common drugs of addiction, namely, amphetamine, methylamphetamine, phethidine, and methadone. The procedure was as follows: Two milliliters of urine was made alkaline with the addition of 0.2 N NaOH. Fifty microliters of chloroform containing a known quantity of internal standard was added and the solution agitated on a "Vortex" for less than 1 min. After centrifugation (1 min), a 1- to 4-μl aliquot of the organic phase was injected directly into the GLC column. Recoveries of the respective drug were good, 78 ± 5, 80 ± 5, 83 ± 3, and 99 ± 1%, and there were no interfering peaks from endogenous components. The method is extremely rapid (less than 15 min) and simple. Kelsey et al. [33] used the following GLC procedure for the analysis of fluphenazine and fluphenazine sulfoxide in the urine of schizophrenic patients. A sample of urine was filtered and a 200-ml aliquot was passed through a 1 × 25-cm Amberlite XAD-2 resin. After urine passage, the column was washed with 100 ml of an ammonium chloride solution buffered at pH 8.5 and then with 50 ml of water. The drugs and urinary pigments were then eluted with 200 ml of methanol. Following removal of the solvents under reduced pressure, the residue was dissolved in a chloroform-methanol solution and then extracted with ammonium hydroxide to remove most of the urinary pigments. The organic phase was evaporated to dryness under nitrogen and the residue trimethylsilylated with bistrimethylsilytrifluoroacetamide (BSTFA) in pyridine to form the trimethylsilyl (TMSi) derivative which was required for GLC analysis using a flame ionization detector (FID). When an electron capture detector (ECD) is used to obtain a high degree of sensitivity, sophisticated methodology may be required as illustrated by the procedure employed by Arnold and Ford [34]. The catechol-containing compounds (e.g., dopa) were isolated from brain tissue which was previously homogenized in cold perchloric acid. An antioxidant (sodium metabisulfate) was added and the homogenate then centrifuged for 30 min. The compounds in the supernatant were extracted by adsorption on activated aluminum oxide at a controlled pH (8.6). After a 5- to 10-min period the aluminum oxide was allowed to settle and the supernatant discarded. The alumina was washed with water four times and then lyophilized. The catechols were removed from the alumina by stirring with 2 N HCl in methanol for 30 min. An aliquot of this solution was then dried under nitrogen and treated with a fluoroacyl anhydride in acetonitrile. Excess acylating agent was removed under nitrogen and the residue dissolved in benzene for injection in the gas chromatograph. In our laboratories and those of others [35] more favorable results are obtained if the excess acylating reagent is hydrolyzed first with water and then with ammonium hydroxide. The derivative can then be extracted into a small volume of organic solvent free of reagent.

B. Derivatization

Derivatization is a key component in GLC analysis for a number of reasons. Some drugs must be converted to more volatile, less polar derivatives to permit their analysis by GLC methods. Other drugs can be chromatographed successfully directly, but it is often advantageous to prepare one of a variety of derivatives to reduce adsorption resulting from interaction of free hydroxyl or amino groups with active sites. This is especially true for analysis at the nanogram level, and facilitates the quantitative analysis of drugs and their often more polar metabolites. Further,

GLC separation of closely related compounds is often improved by derivatization. The introduction of moieties possessing high coefficients of electron affinity provides high sensitivity in work using ECD, often down to the nanogram level or lower. In the preparation of a derivative one must be sure that (a) the derivative is formed quantitatively even in the presence of often massive amounts of extraneous material, (b) no loss occurs as a function of time due to the lack of stability of the derivative or unexpected side reactions, and (c) no abnormal functional group alterations occur during the reaction. The last point is extremely critical when one is employing this technique for structure elucidation.

The derivatization approaches most frequently employed are methylation, acylation, trimethylsilylation, methoxime formation, or any combination of these. We would like to discuss several of these methods.

Methylation is a classic approach for the conversion of polar compounds to less polar derivatives. One of the standard methods for methylation, use of dimethyl sulfate, is employed frequently in GLC methods for the analysis of various drugs. Fiereck and Tietz [17] used this reagent to convert 11 barbiturates and glutethimide from blood samples to their methylated derivatives. The same reagent was employed by Prescott and Redman [36] in their GLC method for the determination of tolbutamide and chlorpropamide in plasma. These two drugs decompose when subjected to GLC at elevated temperature, but the N-methyl derivatives are more stable and can be chromatographed intact using the appropriate column conditions. Simmons et al. [37] deliberately employ conditions (large quantity of glass wool in the flash heater) under which N-methyltolbutamide is quantitatively transformed to N-methyl-p-toluenesulfonamide, and it is this compound which is eluted and quantified in their method for the determination of tolbutamide in serum.

Carboxyl group-containing drugs can be readily converted to methyl esters by reaction with diazomethane; this route has been employed by many investigators. Zacchei and Weidner [38] used this approach to form the methyl ester of probenecid. Ferry et al. [39] converted indomethacin to its ethyl ester using diazoethane in their method for the determination of this drug in plasma and serum. The ethyl ester was chosen so as to distinguish the drug derivative from that of the dealkylated metabolite; diazomethane, in addition to esterifying the parent drug, might convert the phenolic metabolite to indomethacin methyl ester.

Dimethylformamide dimethylacetal has been shown by Thenot et al. [40] to be an excellent reagent for the methylation of fatty acids. This reagent was later shown by the same group to react with amino acids to form N-dimethylaminomethylene methyl ester derivatives [41]. More recently this reagent has been used as a derivatizing agent for the GLC of the barbiturates and glutethimide; these drugs are converted to the corresponding acetals [42]. Methylation of ibuprofen isolated from plasma has been achieved by use of the reagent 1,1'-carbonyldiimidazole [43].

So-called flash or on-column methylation has become a frequently employed method for the conversion of polar drugs such as the barbiturates and diphenylhydantoin to derivatives possessing greatly improved chromatographic properties. The early work of MacGee [44] and Stevenson [45] employed tetramethylammonium hydroxide; the more recently introduced trimethylanilinium hydroxide, first reported by Brochmann-Hanssen and Oke [46], is also widely used. MacGee [47] has used aqueous tetraethylammonium hydroxide as the derivatizing agent in this method for the identification and quantification of barbiturates and glutethimide in blood. Superior results were observed with a "slow" rather than "rapid" injection. Estas and Dumont [48] determined 5,5-diphenylhydantoin and its major metabolite, 5-(p-hydroxyphenyl)-5-phenylhydantoin, in serum by use of tetramethylammonium hydroxide. Hammer et al. [49] developed an assay for the same two compounds but used trimethylanilinium hydroxide as the derivatizing reagent; the authors hold that on-column methylation is preferred over trimethylsilylation because of (a) the stability of the reagent, (b) the instantaneous quantitative methylation in the flash heater, and (c) water need not be excluded from the plasma extract. Perchalski et al. [50] have described a rapid and simultaneous GLC method for the determination of phenobarbital, primidone, and diphenylhydantoin involving on-column methylation with trimethylanilinium hydroxide in methanol. Kowblansky and co-workers [51] utilize flash butylation in their GLC method for the determination of xanthines and barbiturates to preclude any misidentification as a result of methylation. Osiewicz and associates [52] have found that the GLC analysis of phenobarbital using trimethylanilinium hydroxide in methanol results in the formation of both N,N-dimethylphenobarbital and N-methyl-α-phenylbutyramide. Formation of the latter depends on the concentration of the reagent and the period of time the phenobarbital is in the derivatizing solution prior to injection. These authors deliberately employ derivatizing conditions (1.8 M trimethylanilinium hydroxide in methanol) which favor formation of the hydrolysis and subsequent methylation product (N-methyl-α-phenylbutyramide) and use this peak for quantification of phenobarbital. Ervik and Gustavii [53] have employed "extractive methylation" for the conversion of chlorthalidone to its tetramethyl derivative and applied this technique to determination of the drug in plasma. The drug is extracted as an ion pair with tetrahexylammonium hydroxide into methylene chloride in the presence of a large excess of methyl iodide. Ehrsson [54] has reported on the GLC determination of barbiturates following extractive methylation in carbon disulfide.

Trimethylsilylation is a widely employed method of derivatization, and was early applied to a large number of steroids [6]. Although steroids were the first compounds of moderately high molecular weight and structural complexity extensively studied by GLC [5, 6], relatively few papers have appeared on the quantitative GLC determination (as opposed to identification) of steroid drugs, possibly because of the extensive derivatization often required. Derivatization conditions for dexamethasone have been studied by Thenot and Horning [55]; conversion of the keto group at C-20 to the methoxime and trimethylsilylation of at least the hydroxyl group at C-21 are necessary to permit GLC with the side chain intact. The β-hydroxy-keto cyclopentane ring system of the E family of prostaglandins is also converted to the methoxime-TMSi ether to permit successful GLC of these important compounds [56]. Hamberg [57] has investigated the extent of inhibition of PGE_1 and PGE_2 synthesis in man following administration of several drugs by determining [using a GLC-mass spectrometric (MS) assay] the urinary output of the PGE metabolite 7α-hydroxy-5,11-diketo-tetranorprostane-1,16-dioic acid as its dimethoxime-TMSi ether-dimethyl ester. Kelly [58] has suggested that $PGF_{2\alpha}$ be analyzed by GLC as its methyl ester 9,11-butylboronate-15-TMSi ether. The cyclic boronate is formed only with 9,11-cis, and not with 9,11-trans, hydroxyl groups.

No derivatization step was employed in the GLC assay reported by Shah et al. [59] for griseofulvin in the skin of patients. The satisfactory GLC properties of this chlorine-containing antibiotic plus use of the selective and sensitive electron capture detector permits determination of drug levels down to a few nanograms per milligram of skin. Proof that the peak with the retention time of griseofulvin was indeed the antibiotic was obtained by micropreparative GLC and MS analysis of the collected material, which possessed the same mass spectrum as the authentic compound. Derivatization techniques are absolutely necessary for the GLC of most antibiotics. In those cases where derivatization is required, trimethylsilylation is one of the methods of choice for antibiotics ranging in size and complexity from the low-molecular-weight phosphonomycin [60] to the much larger compounds such as paromomycin [61]. The former is readily converted to its di-TMSi derivative, which possesses good GLC properties. The isolation and purification of this natural product were monitored by GLC of the di-TMSi derivative. Figure 1 shows the chromatogram resulting from the analysis of a phosphonomycin concentrate from fermentation sources (before purification by paper chromatography) treated with bistrimethylsilylacetamide (BSA) to form the TMSi derivative; the multicomponent nature of the sample is evident. Preparative paper chromatography of the antibiotic concentrate yielded a product of increased biological activity which, following trimethylsilylation, was subjected to GLC. As can be discerned from Fig. 2, this material is largely free from contamination. Note that the predominant peak in Fig. 2 possesses the same retention time as one of the larger peaks found in the non-paper-chromatographically purified concentrate (see Fig. 1, arrow); the peaks also gave the same mass spectrum. GLC of the BSA-treated crystalline benzylamine salt of phosphonomycin yielded a peak which corresponded in retention time (17 min) to the large peak from the

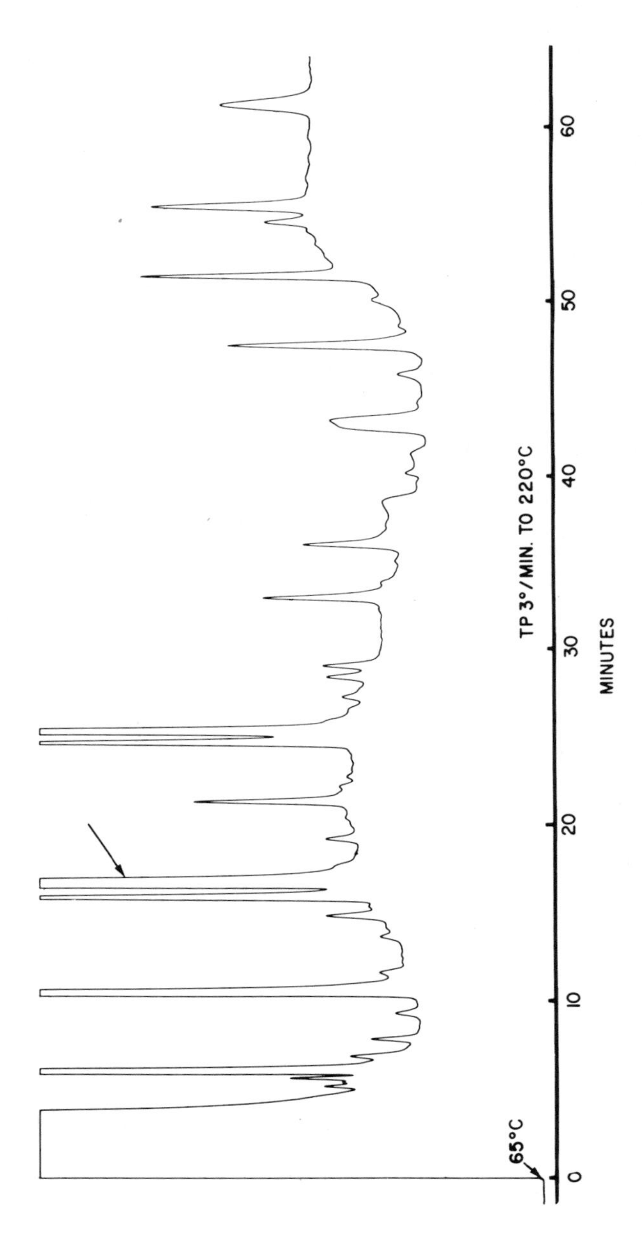

FIG. 1. Chromatogram resulting from analysis of a phosphonomycin concentrate (prior to purification by paper chromatography) treated with BSA. (Reprinted from Shafer et al. [60] with permission. Copyright by Elsevier Scientific Publishing Company.)

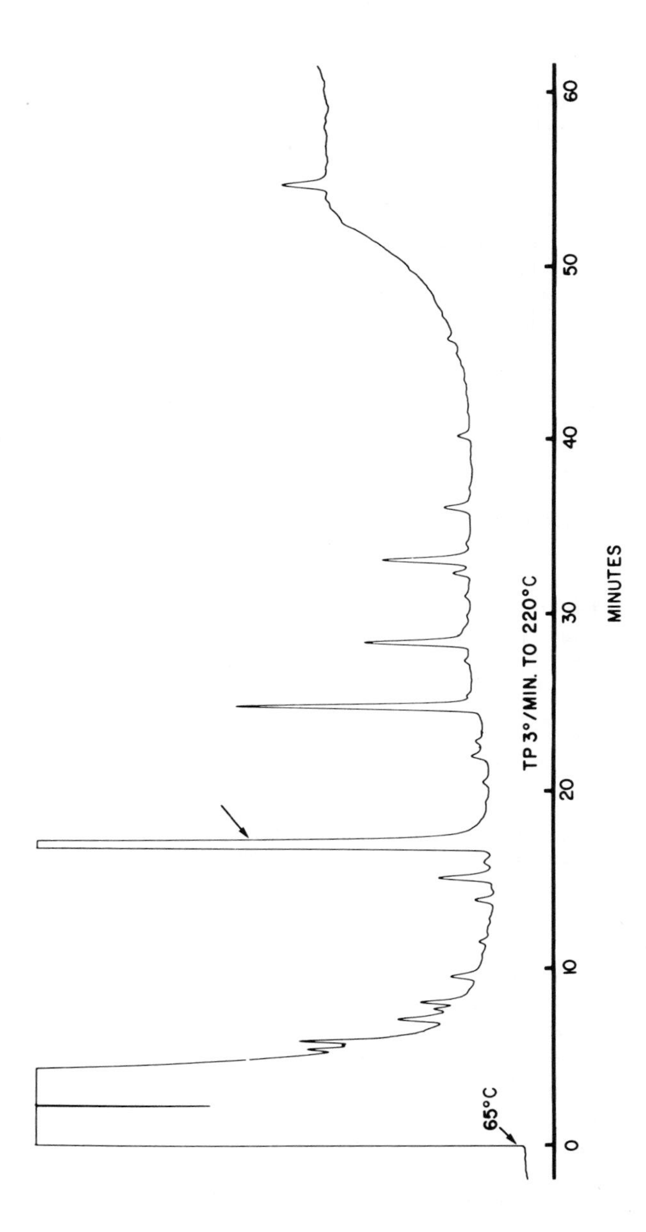

FIG. 2. Chromatogram resulting from analysis of a phosphonomycin concentrate (following purification by paper chromatography; see Fig. 1) treated with BSA. (Reprinted from Shafer et al. [60] with permission. Copyright by Elsevier Scientific Publishing Company.)

sample purified by paper chromatography; combined GLC-MS confirmed the identical nature of the two compounds exhibiting the 17-min retention times.

Aminoglycoside antibiotics possess many more functional groups than phosphonomycin, and persilylation of these compounds forms derivatives possessing molecular weights well in excess of 1000; even with a 1% stationary phase the column temperature employed by Tsuji and Robertson [61, 62] for the GLC analysis of such compounds is approximately 300° C. Combined GLC-MS has been used by these authors to gain insight into the number of TMSi groups actually present on these large molecules. Trimethylsilylated neomycin gave ions in the range of m/e 1300, above the normal mass limit for the type of instrument employed (LKB 9000). A recent paper by Margosis and Tsuji [63] reports on a critical examination to establish the optimal conditions for the GLC analysis of neomycin. The authors emphasize proper column conditions, use of the appropriate amount of reagent (trimethylsilyldimethylamine) and reaction conditions, and prompt assay of samples following their derivatization. Van Giessen and Tsuji [64] have published a GLC method for neomycin in petrolatum-based ointments based on trimethylsilylation. Tsuji and Robertson [65] have also published on the GLC determination of the macrolide antibiotic erythromycin employing trimethylsilylation. The molecular ions found by combined GLC-MS for the TMSi derivatives of erythromycins A, B, and C were m/e 1093, 1005, and 1151, respectively, indicating that all hydroxyl groups were derivatized. Calculations of biopotencies of lots of erythromycin on the basis of GLC data and also by microbiological assay agreed very well. A number of penicillins have been separated and quantified by GLC of their TMSi derivatives [66]. GLC following trimethylsilylation has been found by Brodasky and Argoudelis [67] to be effective in separating closely related members of the celestosaminide family of antibiotics which could not be separated by paper chromatography or TLC. Retention and mass spectral data permitted the characteriziation of these compounds. Margosis [68] has employed a GLC method involving trimethylsilylation with BSA for the analysis of chloramphenicol in bulk material and in several pharmaceutical dosage preparations. Good correlation was observed between the GLC assay and two "official" assays (microbiological; spectrophotometric). A collaborative study on the GLC analysis for chloramphenicol found that the most common source of error involves the derivatization step [69]. The overall results demonstrated the validity of the method, which was proposed for inclusion in the Code of Federal Regulations and recommended for primary compendial usage [69]. Brown and Bowman [70] have recently published a GLC assay for spectinomycin based on the conversion of this antibiotic to its tetra-TMSi derivative using hexamethyldisilazane. A relative standard deviation of 1% was observed with a lower limit of detection (GLC) of 0.1 μg.

Brown [71] has published a GLC method for the assay of cycloheximide, Ia, which involves trimethylsilylation. The antibiotic is initially

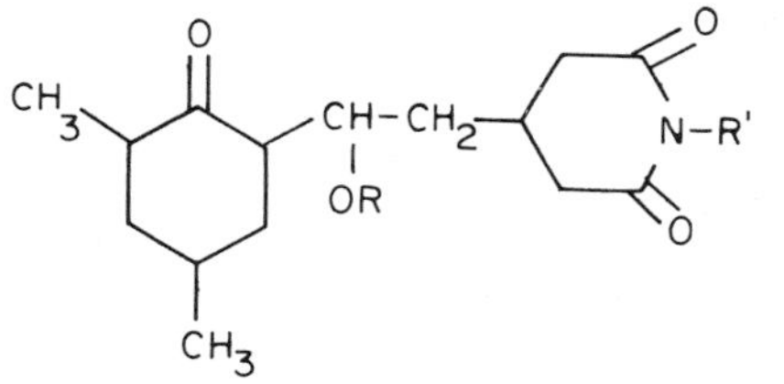

Ia, R = R' = H

Ib, R = TMSi, R' = H

Ic, R = R' = TMSi

converted to a mixture of the O-TMSi and N,O-di-TMSi derivatives (Ib and Ic, respectively). Not only were two peaks observed regardless of derivatization conditions, but Ic was found to undergo an on-column conversion to Ib. The author took advantage of the labile nature of Ic by incorporating addition of the appropriate amount of i-propanol to the trimethylsilylation reaction mixture prior to GLC. This step resulted in quantitative conversion of Ic to Ib, the derivative possessing excellent GLC properties. By this approach, Ia can be quantitatively converted to Ib.

The separation and quantification of tetracyclines as their TMSi derivatives has been achieved by Tsuji and Robertson [72]. The reagent mixture [BSA/trimethylchlorosilane (TMCS)/pyridine, 1:1:2] containing the internal standard is added to the sample; 24 hr at room temperature is required to form the penta-TMSi derivative of tetracycline. Less potent reagents formed unstable derivatives and more potent derivatization agents caused degradation of tetracycline. A greater than 1:1 ratio of BSA to TMCS resulted in incomplete trimethylsilylation, whereas in increase in the proportion of the TMCS to BSA led to degradation of tetracyline, as did reaction temperatures >35° C; at <15° C the rate of derivatization was reduced considerably with no obvious benefit. The system of choice for separation of the tetracyclines was a 6-ft column containing 3% JXR packing; however, several members of this family (e.g., chlorotetracycline and 4-epianhydrotetracycline) failed to separate from each other. A 10% OV-25 stationary phase was effective in separating these two compounds, but columns >2 ft in length could not be used because of unacceptable levels of adsorption and degradation of TMSi tetracycline resulting from the high operating temperature required. Tetracycline hydrochloride standards from the authors' laboratory [72] and the USP, and a variety of lots of powdered drugs from commercial sources, were analyzed for potency by UV, microbiological, and GLC methods. It was concluded that there were no significant differences among the three different approaches. A relative standard deviation of 2.3% was found for the GLC determination of tetracycline. Six months following the appearance of this report the same laboratory published a paper on the determination of tetracyclines by high-pressure liquid chromatography using a 280-nm UV monitor [73]. The high-pressure liquid chromatography approach, unlike the GLC assay, requires no derivatization, and the authors state that the former is the preferred method of analysis for tetracycline.

Since many drugs are not detectable at the nanogram level without the introduction of suitable electron-capturing groups (e.g., those containing one chlorine or at least three fluorine atoms), derivatization is a very important aspect of quantitation at low levels. Heptafluorobutyrylation is one of a number of widely employed reactions for enhancing sensitivity. Kaiser et al. [74] have presented a GLC method for the assay of medroxyprogesterone acetate in plasma. Formation of the 3-enol heptafluorobutyrate permits electron capture detection down to 1 ng/ml plasma. Plasma drug concentrations, as measured by the GLC assay, were 5 to 10 times lower than those found using a radioimmunoassay, suggesting that in the latter drug-related substances other than intact parent drug were being measured. A novel approach to the introduction of electron capture properties has been reported by Burchfield et al. [75] for the drug dapsone, 4,4'-diaminodiphenylsulfone. The free amino groups are diazotized with nitrous acid; replacement of the diazonium groups with iodine atoms results in formation of 4,4'-diiododiphenylsulfone which possesses good GLC properties and can be detected at subnanogram levels.

An imaginative derivatization approach for the GLC assay of sulfonamides such as sulfaquinoxaline (see below) has been reported by Daun [76].

$H_2N-C_6H_4-SO_2N(H)-R$ R = (quinoxalinyl, N, N) in sulfaquinoxaline

Treatment with diazomethane results in methylation of the sulfonamide moiety; the primary amino group is then converted to the heptafluorobutyramide by reaction with the corresponding anhydride. Such a derivative possesses excellent GLC properties and can be determined quantitatively by electron capture techniques at the subnanogram level.

Electron capture properties have been conferred upon prostaglandin $F_{2\alpha}$ by conversion to its pentafluorobenzyl ester; the latter is subjected to GLC as its tri-TMSi ether [77]. An assay utilizing this derivatization approach with a lower sensitivity limit of 12.5 pg of injected ester has permitted the quantitative determination of $PGF_{2\alpha}$ plasma levels in a monkey dosed with this drug. A TLC purification of the solvent extract prior to GLC is employed to increase the specificity of the procedure. Analysis of platelet-free monkey plasma containing added $PGF_{2\alpha}$ by the GLC-EC method, radioimmunoassay, and a GLC-MS assay has demonstrated that the first and second approaches give approximately the same values, whereas with GLC-MS the results are somewhat lower [77].

III. DETECTORS

There are essentially two groups of widely used GLC detectors: (a) nonspecific detectors: thermal conductivity (TCD) and flame ionization (FID), which respond to almost any substance emerging from the column, and (b) selective detectors: electron capture (ECD), thermionic (TID), flame photometric (FPD), and Coulson conductivity (CCD), which respond to only certain types of compounds. The detector response reflects the amount of drug eluting from the column and the observed signal is proportional to the concentration of material. Quantitative analysis depends on a known relationship between sample size and detector response. The greater the linearity, the more accurate are the results.

A. Flame Ionization Detector

The FID is the most frequently employed detector in the analysis of drugs or metabolites, or both, responding to almost every organic compound. Numerous investigators have found the FID to yield dependable performances, good sensitivity (down to a few nanograms of compound injected onto the column), excellent stability, and a wide linear range (10^6-10^8). Frequently employed conditions include a column flow rate of 30 to 60 ml/min, a hydrogen flow of 30 to 50 ml/min, and an air flow of 300 ml/min. The basic components of an FID are a jet and a collector ring; these serve as the electrodes, and a potential of several hundred volts is applied across them. The column effluent (argon, helium, and nitrogen are employed as carrier gases with the FID) is mixed with hydrogen, passes through the jet, and is burned in an atmosphere of air at the tip of the jet, just below the collector ring. A small fraction (0.1%) of the organic molecules are ionized by a mechanism that is not completely understood. A low intensity current flow occurs when no sample components are being eluted (i.e., column background). While sample components are being eluted from the column the ion concentration and hence current flow are greatly increased; the increase in signal is amplified and ultimately displayed on a recorder or sent to an integrator. Detailed discussions of the FID can be found in numerous publications [78, 79]. Hundreds of papers reporting the use of this detector have appeared in the recent literature. Both qualitative and quantitative analyses have been carried out on drugs present in biological systems. Several representative examples of the use of FID in drug analysis are presented.

Antipyrine plasma levels (or half-life) are commonly used to assess the hepatic drug-metabolizing enzyme activity in man. The drug is well absorbed, rapidly and ubiquitously distributed, and extensively metabolized by the liver microsomal enzymes. Prescott et al. [8] devised a rapid GLC method for the estimation of this drug in plasma. The method previously employed [80] involved a change in absorbance following conversion to 4-nitrosoantipyrine and required accurate timing of the reaction. The authors extracted the drug and internal standard from plasma with chloroform and then injected various aliquots into a gas chromatograph equipped with an FID and a U-shaped column packed with 0.5% SE-30 and 0.5% Carbowax 20 M on Gas-Chrom-Q. The recovery of antipyrine from plasma in the range of 5 to 50 μg/ml was 101.5% with a standard deviation of 4.4%. The data were calculated using the standard peak height ratio method.

Berman and Spirtes [81] studied the metabolism of chlorpromazine in rat and rabbit liver microsomal preparations using the Curry [21] extraction and GLC method. The metabolites examined were chlorpromazine N-oxide, monodemethyl chlorpromazine, didemethyl chlorpromazine, chlorpromazine sulfoxide, and the monodemethyl chlorpromazine sulfoxide. The major metabolites in both liver systems were the N-oxide (38-62% of parent) and the monodemethyl drug (28-77% of parent). The measured metabolites accounted for 92 to 104% of the chlorpromazine that was metabolized.

Bertagni et al. [82] studied biliary excretion of the free and conjugated hydroxy metabolites N-methyloxazepam and oxazepam in rats, guinea pigs, and mice following intravenous administration of diazepam and N-demethyldiazepam. The GLC method (FID) of analysis was capable of detecting these compounds (parent drugs and metabolites) at levels not less than 0.15% of the administered dose. Three and four percent (guinea pigs) and 13 and 33% (mice) of diazepam and N-methyldiazepam, respectively, were recovered as conjugated oxazepam. No conjugated metabolites were detected in the bile of rats. None of the species exhibited biliary excretion of the parent drugs or their free hydroxy metabolites.

An improved method for the determination of dibenzepin (antidepressant) and its five N-demethylated metabolites in human urine has been reported by DeLeenheer and Heyndricks [83] using a gas chromatograph equipped with a dual FID. Special attention was paid to the complete characterization of free basic compounds and their assay by GLC using an internal standard. Both qualitative and quantitative data were obtained. The levels of metabolites detected were 2 to 6% of the administered dose. The lower limit of detection was about 1 μg/ml. Williams et al. [84] described a rapid method for the detection and quantitation of therapeutic barbiturate levels in serum using an FID gas chromatograph. Hexabarbitone was employed as the internal standard. As little as 10 μg of barbiturate per 100 ml of serum was determined using a 1-ml serum sample. The authors state that increased sensitivity was obtained when the carrier gas (N_2) was passed over formic acid before entering the GLC column.

B. Electron Capture Detector

One of the major difficulties which besets drug analysis is the low levels of drug and/or metabolites present in the biological sample. Although the FID has proven to have good sensitivity, the amount of substance required for analysis is many times greater than the minimal detectable quantity since only a small portion of the total sample is usually injected. The ECD, developed by Lovelock [85], is extremely sensitive (nanogram to picogram levels) to only certain types of molecules (e.g., compounds containing chlorine, bromine, iodine, or numerous fluorine atoms; polynuclear aromatics, polyunsaturated ketones; anhydrides) and for this reason is more selective than the commonly employed FID. In the ECD the carrier gas is ionized by a radioactive source (β rays; ^{63}Ni or ^{3}H). When a drug or its derivative with an affinity for electrons enters the detector chamber negative ions are formed by the collision of the drug molecule and the free electrons, resulting in a decrease in the background ion current. This negative change in ion current is then observed as a typical positive peak (following reversal of polarity) displayed on the recorder. The ECD is one of the most difficult detectors to use successfully and unless extreme care is exercised to establish proper operating conditions disappointing and misleading results and considerable frustration are inevitable. The reactions involved in the electron-capturing process and therefore detector response are affected not only by the operating conditions of the detector but also by the geometry of the detector itself. These detectors should be cleaned frequently since they tend to decrease in sensitivity with extensive analysis of derivatized biological extracts. Most electron capture detectors suffer from a narrow linear range. The ^{3}H detector has about a 500-fold linear range and relatively low operating temperatures ($<220°$ C), whereas the early pulse or dc-operated ^{63}Ni detector has a smaller linear range, about 50-fold. The more recent ^{63}Ni detectors exhibit a much greater linear range, thereby overcoming the problem of the older detectors, and are stable to a temperature of 350° C. A review article on the ECD has recently appeared [86].

As stated previously, the ECD is extremely sensitive and specific to certain types of molecules. Most drugs are not detected by the ECD at the sensitivity required without prior derivatization to a compound that possesses electron-capturing properties. The functional groups usually derivatized include primary and secondary amino, phenolic, and hydroxyl moieties. Introduction of heptafluorobutyryl-, chloroacetyl-, and pentafluorophenyl-containing noieties is frequently employed. Some typical results obtained with GLC methods utilizing electron capture detectors are presented. Blake et al. [87] discussed rapid screening procedures for drugs which involve (a) directly acylating the amine or hydroxyl sites; (b) converting compounds to species that can be derivatized and then acylated; and (c) acylating aromatic compounds by a rapid Friedel-Crafts acylation. Examples of the three reactions were presented.

Blood and urinary levels of bromopheniramine were determined by Bruce et al. [88] using a gas chromatograph equipped with an ECD and an SE-30 column. Their initial experiments were performed with an FID; however, the procedure was not sufficiently sensitive to quantitate the submicrogram blood levels obtained following the normal human dose (4-8 mg of the antihistamine). In addition, some naturally occurring basic materials interfered with the analysis. In the ECD assay the bromine atom imparted sufficient electron-capturing properties to the molecule, but the tritium foil detector which was employed did not permit a high enough operating temperature to analyze the unaltered molecule. Consequently, the compound was oxidized to the corresponding ketone prior to GLC analysis. However, in this reaction any metabolites which resulted from side-chain metabolic oxidations would also have been determined as parent drug. To preclude this possibility the investigators subjected the samples to prior partition chromatography to separate the known demethylated and didemethylated metabolites. The newer ^{63}Ni detector would have eliminated the need to oxidize the compound to the ketone, and thus the partition chromatographic step would have been unnecessary. Moffat and Horning [89] have reported data on the use of pentafluorobenzaldehyde to increase the sensitivity of detection of the primary amines in the catecholamine series. The sensitivity of detection for the derivatives of the 10 compounds fell within the range of 2.0 to 5.4×10^{-13} mole/sec (FID) and 1.3 to 2.4×10^{-16} mole/sec (ECD). This 1000-fold sensitivity increase is typical when comparing the FID to the ECD. Ten picograms of each derivative was detected by the ECD. Horning et al. [90] have also reported on the GLC separation and increased sensitivity resulting from derivatization of hydroxyl-substituted amines of biological importance including the catecholamines using the ECD. Both the TMSi-N-acetyl and the TMSi-N-heptafluorobutyryl derivatives were prepared. Edwards and Blau [18] have analyzed the low levels of phenylethylamines present in biological extracts. The compounds were chromatographed as their N-dinitrophenyl (DNP), O-TMSi derivatives. The DNP amines were formed with 2,4-dinitrobenzenesulfonic acid and the hydroxyl groups were converted to the TMSi ethers with BSA. The method was employed to study these biologically important amines in brain, liver, plasma, and urine. Levels as low as 20 ng/g tissue have been determined. The determination of chlordiazepoxide (psychotherapeutic agent) plasma concentrations by electron capture gas chromatography was reported by Zingales [91]. The method was developed to detect the low levels of drugs and metabolites following therapeutic doses. The authors state that the previously developed colorimetric, spectrofluorometric, and GLC (FID) methods are not suitable because of time-consuming separation processes and the lack of sensitivity and specificity. In the ECD method the compound is chromatographed in its intact form and is clearly separated from the known demethylated metabolite. No internal standard was used in this procedure.

C. Element-Selective Detectors

Although the great majority of drug analyses by GLC involve either FID or ECD, element-selective detection systems [e.g., alkali metal (Rb) FID] for compounds containing nitrogen (and less frequently phosphorus and sulfur) are increasingly employed [92]. The absolute sensitivity of such detectors is, generally speaking, not much greater than that of a flame ionization detector, but they exhibit several 1000-fold less sensitivity toward non-nitrogen-containing compounds (both sample components and the solvent used for injection), and it is this specificity which can be used to advantage. Since many drugs contain at least one nitrogen atom, whereas most common contaminants from solvents, plasma, and tissue do not, it would appear that this type of detection approach should be particularly useful in drug analysis. Thus, the approach of choice for the determination of dicyclomine in human plasma was reported by Meffin et al. [93] to involve use of a nitrogen-selective flame detector. Quantitative estimation of this drug in plasma at levels down to 1 to 2 ng/ml was possible.

Goudie and Burnett [94] have reported that the selectivity of a nitrogen detector (rubidium chloride) enabled them to employ a simple isolation procedure in a GLC method for phenobarbitone, primidone, and phenytoin in human serum. The serum is extracted with cyclohexane to remove lipids and other extraneous substances, and then reextracted with methylene chloride. The methylene chloride extract is taken to dryness, dissolved in a methanolic solution of trimethylanilinium hydroxide, and an aliquot is analyzed. Breimer and VanRossum [95] have reported on a GLC determination for hexobarbital in plasma at levels down to 50 ng/ml using a nitrogen-selective alkali FID. Not only did they observe greater absolute sensitivity for the drug with the nitrogen detector than with the FID, but in addition, the former is far less affected by the solvent and background peaks in plasma. Because of the great selectivity of the nitrogen detector, plasma extracts (petroleum ether/amyl alcohol, 100/2) can be analyzed directly. No additional time-consuming purification is required, as would be the case with an FID.

Caddy et al. [96] were not able to analyze urinary extracts for amphetamine using an FID because of solvent front interference; however, by switching to a nitrogen-specific alkali FID this problem was eliminated. A rubidium chloride-containing detector with increased sensitivity toward nitrogenous compounds was also employed by James and Waring [97] for the determination of pentazocine levels in rabbit blood. This approach is held [97] to be much more sensitive than an FID method reported earlier [98], but considerably less sensitive than an ECD approach published by Brötell et al. [99]. The latter, however, requires alkylation with pentafluorobenzyl bromide to form the pentafluorobenzyl ether.

A GLC method for the determination of thiopental in plasma using an alkali FID in the nitrogen mode has been reported by Sennello and Kohn [100]. The sensitivity of this detection system toward the drug was found to be similar to that of an FID, but the selectivity was far greater, allowing analysis of relatively crude extracts. Riedmann [101, 102] has summarized the advantages of using a nitrogen-selective FID, namely, minimal sample preparation, response proportional to nitrogen content, great selectivity and specificity, and excellent sensitivity. The performance of element-selective detectors, especially those for nitrogen, is critically dependent on operating parameters; see, for example, Breimer and Van-Rossum [95]. The nitrogen FID detector must be carefully "tuned" to yield the optimal balance between sensitivity and selectivity. Critical factors include the shape and temperature of the flame (determined by the air and hydrogen flow rates) and the area of contact of the flame with the alkali metal surface.

A flame photometric detector operated in the phosphorus-selective mode has been employed by McCallum [103] to measure Δ^9-tetrahydrocannabinol and cannabinol levels in human blood. Rather than converting the drugs to a derivative such as the heptafluorobutyrate for detection by electron capture (e.g., Fenimore et al. [104]), the author transformed the phenols to their diethyl phosphate esters [103]. Less than 1 ng of the Δ^9-THC could be assayed per milliliter of plasma using this approach. The limit of detection for Δ^9-THC as the heptafluorobutyrate using electron capture detection, in contrast, has been reported to be <0.1 ng/ml. Jackson and Reynolds [105] have reported on a GLC method for the determination of cyclophosphamide (a potential defleecing agent) residues in sheep tissues using a phosphorus-sensitive detector. Earlier colorimetric methods proved to be neither sufficiently sensitive nor selective for trace analysis in tissue. The GLC approach is held to be quantitative down to 0.01 ppm in muscle tissue using 10-g samples.

Use of a flame photometric detector in the sulfur-specific mode has been reported by Schirmer and Pierson [106] to allow the measurement of methapyridine in human plasma at concentrations as low as 50 ng/ml. The authors reported that this detection system was not only more sensitive toward methapyridine than on FID, but plasma and urinary extracts showed a very low background, in distinction to that observed with an FID. None (i.e., <50 ng/ml) of the drug was found in the plasma or urine of patients receiving medication. This was ascribed by the authors to rapid metabolism, and a (hydroxylated?) metabolite of the drug was found in urine.

D. Mass Spectrometric Detection

Although GLC methods of analysis are often far more specific than classical approaches (e.g., colorimetric, spectrophotometric), an isolate from a biological source may well contain a compound which possesses the same retention time with a given stationary phase as the drug or metabolite of interest. This can result in a false positive result and lead to erroneous quantitation. Furthermore, the identification of an unknown on the basis of retention behavior alone, even if the reference compound is available for comparison, is open to question. The uncertainty can be reduced significantly by comparison of retention times of derivatives and use of a variety of unrelated stationary phases, but a more definitive approach to positive identification of an unknown and verification of the specificity of a GLC assay involves the direct combination of GLC with MS. The latter technique provides highly characteristic data, and when combined with GLC to allow purification of isolates and to distinguish between isomers and closely related compounds offers a unique opportunity for identification of a compound. Suffice it to say here that in a mass spectrometer the sample molecules are ionized (usually by electron bombardment) and the resulting molecular and fragment ions are separated (in a magnetic or electric field) on the basis of their mass-to-charge (m/e) values. A plot of abundance versus m/e value yields the mass spectrum. Detailed descriptions of the events operative in a mass spectrometer are available [107, 108].

Of course, when dealing with a drug containing a radioactive label, use of a GLC system capable of detecting radioactivity will indicate which components eluted from a column are labeled and hence drug related. This approach will not, however, demonstrate the structure of the labeled compound; the parent compound may have undergone transformation to a metabolite with the same retention time. GLC with radioactivity detection prior to GLC-MS using the same column conditions makes it possible to ascertain which components in an isolate are derived from a labeled drug and hence should be examined by the latter technique. An example of this important approach is seen in Fig. 3. An acidic radioactive metabolite fraction isolated from the urine of a wether lamb dosed with ^{14}C-bis(chloromethyl)sulfone was subjected to trimethylsilylation. GLC with radioactivity detection indicated that most of the injected radioactivity was associated with the large peak possessing the 5-min retention time (Fig. 3). The mass spectrum of the radioactive component was obtained by combined GLC-MS. The derivatized metabolite was identified as the TMSi ester of chloromethanesulfinic acid [109].

When the use of radioactive drugs is not desirable or possible in certain instances, the combination of GLC and MS then becomes the method of choice for characterizing compounds as drug related. The approach is particularly effective if the drug-related compounds contain chlorine or bromine, which yield characteristic and readily recognized isotope clusters. Mass spectrometrically recognizable cluster patterns can be artificially created by labeling a drug with stable isotopes, e.g., carbon-13, nitrogen-15, or deuterium.

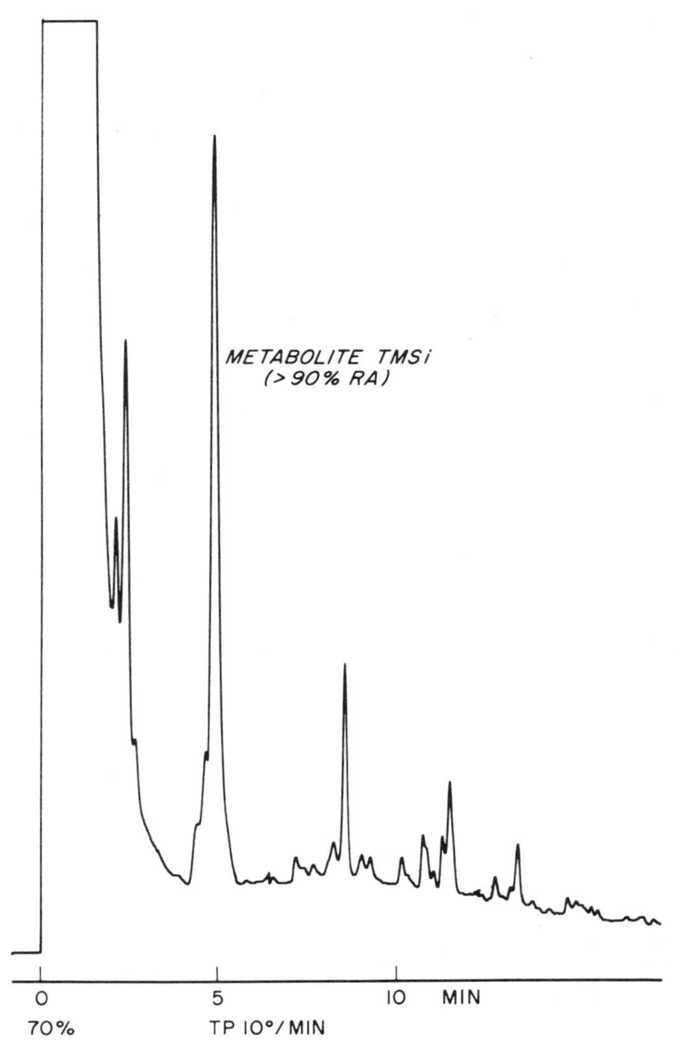

FIG. 3. Chromatogram resulting from analysis of trimethylsilylated metabolite fraction from urine of a lamb dosed with ^{14}C-labeled bis(chloromethyl)sulfone. More than 90% of the available injected radioactivity was associated with the large peak possessing a retention time of 5 min. (Reprinted from Wolf et al. [109] with permission. Copyright by the American Chemical Society.)

The combination of GLC and MS for identification purposes is used in many areas. For example, Nakamura et al. [110] have applied this technique to identify and quantify heroin in illicit samples. A minimal amount of sample workup was necessary because of the ability of the GLC column to separate the various components. A mass spectrum (obtained from a component possessing the appropriate retention time) which is identical to that of a reference sample is a strong piece of court evidence, as it is much less equivocal than a retention time.

Fischer and Ambre [111] employed GLC-MS to demonstrate that with OV-1 both glutethimide and a metabolite were eluted at the same retention time to give a symmetrical peak, resulting in an overestimation when the drug level of urine was assayed using this nonselective stationary phase. Two peaks were observed when the analyses were carried out using Carbowax 20 or OV-225 columns. Using these stationary phases it proved possible to assay urinary levels for both the parent drug and the metabolite, identified by its GLC-MS properties as α-phenyl-α-glutaconimide. It is clear that retention time and even peak shape are no guarantee of GLC component identity or homogeneity, and that GLC-MS is a superior method for the characterization of GLC effluent components.

Another example of using GLC-MS techniques to validate a GLC assay for a drug is from the paper of Cala et al. [112] on the determination of pyrimethamine in tissue. The mass spectrum of pyrimethamine (Fig. 4) is dominated by a very characteristic and intense cluster of ions in the region

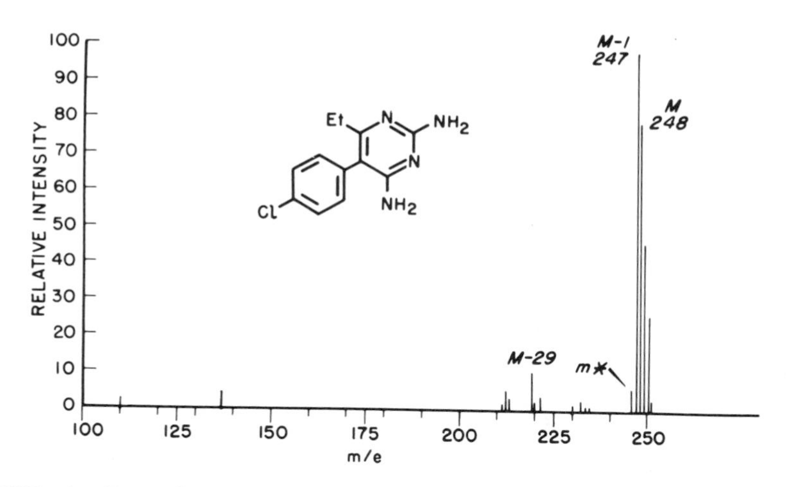

FIG. 4. Partial mass spectrum of pyrimethamine obtained at 70 eV. (Reprinted from Cala et al. [112] with permission. Copyright by the American Chemical Society.)

of the molecular ion (m/e 248). Thus, rather than scanning a complete spectrum to identify as authentic drug the compound eluted from the column at the retention time of pyrimethamine, a reduced scan including only the molecular ion region should be sufficient. Further, the instrument can be set to scan repetitively over a small mass range, and for a known quantity (20 ng) of pyrimethamine the response seen in Fig. 5 (lower) was observed. The result observed when injection of an aliquot (held to contain 20 ng, on

REPETITIVE PARTIAL MASS SCAN
m/e 247-250

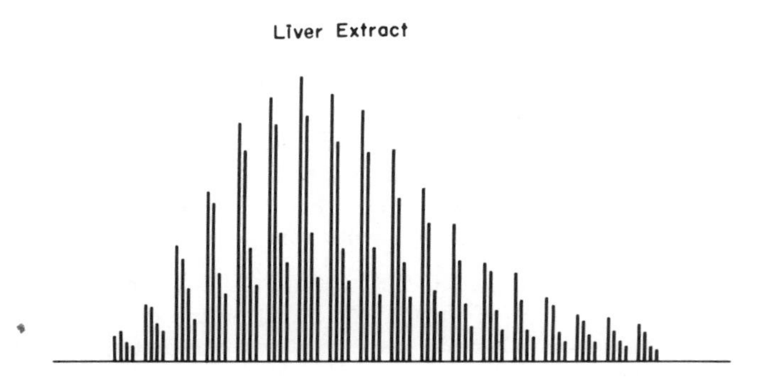

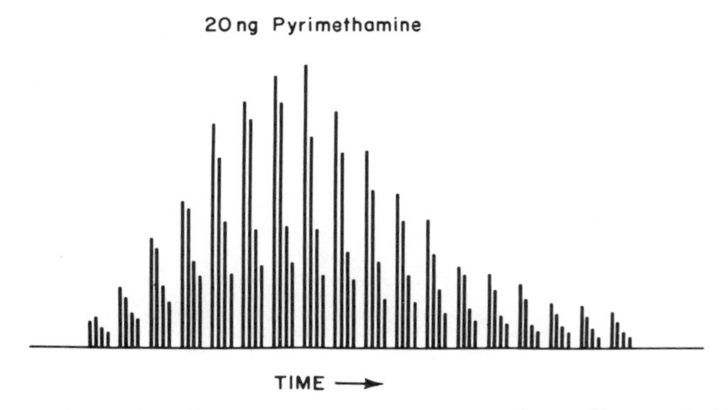

FIG. 5. Mass spectrometric responses resulting from repetitive scanning from m/e 247 to 250 during elution from the GLC column of 20 ng of pyrimethamine (lower) and an aliquot of a 1-day postdose liver extract determined by a GLC method to contain 20 ng of drug. (Reprinted from Cala et al. [112] with permission. Copyright by the American Chemical Society.)

the basis of a GLC assay; see Sect. VII) of a tissue isolate was subjected to repetitive scan GLC-MS is seen in the upper portion of Fig. 5. Not only are the two envelopes of signals virtually identical in a qualitative sense, but the areas of the envelopes are also identical, demonstrating the validity of the GLC assay. Baczynskyj et al. [113] have reported on a method for quantitation of drugs by repetitive scanning over a narrow mass range using deuterated analogs of the drugs as internal standards and computerization of the data.

Rather than following a small segment of the mass region with time, the mass spectrometer can readily be set to monitor a single ion. If the chosen ion is intense and the compound possesses satisfactory GLC properties, especially with respect to adsorption, this approach can result in very great sensitivity (in the subnanogram region). Mirocha et al. [114] have used this approach for the identification of minute amounts of diethylstilbestrol (DES) in swine feedstuff. The extract was so impure that it was impossible to detect DES (as its TMSi ether derivative) on the total ion current chromatogram (analogous to FID chromatogram; Fig. 6). When the mass spectrometer was set to monitor the molecular ion of the derivative (m/e 412) the result seen in Fig. 7 was obtained. Signals of m/e 412 for both the cis and the trans isomers were observed, as required, at the appropriate retention times. Confirmation of the presence of DES in the extract was obtained by injecting an aliquot of the isolate in a solution of trimethylanilinium hydroxide in methanol and monitoring the molecular ion of the dimethyl ether, m/e 296; both the cis and the trans isomers were observed. Although quantitation of the results was not attempted, it probably could have been if desired.

Use of an accelerating voltage alternator, introduced by Sweeley et al. [115], allows several ions to be followed simultaneously; this is sometimes referred to as multiple ion detection. The mass spectrometer can thus be employed as a selective or ion-specific GLC detector, as discussed by Brooks and Middleditch [116], or the gas chromatograph can be used as an inlet system of high resolving power for the mass spectrometer so as to obtain complete spectra. The former approach is currently in wide use in drug analysis and other areas for both identification and quantification. It has been well described by Holmstedt and co-workers [117, 118], who have coined the term mass fragmentography to describe such a technique and applied it to the quantitation of chlorpromazine and its metabolites. Other recent and pertinent articles are those by Jenden and Cho [119] and Watson [120]. Mass fragmentography is far more sensitive than GLC using an FID. Fanelli and Frigerio [121] have published on FID and mass fragmentographic assays for piribedil in plasma; using the same GLC conditions, the FID and mass fragmentography sensitivity limits were 250 and 10 ng/ml, respectively.

The sensitivity and selectivity of mass fragmentography make this an attractive means for demonstrating the validity of GLC assays. For example,

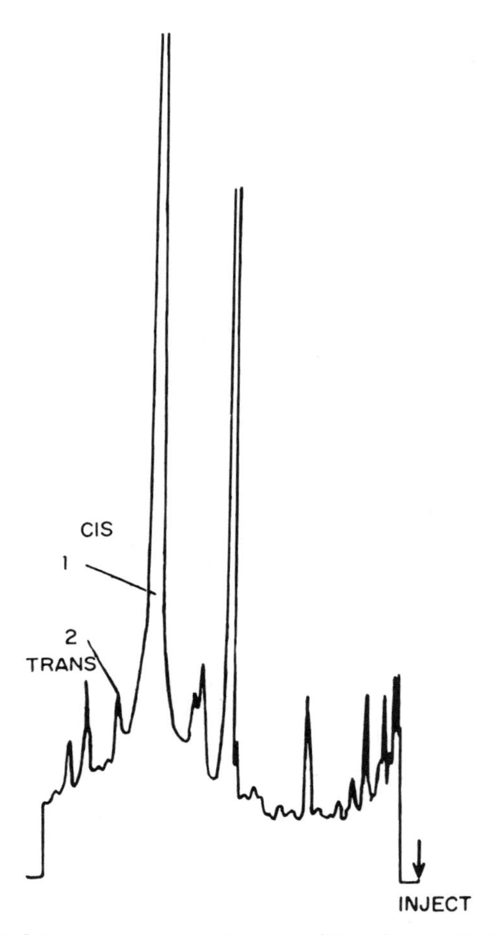

FIG. 6. Total ion current tracing resulting from GLC-MS analysis of trimethylsilylated isolate from contaminated feed. Numeral 1 indicates the retention time of the di-TMSi derivative of cis-DES (which is coeluted with a far larger quantity of another compound). Numeral 2 indicates the presence of a small peak possessing the retention time of the trans isomer. (Taken from Mirocha et al. [114] with permission. Copyright by the American Chemical Society.

this technique has been employed by Palmer and Kolmodin-Hedman [122] to verify an electron capture GLC method for p,p'-DDE in human plasma at the 10- to 60-ng/ml level. Mass fragmentography has found application in studying pharmacokinetics of drugs in humans. It was originally thought that the hypnotic drug methaqualone was rapidly absorbed and eliminated in humans. Data obtained using a GLC method with FID sensitivity adequate to follow plasma levels for up to 8 hr postdose supported this position [123]. Another study using a more sensitive GLC assay (also FID) indicated that plasma levels dropped only slowly during the period 8 to 22 hr postdose [124]. The greater sensitivity of the mass fragmentographic approach enabled Alvan et al. [125] to follow plasma levels for more than 4 days postdose and obtain the data necessary to ascertain plasma kinetics (which suggest a two-compartment open model). The initial rapid plasma concentration decrease (α slope) is followed by a slow β-phase elimination. Cho and co-workers [126] employed mass fragmentography to study the pharmacokinetics of amphetamine and phentermine in rats; the deuterium-labeled analogs were used as internal standards to establish plasma and brain levels of these drugs. The sensitivity of these GLC-MS isotope dilution assays [127] is such as to make possible the assay of very small volumes of biological fluid. For example, Rane et al. [128] have studied the plasma disappearance of transplacentally transferred diphenylhydantoin (DPH) in the newborn. Their mass fragmentographic approach, using deuterium-labeled

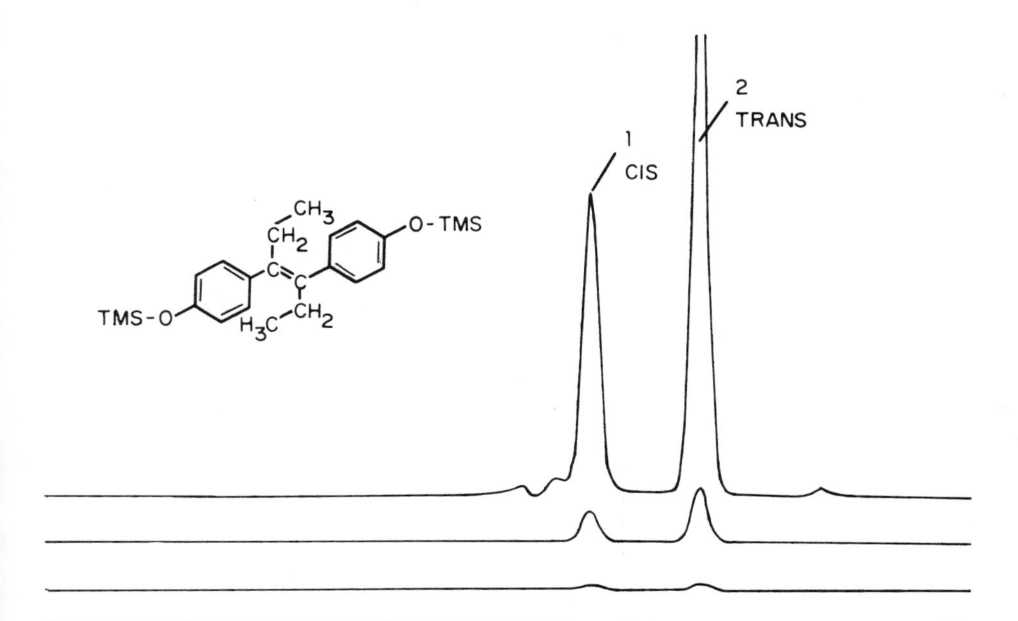

FIG. 7. Mass fragmentogram resulting from monitoring the ion of m/e 412 (molecular ion of the di-TMSi derivative of DES) during GLC-MS of trimethylsilylated isolate (see Fig. 6). The numerals 1 and 2 refer to the cis and trans isomers of the derivative. (Reprinted from Mirocha et al. [114] with permission. Copyright by the American Chemical Society.)

DPH as internal standard, required only 100 μl of plasma to give precise determinations of drug down to 0.01 μg/ml. The wide applicability of this analytical technique is further emphasized in the work reported by Agurell and associates [129]. These authors determined plasma levels of Δ^1-tetrahydrocannabinol in humans who smoked cigarettes containing 10 mg of this marijuana constituent; peak levels (12-26 ng/ml) were found to occur within 10 min of smoking one cigarette, and decreased rapidly thereafter.

This chapter is surely not the place to expound on the value and application of GLC-MS techniques to various aspects of drug analysis. However, it seems to the authors that it is impossible to think of or discuss GLC without mentioning its combination with MS. It is not our intention to set one method on a higher pedestal than the other, for each plays its own important role.

IV. DATA INTERPRETATION

Data interpretation on the individual components emerging from GLC columns can yield both qualitative and quantitative information. The qualitative aspects of GLC have been reported in detail [6, 101, 130, 131] and consequently only the salient points will be discussed. Qualitative analysis or identification is based primarily on the retention time (or volume) or more often on the retention time relative to that of a reference standard. The usual identification is the result of a comparison of the retention data of the unknown compound with those obtained from the authentic reference compounds analyzed under identical conditions. The Kovats retention index [132, 133] is a most useful system for the standardization of the retention data, as is the "methylene unit" approach [134]. These values are independent of changes in GLC operating parameters and provide structural information when one determines the difference between the retention indices (or methylene unit values) on polar and nonpolar columns. Additional information may be gained by analyzing the sample with selective and nonselective detectors. This technique may be used to differentiate between compounds that contain N, S, P, and halogens. Finally, identification can be made by comparing the retention data of derivatives of the unknown with the data obtained on the parent drug and the derivatives of the reference compounds.

Since the quantitative aspects of drug and metabolite analysis comprise the most relevant parameter in the monitoring of blood levels at therapeutic doses, we would like to discuss the methods of calibration, peak measurement, and computer data acquisition and analysis. Some examples of quantitative results are also presented. In the GLC analysis of drugs the areas under the eluting peaks are dependent on the amounts of emerging sample components; therefore, those peak areas after proper calibration can be related quantitatively to the amount of the individual components present. The two methods of calibration commonly used are (a) external calibration and (b) internal standardization. External calibration involves the construction of a calibration curve following the injection of known amounts of the drug to be analyzed. The peak area (or peak height) data are plotted versus weight of the compound injected. The method is simple but requires the injection of precise amounts of sample, and the detector sensitivity must remain constant in order to obtain the results from the calibration curve accurately. A drifting baseline during a series of analyses is suggestive of detector instability. External calibration has produced good results using the FID when the detector is very stable; however, with ECD difficulties may arise owing to instability of the detector. To ensure precise injections of the sample the syringe should first be filled with a small amount of solvent followed by an air bubble and then the precise volume of the sample. Zingales [91] has used this approach for the assay of chlordiazepoxide and stated that an internal standard was unnecessary because of the short retention time of this drug and because standardization of the instrument was easily accomplished. Recovery of the drug added to control plasma in the range of 0.03 to 10 μg was about 90 ± 3%.

The internal standardization technique requires the injection of samples containing known weight ratios of the compound of interest and the internal standard (I.S.). The area (or peak height) ratios are plotted against the known weight ratios. In sample analysis a precise quantity of the internal standard is added to the unknown sample and the mixture chromatographed. The observed area ratios are then used to determine the weight ratio of the unknown sample. For best performance the internal standard should be chemically similar to the drug being analyzed, so that it will behave chromatographically similar to the drug analyzed; added to the initial biological specimen (plasma, urine, etc.) before any manipulation, e.g., solvent extraction or TLC (this compensates for losses caused by inaccurate and nonquantitative transfers); and cleanly separated from all other peaks in the chromatogram. This method compensates for small changes in operating conditions and for inaccuracies in injection volumes (which are highly troublesome in the external calibration method). McNair and Bonelli [131] discuss the quantitation aspects of analysis in detail. The best calibration curves are obtained when the appropriate weight ratios are added to the initial biological sample and then carried through the entire procedure. The standard curve should cover the entire range used in the analysis so that any nonlinearity in the system will have a negligible effect on the results. A typical calibration curve used in the analysis of probenecid [38] using an FID is presented in Fig. 8. A linear relationship is observed when the peak height ratio is plotted versus the weight ratio. The curve was obtained following the addition of known weight ratios of probenecid and internal standard to control plasma and then carried through the entire procedure. A summary of the recovery results of 0.2 to 60 μg of drug added

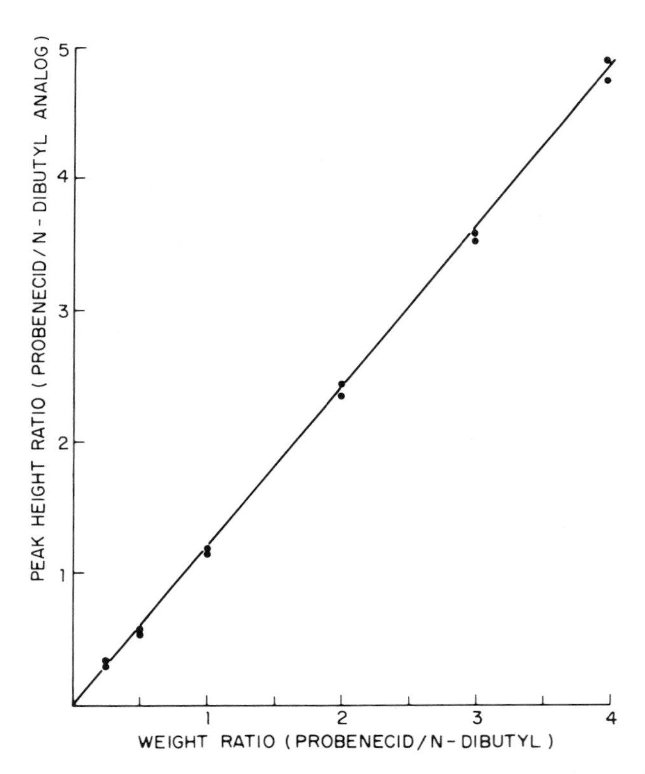

FIG. 8. Relationship between peak height ratio (probenecid/N-dibutyl analog) and weight ratio (probenecid/N-dibutyl analog). Five microliters was injected in each case out of 100 μl following recovery from plasma. (Reprinted from Zacchei and Weidner [38] with permission. Copyright by the American Pharmaceutical Association.)

to control dog plasma is presented in Table 1. The mean recovery was 100.5 ± 6.4% over the entire concentration range. Zacchei and Weidner [135] have used a similar technique with an ECD in the quantitation of a novel antiarrhythmic drug (MK-251). Figure 9 shows a plot of the peak height ratio (drug/I.S.) versus weight ratio. The amount of drug added to the initial tube was in the 5- to 100-ng range and the amount of I.S. added was 50 ng. Typical recovery data are presented in Table 2 and Fig. 10. The recovery of drug added to control dog plasma was 102.5 ± 13.5% over the entire range for a period of several months.

Either peak height or peak area measurements may be employed in quantitation. Ball et al. [136-138] discuss the precision of the methods and the following conclusions can be presented. Peak height measurements

TABLE 1

Recovery of Probenecid from Plasma[a]

Amount added (μg/ml)		Amount recovered[b]						Mean ± S.D.
60.0	A	62.1	61.3	59.2	59.3	61.1	61.7	60.78 ± 1.24
	B	103	102	99	99	102	103	
45.0	A	44.8	44.5	45.5	46.1	45.6	44.7	45.20 ± 0.63
	B	99	99	101	102	101	99	
30.0	A	29.9	28.9	31.0	29.6	29.0	29.9	29.72 ± 0.76
	B	100	97	103	99	97	100	
15.0	A	15.0	15.4	15.4	15.1	15.0	15.2	15.18 ± 0.18
	B	100	102	102	101	100	101	
7.5	A	7.25	7.17	7.69	7.38	7.24	7.36	7.34 ± 0.18
	B	97	96	103	98	97	98	
5.0	A	4.83	5.20	4.82	5.28	4.68	5.05	4.98 ± 0.24
	B	97	104	96	106	94	101	
1.0	A	0.95	1.10	0.91	0.89			0.96 ± 0.09
	B	95	110	91	89			
0.5	A	0.55	0.53	0.46	0.48			0.51 ± 0.04
	B	110	106	92	96			
0.25	A	0.30	0.21	0.27	0.27	0.30		0.27 ± 0.03
	B	120	84	108	108	120		

[a]From [138]. Copyright by the American Pharmaceutical Association.

[b]Values in the A rows represent micrograms recovered; values in the B rows represent percent recovery.

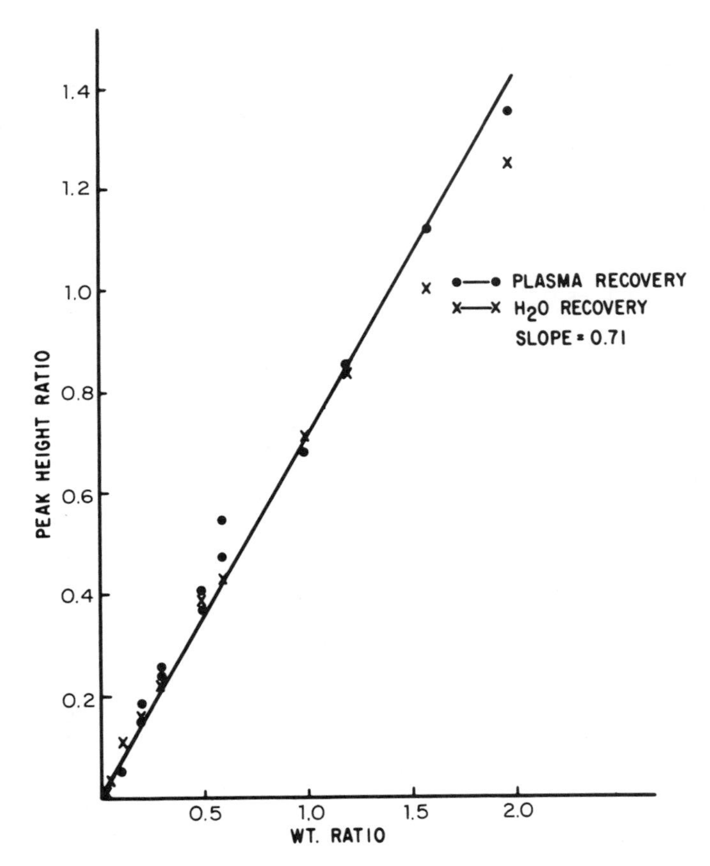

FIG. 9. Relationship between peak height ratio (trifluoroacetylated drug/trifluoroacetylated I.S.) and weight ratio (drug/I.S.). Five microliters was injected in each case out of 100 to 500 μl following appropriate recovery (water or plasma). (Reprinted from Zacchei and Weidner [135] with permission. Copyright by the American Pharmaceutical Association.)

are faster and inherently more precise than the peak area measurements, since fewer manipulations are required. For incompletely resolved peaks, the adjacent peak contributes significantly more to the peak area than it does to the height of the measured peak. Peak height measurements of narrow peaks on sloping baselines may be as much as an order of magnitude in error when comparing the area measurements. However, peak height determinations are more sensitive to column overload, extensive tailing, and small changes in operating conditions (column temperature and carrier flow rate). Consequently, precise control of the operating parameters

TABLE 2

Recovery of Drug from Dog Plasma and Water[a]

Amount added (ng/ml)		H_2O	Amount recovered[b]						Mean ± S.D. of plasma recovery
			Plasma						
100	A	87.0	93.8	99.4	—[c]	105.6	99.4	109.2	101.5 ± 6.0
	B	87	94	99	—	106	99	109	
80	A	72.9	78.3	76.1	81.6	91.2	81.6	79.6	81.4 ± 5.2
	B	91	98	95	102	114	102	99	
60	A	60.1	59.4	66.5	70.6	77.5	68.6	—	68.5 ± 6.6
	B	100	99	111	118	129	114	—	
50	A	50.3	47.3	56.2	59.0	43.9	51.4	47.2	50.8 ± 5.8
	B	101	95	112	118	88	103	94	
30	A	30.6	32.9	35.7	35.0	37.7	35.0	—	35.3 ± 1.7
	B	102	110	119	117	125	117	—	
25	A	28.1	25	28.8	21.9	24.7	25.4	25.4	25.2 ± 0.9
	B	112	100	115	88	99	102	102	
15	A	16.6	16.6	13.7	18.5	15.1	15.8	11.9	15.3 ± 2.3
	B	111	111	91	123	101	105	79	
10	A	11.6	10.6	9.6	11.7	9.6	10.3	7.8	9.9 ± 1.3
	B	116	106	96	117	96	103	78	
5	A	10.1	3.7	3.4	5.5	5.5	4.8	3.5	4.4 ± 1.0
	B	202	74	69	110	110	96	70	

[a]From [135]. Copyright by the American Pharmaceutical Association.

[b]Values in the A rows represent nanograms recovered; values in the B rows represent percent recovery.

[c]No sample analyzed.

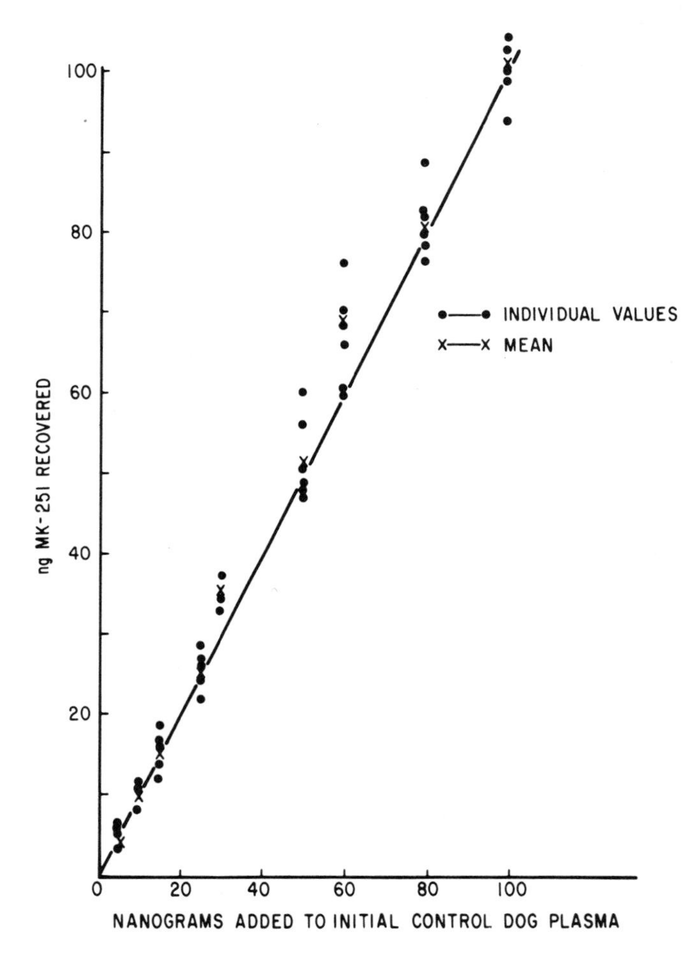

FIG. 10. A typical recovery curve where various amounts of drug were added to control dog plasma. The curve shows the individual values and the mean for each concentration. Five samples were run at each concentration.

must be maintained if one is to take advantage of the greater precision. The authors investigated four methods of area calculations: triangulation, cut and weigh, planimetry, and height-width. Triangulation was shown to be the least precise method and not recommended. The cut and weigh method suffers from variations in paper thickness and moisture content. The same degree of precision (~0.5%) was obtained for peaks of large areas from either the height-width or planimetry measurements. The height-width method is preferred since the error involved in measuring smaller areas is less than the planimetry method.

The aforementioned methods are time-consuming and for work involving large numbers of samples they are usually replaced by some automatic method of determining the detector response directly from the electrometer. These automatic systems range from simple integrators to the use of on-line computers. Many of the vendors of gas chromatographs provide equipment which not only produce a chromatogram but also print a digital record of the integration of the area under each peak, the time of each peak, and the total area for the run. Some systems even produce a computer-compatible punched tape record of the GLC analysis and a typewritten record of the integration data for each sample via a teletype unit. Automation of the heavy routine analytical load of multichromatograph laboratories has been successfully accomplished by many different types of computers [139]. Dessy and Titus [140] discuss the problems involved in computer interfacing to gas chromatographs. They used as a working example the development of an automated gas-chromatographic package servicing eight gas chromatographs. Operator problems and system reliability with respect to GLC automation were discussed by Deans [141]. He described a modular automatic system and discussed the credibility of analysis, factors affecting credibility, operator errors, instrument malfunctions, and, most importantly, reliability of analysis. Craven et al. [142] describe an on-line computer-GLC data handling system which monitors up to 30 chromatographs simultaneously. The system utilizes the IBM 1130 computer with 8K word core storage and 500K disk storage. The system has software provisions for area reallocation between merged peaks and compensation for peak integration on tailing peaks or baseline drift. The system calculates relative retention times (using up to three reference peaks) and retention indices on request. Analysis can be calculated on area percent, mole percent, or weight percent by internal or external normalization. The results are generated at the termination of the analysis and can be stored for later summary or retention index reporting. This approach permits both the reception of GLC data "on the fly" and the output of the information moments after terminating the analysis.

The computer has eliminated the routine calculation in GLC and provides integrations and calculations which are more reproducible than can be obtained manually. The data are obtained immediately and documented in a precise and presentable form. The large dynamic range of the interface eliminates the need to attenuate the signal from the chromatograph and to permit the accurate integration of peaks which go off scale.

V. PHARMACEUTICAL PREPARATIONS

The qualitative and quantitative analysis of drug dosage forms, although possibly lacking the glamor and acclaim of determining plasma levels in patients at the nanogram-per-milliliter level, is of critical importance to the pharmaceutical industry and allied professions. Indeed, the safety and validity of studies involving human and other species are dependent on the administration of preparations of known properties. GLC methods for the analysis of pharmaceutical preparations have proved useful with a large variety of compounds. Several recent papers describe analyses for two of the oldest and most widely used drugs, aspirin and salicylic acid. Watson et al. [143] have reported on the rapid and simultaneous determination of these two compounds in admixtures and in aspirin tablets by GLC of their methyl esters. Derivatization is achieved by addition of diazomethane to a tetrahydrofuran solution of the powdered sample. The reaction is carried out for <5 min at -15° C to preclude formation of methyl O-methoxybenzoate, the internal standard. Since the level of salicylic acid in aspirin tablets is very low, the methyl salicylate peak is much smaller than the peak for methyl O-acetoxybenzoate. Thus, when using the same attenuation setting for both peaks the latter will be far off scale; an electronic digital integrator is employed to achieve quantitation. The same laboratory has developed an assay based on this approach (methylation; methyl O-methoxysalicylate internal standard) for salicylic acid in analgesic preparations containing aspirin, phenacetin, caffeine, and propoxyphene or codeine [144]. The method was employed to study aspirin stability in these commercial multicomponent formulations and to elucidate the nature and causes of the degradation. Another GLC method for the determination of aspirin and salicylic acid (as their TMSi derivatives) in solid dosage forms has been reported by Patel and co-workers [145]. Traces of acetylsalicylsalicylic acid (<0.1%) were also found in a majority of the samples investigated. Temperature programming was employed because of the great differences in volatility, and hence isothermal retention times, for the three compounds. This technique is one of the strengths of analysis by GLC, for it permits the analyst to obtain profiles [146, 147] on mixtures of compounds (synthetic or naturally occurring) with widely differing molecular weights or polarities, or both. The analysis is not limited to a few closely related substances, and by scanning for a wide range of compounds unexpected components may be discovered in samples.

A GLC assay for vitamins D_2 and D_3 in the presence of vitamin E (α, γ, δ-tocopherols) and vitamin A in dosage forms has been described by Fetter and associates [148]. These fat-soluble vitamins are isolated in their alcohol forms (saponification followed by extraction with toluene) and converted to their propionates by reaction with propionic anhydride. Analyses are carried out on SE-52 with samples containing vitamin D_2, and on SE-30 with samples containing vitamin D_3. Vitamin A propionate undergoes a thermal degradation in the injector zone and the pyrolysis products elute in the solvent front. The vitamin D propionates undergo a thermally induced rearrangement to mixtures of the pyro and isopyro forms, which are separated from the tocopherol propionates and quantified (trioctanoin internal standard).

Mixtures of the equine estrogens are found in several pharmaceutical preparations, normally as their sulfate esters. These compounds are usually separated by GLC in derivatized form following hydrolysis. McErlane [149] advocated using a combination of TMSi and methoxime-TMSi derivatives; however, Schroeder et al. [150] have recommended analysis of the free phenols. Roman and co-workers [151] have reported the analysis of esterified estrogens in tablets employing a TMSi ether approach.

Karkhanis and co-workers [152] have published a GLC method for the determination of resorcinol monoacetate which can be adapted to the quality control of the drug in dermatological creams and lotions. Previously published methods of analysis for resorcinol monoacetate have not been applied to dermatological products because of the complex nature of the preparations. The approach reported by Karkhanis et al. [152] involves acetylation of the sample and benzene extraction of the diacetates of resorcinol and orcinol (added at the beginning of the assay as an internal standard).

A GLC method for the analysis of iodochlorhydroxyquinoline and related halogenated 8-hydroxyquinolines as their TMSi derivatives has been reported by Gruber et al. [153]. According to these authors previous approaches have shared a common failing, namely, the inability to distinguish impurities (halogenated 8-hydroxyquinolines) from each other and from iodochlorhydroxyquinoline. GLC methods were not successful until trimethylsilylation (with the reagent trimethylsilylimidazole) was utilized. Using this derivatization approach the closely related halogenated 8-hydroxyquinolines were found to be readily separable by GLC, and the authors [153] suggest that their procedure is a significant improvement over the then currently official (USP XVIII; NF XIII) method for iodochlorhydroxyquinoline-hydrocortisone cream.

The determination of drugs as their hydrochlorides by GLC can be carried out directly or indirectly. An example of the former approach is the GLC analysis for meclizine hydrochloride in tablet formulations reported by Wong et al. [154]. Pulverized tablets are extracted with chloroform containing an internal standard, and an aliquot of the resulting solution injected. Since meclizine hydrochloride yields a peak with the same retention as the free base, it is evident that HCl is cleanly eliminated from the salt via a thermal conversion. On the other hand, Sennello [155], in his work on the determination of methamphetamine hydrochloride in sustained-release tablets, observed that the assay results were not reliable unless the drug was converted to its free base (using KOH) prior to GLC.

The need for such a chemical conversion should be established in each individual case.

A simple but specific GLC method for determining the purity of phenylbutazone raw material and solid dosage forms of this drug has been published by Watson et al. [156]. The GLC procedure is held to be much less tedious than the earlier UV spectrophotometric method which, although direct, lacks specificity. The basis for the GLC assay is that whereas phenylbutazone is eluted as a single symmetrical peak, each degradation product of the drug undergoes thermally induced breakdown in the flash heater zone of the chromatograph to yield a reproducible pattern of peaks. The latter serve to indicate the extent of decomposition present in the drug formulation. Some of these decomposition products absorb at 264 nm and thus interfere with the UV method. They do not possess the same retention time as the drug and are identified using the GLC approach.

A GLC method for the analysis of L-dopa and closely related amino acids in capsules and tablets of this drug has been published by Chang and associates [157]. Trimethylsilylation was employed to convert these highly polar compounds to nonpolar derivatives suitable for GLC analysis. The method employs α-methyldopa (Aldomet) as internal standard for L-dopa; hence an assay for the former drug could probably be readily established using L-dopa as internal standard.

Although not a drug, sorbitol is a frequently encountered ingredient of pharmaceutical preparations. Simple but specific and reliable assays for this hexahydric alcohol in the presence of other substances including mannitol have recently been reported by two laboratories [158, 159]. Both approaches involve GLC of the hexacetyl derivatives, with good separation of sorbitol and mannitol hexaacetates, in contradistinction to results with the corresponding TMSi derivatives. Each group of authors stated that their approach was superior to the then official USP procedure (non-GLC) for sorbitol on the basis of ease and speed; furthermore, the GLC approaches allowed the simultaneous determination of both sorbitol and mannitol.

A GLC method for the analysis of suspensions containing ephedrine, phenobarbital, and theophylline (frequently used in combination for sedation and bronchodilation) has been reported by Schultz and Paveenbampen [160]. A pH-dependent extraction procedure was employed to separate ephedrine from phenobarbital and theophylline; the two fractions are then subjected to GLC analysis with internal standards on an OV-17 column at two different temperatures. Recoveries, mean standard deviations, and coefficients of variation were eminently satisfactory.

Alkaloids, among the first compounds of considerable molecular weight and structural complexity to be separated by GLC [161], are frequently found as components of pharmaceutical preparations. A number of GLC methods have been developed for their analysis in dosage forms and related samples. Sondack and Koch [162] have reported the GLC determination of strychnine and brucine in liquid and tablet formulations. A simple isolation and extraction procedure is employed. The sample is adjusted to alkaline pH, extracted with chloroform, and the extract taken to dryness. The residue is redissolved in a solution containing the internal standard and analyzed. This method is considerably less complicated than that described in NF XI for assay of nux vomica tincture for strychnine alone (any brucine must be destroyed, or it would interfere with the strychnine determination). Whereas this GLC assay is carried out without derivatization, a method for the determination of ergonovine maleate in pharmaceutical preparations employing brucine as internal standard involves use of trimethylsilylation [163]. The GLC approach allows both quantitation of the drug and also estimation of the extent of degradation (as determined by earlier colorimetric and TLC procedures, respectively) of this labile compound. Degradation always occurs during sample workup, and since the GLC assay involves less handling of the sample than other methods, less degradation is encountered with the former. Replicate GLC assays of 0.1-mg/tablet samples gave a relative standard deviation of ±3% with a relative error of -2%.

Grady and Zimmerman [164] have reported on a collaborative study of a GLC assay for atropine and scopolamine in pharmaceutical preparations. The simple, rapid procedure involves adjusting the preparation to a basic pH, adding the internal standard, and extracting with methylene chloride. The organic phase is reduced in volume and an aliquot is analyzed by GLC. No derivatization steps are involved. Accuracy, reliability, sensitivity, specificity, and precision were found to be satisfactory. Poorly prepared and conditioned GLC columns have been implicated in the excessive tailing and on-column dehydration of belladonna alkaloids such as atropine and scopolamine. Such phenomena are not observed in this assay, and the authors credit the column system employed (3% OV-17) and a special curing (or bake out) procedure for their success.

A GLC method for the specific and accurate determination of scopolamine and its acidic and basic degradation products has been reported by Windheuser et al. [165]. This approach uses trimethylsilylation and temperature programming; these techniques are required because of the polar nature of the degradation products and their wide range of retention times.

Santoro and co-workers [166] have published a GLC method for the determination of hyoscyamine sulfate, atropine sulfate, and scopolamine hydrobromide in pharmaceutical preparations also containing chlorpheniramine and phenylpropanolamine hydrochloride. The first two compounds (as their free bases) coelute from the OV-17 column used in the analysis, which employs homatropine as internal standard. In another publication Grabowski and associates [167] describe their GLC method for the anlysis of homatropine methyl bromide in tablets and elixirs also containing other tropine derivatives. The homatropine methyl bromide is hydrolyzed to mandelic acid, which is trimethylsilylated and quantified. No interference has been observed from the usual components of tablets and elixirs.

Greenwood and Guppy [168] have described a direct GLC method for the analysis of a large number of basic nitrogenous drugs in a variety of pharmaceutical preparations. This approach, which employs an OV-17 column operated under isothermal conditions between 200° and 270° C, is held to be much more specific than classical methods and can be used for routine quality control work. The paper contains numerous references.

GLC methods can also be employed for the determination of drugs in non-FDA-approved dosage forms. For example, Moore and Bena [169] have published a paper on the assay of heroin, and also the commonly employed diluent quinine hydrochloride, in illicit preparations. In addition to being faster and more readily applied to small samples than a spectrophotometric method, the GLC assay also compared favorably with respect to accuracy. Washings from paraphernalia such as cookers, syringes, and strainers employed for heroin administration can be assayed for the drug and also for the related compounds cocaine and 0^6-monoacetylmorphine, the latter a hydrolysis product of heroin. GLC has been recommended for the identification of the street drug LSD in the presence of its N-methyl-N-n-propyl isomer [170]. The authors were not able to separate the isomers by TLC. LSD itself possesses poor GLC properties, but adequate behavior is found for its TMSi derivative. A column containing textured glass beads coated with 0.25% OV-17 was employed for the separation. The sensitivity of the method is not high, since sample equivalent to 50 to 100 μg of LSD is required per analysis. A single-pH extraction procedure for detecting drugs of abuse in urine via TLC has been reported by Stoner and Parker [171]. It might also prove useful to examine the extract by GLC.

Segelman and Sofia [172], using a GLC procedure, have found that boiled marijuana is enriched significantly in (-)-trans-Δ^9-tetrahydrocannabinol content. Since Δ^9-THC is probably the major psychoactive constituent of marijuana, this may explain the increased potency (on a unit mass basis) of the water-extracted and dried plant material. A GLC assay has been employed to determine the cannabinoid (Δ^9-THC, cannabidiol, and cannabinol) content of marijuana plants grown in Mississippi from seeds from many parts of the world [173]. The cannabinoid levels were used as the criterion for dividing the marijuana into two chemical phenotypes.

VI. HUMAN HEALTH DRUGS

Cornish et al. [174] presented a review of pharmaceuticals and related drugs in 1973. The information is arranged in tabular form listing the classes of compounds, method and subject of analysis, and the appropriate reference. The review covers the period from June 1970 to June 1972 and more than 2500 references are reported. Gochman and Young [175] have reviewed the qualitative and quantitative methods of analysis of substituents encountered in clinical chemistry. In their review, which covers 1971 and 1972, reference is made to the analysis of therapeutic drugs by a variety of techniques with special emphasis on GLC. Vesell and Passananti [176] in their review article discuss the utility of such clinical chemical determinations of drug concentrations in biological fluids. The authors stress the necessity of determining blood levels of a drug in an effort to prevent possible drug toxicity. A list of the common drugs analyzed, the method, and the approximate serum concentrations is presented. Reid et al. [177] describe a simple, rapid, and precise method for the preliminary resolution and quantitation by GLC of toxic drug substances present in human body fluids. A two-component (1% SE-30 and 1% QF-1) liquid phase and temperature programming are employed. The sample components are identified by comparison of their retention times with those of standards. The data can usually be reported to the physician in less than an hour. They have found a useful correlation between total barbiturate, state of anesthesia, and potency of barbiturates. Their technique is applicable to only those drugs that are extractable from the biological fluids using chloroform.

Another systematic approach to the application of GLC to toxicity was presented by Jain and Kirk [178, 179]. Procedures for the extraction and quantitation of alkaloids and antihistamines are discussed. Most of their studies dealt with high concentrations of drug in the blood. An FID was used for the analysis of barbiturates, alkaloids, tranquilizers, and antihistamines. Polyester Hi-Eff-8B was used as stationary phase for these studies. Cardini et al. [180] described a GLC method used to analyze pharmaceutical preparations of psychotropic drugs. A number of monoamine oxidase inhibitors and other stimulating psychotropic drugs were separated and analyzed. The primary goal of these studies was to determine the purity of various pharmaceutical preparations and for the separation and identification of mixtures. Some of the drugs analyzed, specifically the hydrazines, undergo pyrolysis under the GLC conditions employed. The gas phase analysis of human drugs and metabolites using "metabolic profiles" has been reported by Horning and Horning [23, 24, 146]. These authors discussed multicomponent analyses for three types of urinary constituents: steroids, acids, and drugs and their metabolites. The techniques involved GLC, GLC-MS, and computerization. In most instances derivatives were prepared to increase volatility and thereby provide GLC separation and subsequent characterization.

The high degree of specificity and sensitivity of GLC analyses can be illustrated in the analysis of several drugs. Published methods for the determination of amitriptyline, a widely used antidepressant, have been reported [181-185]; however, they lacked either the sensitivity (FID) or specificity (UV detection) necessary for the determination of plasma levels following administration of therapeutic doses in the presence of nortriptyline, a known metabolite. Hucker and Stauffer [1] developed a method for the measurement of steady-state levels of amitriptyline and nortriptyline in human plasma. The drugs were extracted with heptane-3% isoamyl alcohol from the samples at a basic pH and back-extracted into 0.1 N HCl. After subsequent extraction into ether, the drugs were analyzed on a 1.5% OV-17 column. The method possesses sufficient sensitivity to detect concentrations as low as 25 ng/ml. Normal amitriptyline levels are on the order of 40 to 90 ng/ml following doses of 75 mg/day.

Diphenylhydantoin (DPH) is a drug widely used in the treatment of grand mal epileptiform seizures, and numerous GLC methods for the quantitation of this drug have been reported in the literature [44, 50, 186-190]. The results obtained from GLC analyses were compared with those obtained following spectrophotometric techniques; in general, the GLC methods exhibited increased sensitivity and specificity. The GLC methods employed either no derivatization [186, 190] or derivatization with tetramethylammonium hydroxide [44, 50, 187] or trimethylanilinium hydroxide [188, 189].

VandenHeuvel et al. [191] have reported a GLC method for hydrochlorothiazide in biological fluids. The drug is converted to its tetramethyl derivative; an ECD is employed for assay at the 0.05-μg/ml level.

A GLC method for carbamazepine, another anticonvulsant drug, was reported by Friel and Green [192]. The drug, following appropriate extraction procedures, was injected directly into a 10% OV-7 column. Adsorption and decomposition problems that were discussed by Meijer [193] and Street [194] were minimized by masking and neutralization of the column solid support and by use of a solvent (carbon tetrachloride) with a low dielectric constant.

A. Pharmacokinetics

The role of GLC in pharmacology and toxicology as related to pharmacokinetic analysis of a drug was described by VanRossum et al. [195]. Pharmacokinetic parameters may be determined accurately only if the entire concnetration curve is measured. This implies that a fairly large number of blood samples should be analyzed while keeping sample size at a minimum. A GLC method has advantages resulting from the inherent sensitivity and specificity as compared to colorimetric or spectrophotometric techniques. The drug plasma-time curve is a time-dependent function which is a reflection of what the body does to a drug. The data provide insight into the kinetics of drug metabolism, excretion, and retention, thereby permitting an evaluation of specific dosing regimens to be followed to provide the necessary pharmacological effect.

Kaiser and Van Giessen [196] described a simple, rapid, sensitive, and specific GLC method for the determination of ibuprofen (an orally active antiinflammatory drug) in the human plasma. The procedure was successfully applied to drug absorption studies in man. The method is sensitive to 0.5 μg ibuprofen/ml plasma with an average recovery of 94.8 ± 6.6% (S.D.). GLC-MS analysis confirmed the specificity of the method. The authors were subsequantly able to evaluate the pharmacokinetics and drug bioavailability from various dosage formulations. In the same year Kaiser and Glenn [197] did a correlation study with plasma ibuprofen levels and the biological activity. Their data indicated that plasma drug concentrations in normal and polyarthritic rats were dose related. The logarithm of biological activity, expressed as percent inhibition of polyarthritis (% IPA)/(100 - % IPA), was related to the logarithms of (a) administered dose (mg/kg), (b) plasma drug concentrations (μg/ml) at 2 hr postdrug, and (c) the average plasma drug concentrations in a dosage interval at the equilibrium state. No differences were observed, however, in the rate of elimination of drug from the normal rats or polyarthritic rats.

Yashiki et al. [198] reported on the pharmacokinetics and metabolic fate of 5-n-butyl-1-cyclohexyl-2,4,6-trioxoperhydropyrimidine (BCP) in man. A GLC procedure was developed which provided greater sensitivity and specificity than the previously used UV method [199]. A chloroform extract was taken to dryness, subjected to trimethylsilylation, and subsequently analyzed on a 2% neopentyl glycol succinate column. The method possessed a sensitivity limit of 0.1 μg of BCP. A comparison of the results obtained from the GLC method and the UV method was presented; in most instances the UV method gave higher values, suggesting possible interference.

Analytical and pharmacokinetic studies on the butyrophenones (haloperidol and trifluperidol) were described by Marcucci et al. [200]. The butyrophenones represent a class of powerful neuroleptics which are used for the treatment of mental disease. The studies were undertaken to provide some insight into the absorption, excretion, distribution, and metabolism of this class of drugs. The disposition of haloperidol and trifluperidol was extensively investigated in rats. The GLC method was based on the electron capture response of these drugs employing a ^{63}Ni detector. No derivatization was required. Increased sensitivity was observed with an increase in the number of halogen atoms on the butyrophenone. Specificity of analysis was confirmed by GLC-MS techniques. Using the electron capture GLC method (3% OV-17), Marcucci and co-workers [200] demonstrated the rapid accumulation of haloperidol in the brain. The rat brain/blood ratio 1 min after injection of the drug was 6.6. No drug (≤0.02 μg/ml) was detected in the blood after 1 hr; however, the brain concentration at 5 hr was still very high (≃0.04 μg/g). Similar results were obtained with trifluperidol; however, unlike haloperidol, trifluperidol did accumulate in the adipose tissue. When the neuroleptic activity of the two drugs was evaluated in the amphetamine antagonism test on rats, it was shown that the brain concentration of the drugs must be greater than 30 ng/g to inhibit completely the effects of amphetamine.

VII. ANIMAL HEALTH DRUGS

The determination of residue levels in tissue to be consumed by humans is a necessity in bringing a new animal health drug onto the market, and GLC methods are frequently employed to obtain the required data. Cala et al. [112] have recently described the successful application of GLC combined with electron capture detection to the analysis of the coccidiostat pyrimethamine (2,4,-diamino-5-p-chlorophenyl-6-ethylpyrimidine) in chicken tissue with an assay sensitivity limit of 0.05 ppm. Although pyrimethamine would appear to be readily susceptible to adsorption during GLC, a linear relationship was observed between quantity injected and detector response from 20 ng down to 2 ng without derivatization. The chromatogram resulting from injection of 10 ng of pyrimethamine is shown in Figure 11. Analysis of 1% of the extract from a 2-g sample of liver (3 days postdose) gave the chromatogram shown in Fig. 12. Specificity of

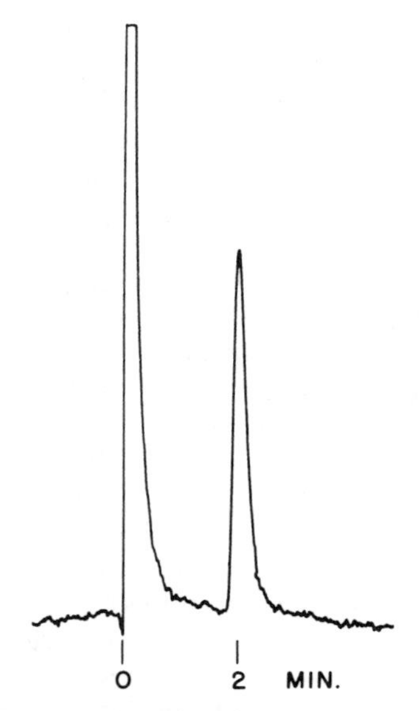

FIG. 11. Chromatogram resulting from analysis of 10 ng of pyrimethamine using an electron capture detector. (Reprinted from Cala et al. [112] with permission. Copyright by the American Chemical Society.)

the method was demonstrated in studies using ^{14}C-labeled drug and combined GLC-MS (Sect. III D).

Rafoxanide [3'-chloro-4'-(4-chlorophenoxy)-3, 5-diiodosalicylanilide], in contrast to pyrimethamine, must be derivatized in order to undergo GLC analysis [201]. This anthelmintic was found to form a di-TMSi derivative, but the high molecular weight (769) of this compound resulted in long retention times under column conditions suitable for steroids. The four halogen atoms should provide for very good electron capture sensitivity, but on a 2-ft column the lower practical detection limit was 1 ng of drug per injection. Reduction of the column residence time of this compound was found to result in increased detector response (peak area), suggesting on-column sample loss and degradation (transfer of a TMSi group from the sample molecule to the packing). A column containing only 4 in. of packing was then chosen for the GLC assay, and a measurable response was obtained from injection of as little as 0.2 ng of drug (as its derivative). This system was adequate for separation from the solvent front; no metabolites were present to interfere. The left panel of Fig. 13 shows the chromatogram resulting from GLC on the 4-in. column of 1 ng of Rafoxanide (as its di-TMSi derivative). The

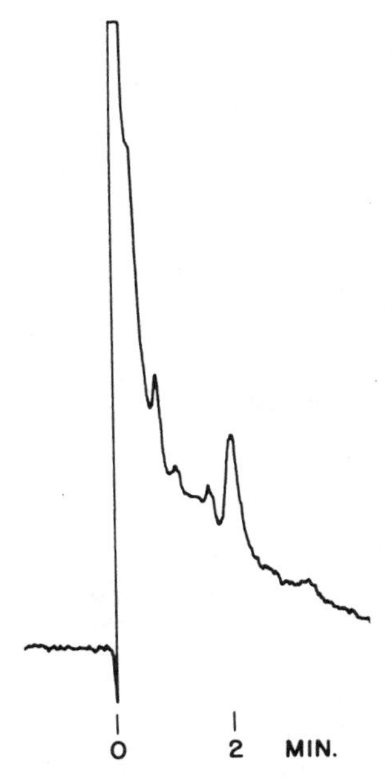

FIG. 12. Chromatogram resulting from the analysis of an isolate from liver of a chicken 3 days following administration of pyrimethamine; the drug residue was found to be 0.2 ppm. (Reprinted from Cala et al. [112] with permission. Copyright by the American Chemical Society.)

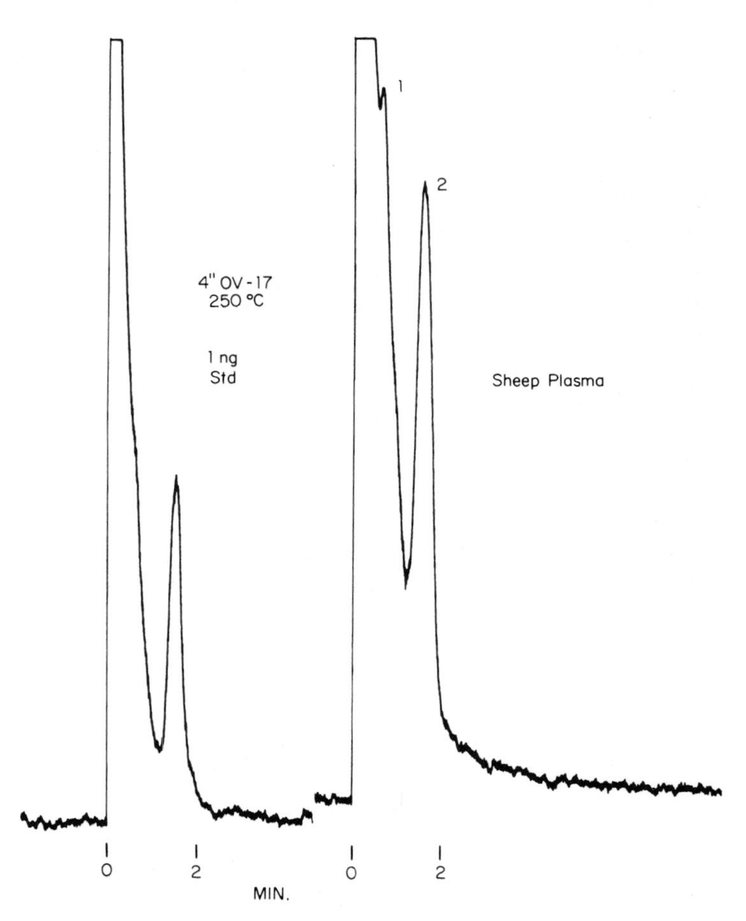

FIG. 13. Left panel: Chromatogram resulting from analysis of 1 ng of rafoxanide as its di-TMSi derivative. Right panel: Chromatogram resulting from analysis of 1/50 of a trimethylsilylated isolate from 1 ml of plasma from a sheep dosed with rafoxanide. Peak 2 corresponds to the di-TMSi derivative of rafoxanide; peak 1 results from the di-TMSi derivative of desiodo-Rafoxanide (produced by photolysis during sample workup.) (Reprinted from Talley et al. [201] with permission. Copyright by the American Chemical Society.)

right panel shows the chromatogram resulting from assay of a trimethylsilylated plasma extract from a sheep treated with Rafoxanide. The small peak designated 1 is caused by the di-TMSi derivative of desiodo-rafoxanide, which can arise via a photolytic reaction and is not a metabolite.

The accurate determination of low tissue residue levels of DES, both free and conjugated, in drug-treated animals intended for human consumption is of great importance from a regulatory and legal viewpoint because of the toxic nature of this chemical. One proposed GLC method involves dissolution of the final isolate in 10 μl of ethanol and injection of this solution in its entirety. Since detection is by electron capture, but no derivatization step is involved, the authors [202] speculate that detection "... is dependent upon the formation of a derivative of DES on the column." It would not be surprising, however, if a compound possessing the structure of DES would exhibit a sufficient electron capture affinity to be detected at the levels reported in this paper (minimum of 20 ng injected). The method of Coffin and Pilon [203] also employs electron capture detection, but the final isolate is trifluoroacetylated prior to GLC, and a 0.4% aliquot of the final sample is injected. This is sufficient to allow determination of DES at the 2- to 10-ppb level. DES exists as an equilibrium mixture of the cis and trans isomers in solution, and derivatization followed by GLC yields two well-separated peaks; the sums of the two areas are used for quantitation. As these authors indicate, this is a very useful feature in assaying for DES, as a doublet must be observed on the chromatogram if DES is indeed present in the sample.

Donoho et al. [204] have published a paper on the GLC determination of DES (and its glucuronide) in cattle tissue with a sensitivity of 1 ppb using the dichloroacetyl derivative and electron capture detection. Peaks arising from both cis and trans isomers are observed, but generally only the latter is quantified. Prior to the assay of tissue isolates the column is "primed" by injection of standards until successive injections of the same sample yield reproducible peak heights. With very low level DES isolates from tissue the authors noted a "potentiation" effect: A known amount of derivatized DES from control tissue samples gave a greater response than the same amount of reference material injected without tissue extract material. This effect was held to be caused by tissue components deactivating the column or eliciting an anomalous detector response, but not to an interfering GLC peak. Verification (confirmation) of assays for controversial compounds such as DES is surely an area of applicability for the specificity and sensitivity of GLC-MS techniques.

VIII. METABOLISM

There are far more literature reports on the use of GLC methods for assay purposes involving parent drug than for metabolite identification. The following few examples illustrate the use of GLC in this latter aspect of drug analysis. Hucker et al. [205] studied the metabolism of 1-(p-chlorobenzoyl)-5-dimethylamino-2-methylindole-3-acetic acid and utilized GLC techniques to demonstrate that a major urinary metabolite of this drug in the human and several other species is p-chlorohippuric acid. The methylated acidic metabolite exhibited a retention time identical to that of methyl p-chlorohippurate, and cochromatography of the methyl esters of the reference standard and the isolate gave a single peak with enhanced intensity. Hucker and Hoffman [206] also employed GLC in a study of the metabolic fate of 1-p-chlorobenzylidenyl-5-methoxy-2-methylindene-3-acetic acid. GLC was employed by Berman and Spirtes [81] to study the metabolism of chlorpromazine by rabbit microsomes. A typical chromatogram contained peaks from six drug-related compounds: chlorpromazine, mono- and didemethyl chlorpromazine, chlorpromazine sulfoxide, monodemethyl chlorpromazine sulfoxide, and chlorpromazine N-oxide. GLC techniques have been used by Dinovo and co-workers [207] in a study of thioridazine and mesoridazine (the S-oxide of the former) metabolism in humans. An apparently drug-related unknown component was found to be present in plasma. This compound was shown not to possess GLC (or TLC) behavior of any of the six known metabolites of thioridazine. The metabolic fate of chlorpropamide, [p-chlorophenyl)sulfonyl]-3-propylurea, in man and the rat was studied using GLC methods [208]. The 2- and 3-hydroxypropyl metabolites were quantified as the dibenzoyl derivatives of the respective metabolite hydrolytic products (the isomeric hydroxypropylamines). De Leenheer and Heyndrickx [209] identified and quantified the human urinary sulfoxide and desmethyl metabolites of methotrimeprazine by GLC. The GLC system was also used preparatively to permit the isolation of these compounds in a sufficient state of purity for infrared and mass spectrometric examination. These authors carried out similar studies with the urine of patients treated with dibenzepine [83]. GLC analysis of acetylated urinary extracts permitted the identification and quantification of parent drug and free metabolites. The structures of the metabolites were confirmed by infrared and mass spectrometry of the compounds isolated by preparative GLC. Walkenstein et al. [210] utilized GLC methods for the assay of parent drug and metabolites in a study of the metabolism of 1,4-dihydro-2,6-dimethyl-4-(2-trifluoromethylphenyl)-3,5-pyridinodicarboxylic acid diethyl ester.

GLC techniques played a key role in the work of Hucker et al. [211] which demonstrated that phenylacetone oxime is an intermediate in the oxidative deamination of amphetamine by a rabbit liver microsomal system. GLC analysis of an extract of the incubation mixture showed four peaks with retention times (SE-30) of 1.5, 1.6, 3.5, and 3.8 min. Authentic phenylacetone oxime was found to exhibit a retention of 3.8 min. Furthermore, GLC of the extract following trimethylsilylation indicated the disappearance of the 3.8-min peak and a new peak of retention time 5.0 min, the retention time of the TMSi derivative of phenylacetone oxime. These GLC data strongly suggested that the 3.8-min component in the isolate was phenylacetone

TABLE 3

Retention Behavior for Zearalenones

Compound	Relative retention time[a]	
	SE-30[a]	QF-1[b]
Drug	0.99	5.40
Metabolite	0.81	—
6'-Ketone	0.82	—
Metabolite TMSi	1.01	4.52
6'-Ketone TMSi	1.01	4.48
Metabolite OMO	1.03	—
6'-Ketone OMO	1.04	—
Metabolite OMO/TMSi	1.29	2.37
6'-Ketone OMO/TMSi	1.27	2.34

[a]Cholestane = 1.00.

[b]Dimethylpolysiloxane stationary phase.

[c]Fluoroalkylpolysiloxane.

oxime. The component was collected (preparative GLC) and examined by direct probe mass spectrometry; the mass spectrum was identical to that of phenylacetone oxime. By use of trimethylsilylation and two columns of differing selectivity, the 1.5- and 1.6-min retention time components were identified as phenylacetone and phenyl-2-propanol. GLC was also employed by Beckett et al. [212] in their in vitro (guinea pig liver microsomal fractions) metabolism study of fenfluramine.

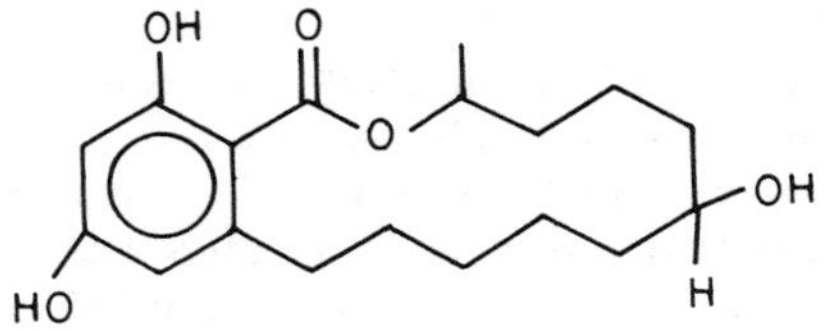

GLC techniques have been employed to identify a major ovine urinary metabolite of the 6'-alcohol from 1',2'-dihydro-(-)-zearalenone (see below) [2]. The pertinent retention data are presented in Table 3. Comparison of the metabolite and the reference compound under a variety of derivatization and GLC conditions resulted in excellent retention data correlation. The appropriate retention time shifts following exposure to the different derivatization conditions comprise an example of GLC functional group analysis at the microgram level. The structure of the metabolite was later confirmed by combined GLC-MS, a desirable if not mandatory approach for metabolism studies.

Gas-liquid radiochromatographic examination of metabolite fractions arising from radiolabeled drugs demonstrates which of the components eluted during GLC analysis are drug related, and hence worthy of further consideration. A well-planned metabolism study should involve use of radiolabeled drug (if possible), gas-liquid radiochromatography, and combined GLC-MS [213].

IX. THERMAL CONVERSIONS OF DRUGS

Although there is far less thermal degradation of organic compounds during GLC at column temperatures >200° C than was originally expected by many persons [214], there are well-documented examples of thermally induced transformations often analogous to those found in classical organic chemistry. Thus, the analyst should not be surprised if structural alterations occur during GLC. When they do occur they may initially be puzzling, but once recognized can yield valuable structural information. Sometimes one can take advantage of them to solve otherwise difficult problems. Several assay procedures for drugs utilizing on-column rearrangements have been reported. Dell et al. [215] have described a GLC method for the determination of bufexamac (4-butoxyphenylacetohydroxamic acid) based on the Lossen rearrangement of the TMSi derivative to the corresponding isocyanate. This is analogous to the earlier report by Vagelos et al. [216] who converted acyl coenzyme A compounds to the corresponding acetylated hydroxamates; GLC of these derivatives resulted in elution of the isocyanates. A GLC method for the determination of bumetanide (3-n-butylamino-4-phenoxy-5-sulfamylbenzoic acid) in urine involves on-column methylation with a concomitant Smiles rearrangement [217]. An assay for tolazamide [1-(n-butyl)-3-p-chlorobenzenesulfonylurea] in plasma is based on quantification of the p-toluenesulfonamide peak resulting from a thermally induced rearrangement [218].

Street [219] has reported on the use of an on-column Hofmann degradation for the identification of drugs containing tertiary amine groups. The drug is chromatographed both before and after exhaustive alkylation with methyl iodide and formation of the quaternary ammonium hydroxide; the latter is injected into the chromatograph where it decomposes to give an olefinic derivative(s) of characteristic retention time(s). This work is an extension of that by Hucker and Miller [182], who demonstrated that on-column Hofmann degradation with tertiary amine drugs can result in improved separation and also be employed in quantitation. Two on-column reactions have been observed to occur during the GLC of long-chain alkyl trimethylammonium bromides [220]. Conversion to tertiary amine and alkyl halide is the major route, but Hofmann degradation also occurs. Thermal elimination during GLC has also been observed with some N-oxides of drugs. Sullivan et al. [221] have reported that injection of methadone N-oxide results in elution of 4,4-diphenyl-5-ketoheptene-2. Chlorpromazine N-oxide undergoes a Cope elimination reaction when subjected to GLC to form 10-allyl-2-chlorophenothiazine [81, 222]. The TMSi derivatives of morphine and morphine N-oxide exhibit identical retention times on three columns of widely differing partitioning properties [223], suggesting conversion of the N-oxide to the parent alkaloid under the conditions employed. Since N-oxides are possible metabolites of tertiary amine drugs, examination of the former by GLC techniques may fail to give direct evidence for their presence in an isolate. Certain sulfate conjugates are also thermally unstable, undergoing on-column conversion to the parent alcohol or olefin [224, 225]. Oxazepam has been shown by two groups of workers [226, 227] to undergo a thermally induced alteration to 6-chloro-4-phenylquinazoline carboxaldehyde. This thermolysis appears to be a general reaction with 3-hydroxy-1,4-benzodiazepin-2-ones.

Westley et al. [228] have reported a GLC assay for lasalocid in solid preparations and fermentation broths based on the quantitative thermolytic conversion of the antibiotic to its retroaldol ketone. The reaction proceeds smoothly and instantaneously upon injection into the chromatograph at 300° C; the relative standard deviation for the pyrolysis assay was found to be 0.87 (lasalocid sodium salt) and 3.7% for samples of fermentation broth.

Carbamazepine undergoes a partial decomposition to yield iminostilbene and 9-methylacridine when a methanolic solution of the drug is examined by GLC [227]. A metabolite of this drug, the 10,11-epoxide, rearranges during GLC to yield 9-acridinecarboxaldehyde [227, 229]; another metabolite, 10,11-dihydroxycarbamazepine, undergoes a pinacol rearrangement on OV-17 (but not on SE-30!) to yield the same aldehyde [227]. Reliance on retention time only would clearly not differentiate between these two metabolites; indeed, even their mass spectra (obtained by combined GLC-MS) would be identical. Combined GLC-MS is the most useful approach for identifying GLC components, but an on-column transformation will, of course, result in misleading MS information. It is for this reason that both GLC- and direct probe-MS examinations of drugs and metabolites are highly desirable.

Carbamates are notoriously susceptible to thermal degradation. GLC of cambendazole [2-(4-thiazolyl)-5-isopropoxycarbonylaminobenzimidazole] at 240° C results in the elution of two components, the parent drug and a compound of shorter retention time, shown to be the corresponding isocyanate [230]. Rather than reducing the extent of isocyanate formation, trimethylsilylation facilitates the rearrangement, for GLC following derivatization results in the elution of only one compound, the isocyanate. Formation of the isocyanate is temperature dependent. GLC at 180° C of a cambendazole metabolite of much lower molecular weight but still possessing the isopropoxycarbonylamino side chain either free or as the TMSi derivative, results in elution of the intact compounds rather than rearrangement products [230]. Thermal decomposition of meprobamate caused a GLC assay for this carbamate drug to be replaced by a colorimetric assay [231]. The GLC method of Martis and Levy [232] for meprobamate avoids the degradation; the drug is hydrolyzed to 2-methyl-2-propyl-2,3-propanediol, which is analyzed as its di-TMSi derivative. Rabinowitz et al. [233] have reported that by judicious choice and careful control of injector port and column temperatures they have developed a specific and rapid direct GLC assay for meprobamate with little, if any, degradation.

Another interesting example of functional group changes during analysis involves ronidazole (1-methyl-5-nitroimidazol-2-ylmethyl carbamate). When chromatographed in the underivatized state at 140° C it is the corresponding alcohol, and not the parent drug, which is eluted from the column [234]. Trimethylsilylation followed by GLC results in elution of the TMSi ether of the alcohol, and it was initially assumed that an on-column conversion of the TMSi derivative of the carbamate to the TMSi ether of the alcohol had occurred. Additional data demonstrated that the TMSi ether of the alcohol was formed from the parent drug in the derivatization medium rather than during GLC, even though reaction conditions were rather mild (BSTFA/pyridine, 4:1, 65° C, 15 min). As the alcohol is a metabolite of ronidazole [235], GLC (or GLC-MS) data alone would not distinguish between the parent drug and its metabolite in an isolate. Other examples of structural changes during derivatization include the formation of the diheptafluorobutyramide of dapsone (p,p'-diaminodiphenylsulfone) by treatment of p-ureido-p'-aminodiphenylsulfone with heptafluorobutyric anhydride [234], the reaction of some steroid ketones during trimethylsilylation to form oxysilylation products (which requires introduction of an oxygen atom) [236], and aromatization of the A ring of norethynodrel during trimethylsilylation with trimethylsilylimidazole [237]. Since very reactive derivatizing agents are commonly employed in connection with GLC work, it is no wonder that unforeseen chemistry occasionally takes place.

X. ASSESSMENT

The great increase during the past few years in the percentage of papers presented at symposia (e.g., the Advances in Chromatography series) and published in journals such as Analytical Chemistry, Journal of Chromatography, Journal of Chromatographic Science, and Journal of Pharmaceutical Sciences that are concerned with the analysis (especially in biological samples) of drugs and metabolites by GLC attests to the general acceptance of these methods by an ever increasing number of workers. Just as the development of reliable thin-film packings and readily mass-produced sensitive detectors in the early 1960s started the ultimate wave of GLC assays, it is likely that several developments of the early 1970s will further increase the attractive nature of GLC-related methodology. High-resolution, thermostable capillary columns possess superb resolving power [96, 238-242], offering the opportunity to distinguish between compounds differing only slightly in structure, but possibly greatly in biological activity. Further, the combination of such columns with picogram detection systems of high selectivity based on mass spectrometry [243] suggests assay systems not dreamed of less than a decade ago. The combination of GLC and MS should be considered as a totally integrated system, and eventually MS-based detection for GLC separations will probably be little more extraordinary than element-selective detectors are today. Indeed, GLC-MS in combination with a computer system as a totally integrated system in drug analysis is already at hand [113, 244-247].

REFERENCES

1. H. B. Hucker and S. C. Stauffer, J. Pharm. Sci., 63, 296 (1974).
2. W. J. A. VandenHeuvel, Sep. Sci., 3, 151 (1968).
3. A. T. James and A. J. P. Martin, Biochem. J., 50, 679 (1952).
4. A. T. James, A. J. P. Martin, and G. H. Smith, Biochem. J., 52, 238 (1952).
5. W. J. A. VandenHeuvel, C. C. Sweeley, and E. C. Horning, J. Am. Chem. Soc., 82, 3481 (1960).
6. E. C. Horning, W. J. A. VandenHeuvel, and B. G. Creech, in Methods of Biochemical Analysis, Vol. 11 (D. Glick, ed.), Wiley (Interscience), New York, 1963.
7. M. Riedmann, Xenobiotica, 3, 411 (1973).
8. L. F. Prescott, K. K. Adjepon-Hamoah, and E. Roberts, J. Pharm. Pharmacol., 25, 205 (1973).
9. T. Walle and H. Ehrsson, Acta Pharm. Suec., 8, 27 (1971).
10. D. E. Case, J. Pharm. Pharmacol., 25, 800 (1973).
11. H. E. Sine, M. J. McKenna, T. A. Rejent, and M. M. Murray, Clin. Chem., 16, 587 (1970).
12. L. B. Foster and C. S. Frings, Clin. Chem., 16, 177 (1970).

13. R. J. Flanagan and G. Withers, J. Clin. Pathol., 25, 899 (1972).
14. J. B. Keenaghan, Anaesthesiology, 20, 110 (1968).
15. J. E. O'Brien, W. Zazulak, V. Abbey, and O. Hinsvark, J. Chromatogr. Sci., 10, 336 (1972).
16. J. Ramsey and D. B. Campbell, J. Chromatogr., 63, 303 (1971).
17. E. A. Fiereck and N. W. Tietz, Clin. Chem., 17, 1024 (1971).
18. D. J. Edwards and K. Blau, Anal. Biochem., 45, 387 (1972).
19. G. P. Beharrell, D. M. Hailey, and M. K. McLaurin, J. Chromatogr., 70, 45 (1972).
20. J. A. F. DeSilva, N. Munno, and R. E. Weinfeld, J. Pharm. Sci., 62, 449 (1973).
21. S. H. Curry, Anal. Chem., 41, 1251 (1968).
22. L. C. Bailey and A. P. Shroff, J. Pharm. Sci., 62, 1274 (1973).
23. M. G. Horning, Biomed. Appl. Gas Chromatogr., 2, 53 (1968).
24. E. C. Horning and M. G. Horning, Clin. Chem., 17, 802 (1971).
25. M. G. Horning, P. Gregory, J. Nowlin, M. Stafford, K. Lertratanangkoon, C. Butler, W. G. Stillwell, and R. M. Hill, Clin. Chem., 20, 282 (1974).
26. N. Weismann, M. L. Lowe, J. M. Bethe, and Q. Demetriore, Clin. Chem., 17, 875 (1971).
27. D. Sohn, J. Simon, M. Hanna, and G. Ghali, J. Chromatogr. Sci., 10, 294 (1972).
28. V. P. Dole, A. Crowther, J. Johnson, M. Monsalvatge, B. Biller, and S. S. Nelson, N.Y. State J. Med., 72, 471 (1972).
29. S. J. Mule, J. Chromatogr., 39, 302 (1969).
30. E. Watson, P. Tramell, and S. M. Kalman, J. Chromatogr., 69, 157 (1972).
31. S. C. Hoffsommer, J. Chromatogr., 51, 243 (1970).
32. J. A. F. DeSilva, F. Munno, and N. Strojny, J. Pharm. Sci., 59, 201 (1970).
33. M. I. Kelsey, A. Keskinger, and E. A. Moscatelli, J. Chromatogr., 75, 294 (1973).
34. E. L. Arnold and R. Ford, Anal. Chem., 45, 85 (1973).
35. T. Walle and H. Ehrsson, Acta Pharm. Suec., 7, 389 (1970).

36. L. F. Prescott and D. R. Redman, J. Pharm. Pharmacol., 24, 713 (1972).

37. D. L. Simmons, R. J. Ranz, and P. Picotte, J. Chromatogr., 71, 421 (1972).

38. A. G. Zacchei and L. Weidner, J. Pharm. Sci., 62, 1972 (1973).

39. D. G. Ferry, D. M. Ferry, P. W. Moller, and E. G. McQueen, J. Chromatogr., 89, 110 (1974).

40. J.-P. Thenot, E. C. Horning, M. Stafford, and M. G. Horning, Anal. Lett., 5, 217 (1972).

41. J.-P. Thenot and E. C. Horning, Anal. Lett., 5, 519 (1972).

42. V. S. Venturella, V. M. Gualario, and R. E. Lang, J. Pharm. Sci., 62, 662 (1973).

43. D. G. Kaiser and G. J. Van Giessen, J. Pharm. Sci., 63, 219 (1974).

44. J. MacGee, Anal. Chem., 42, 421 (1970).

45. G. W. Stevenson, Anal. Chem., 38, 1948 (1966).

46. E. Brochmann-Hanssen and T. O. Oke, J. Pharm. Sci., 58, 370 (1969).

47. J. MacGee, Clin. Chem., 17, 587 (1971).

48. A. Estas and P. A. Dumont, J. Chromatogr., 82, 307 (1973).

49. R. H. Hammer, B. J. Wilder, R. R. Streiff, and A. Mayersdorf, J. Pharm. Sci., 60, 327 (1971).

50. R. J. Perchalski, K. N. Scott, B. J. Wilder, and R. H. Hammer, J. Pharm. Sci., 62, 1735 (1973).

51. M. Kowblansky, B. M. Scheinthal, G. D. Cravello, and L. Chafetz, J. Chromatogr., 76, 467 (1973).

52. R. Osiewicz, V. Aggarual, R. M. Young, and I. Sunshine, J. Chromatogr., 88, 157 (1974).

53. M. Ervik and K. Gustavii, Anal. Chem., 46, 39 (1974).

54. H. Ehrsson, Anal. Chem., 46, 922 (1974).

55. J.-P. Thenot and E. C. Horning, Anal. Lett., 5, 905 (1972).

56. B. Samuelsson, M. Hamberg, and C. C. Sweeley, Anal. Biochem., 38, 301 (1970).

57. M. Hamberg, Biochem. Biophys. Res. Commun., 49, 720 (1972).

58. R. W. Kelly, Anal. Chem., 45, 2079 (1973).

59. V. P. Shah, S. Riegelman, and W. L. Epstein, J. Pharm. Sci., 61, 634 (1972).

60. H. Shafer, W. J. A. VandenHeuvel, R. Ormond, F. A. Kuehl, Jr., and F. J. Wolf, J. Chromatogr., 52, 111 (1970).

61. K. Tsuji and J. H. Robertson, Anal. Chem., 42, 1661 (1970).

62. K. Tsuji and J. H. Robertson, Anal. Chem., 41, 1332 (1969).

63. M. Margosis and K. Tsuji, J. Pharm. Sci., 62, 1836 (1973).

64. B. Van Giessen and K. Tsuji, J. Pharm. Sci., 60, 1068 (1971).

65. K. Tsuji and J. H. Robertson, Anal. Chem., 43, 818 (1971).

66. C. Hishta, D. L. Mays, and M. Garofalo, Anal. Chem., 43, 1530 (1971).

67. T. F. Brodasky and A. D. Argoudelis, J. Antibiot., 26, 131 (1973).

68. M. Margosis, J. Chromatogr., 47, 341 (1970).

69. M. Margosis, J. Pharm. Sci., 63, 435 (1974); J. Chromatogr. Sci., 12, 549 (1974).

70. L. W. Brown and P. B. Bowman, J. Chromatogr. Sci., 12, 373 (1974).

71. L. W. Brown, J. Agric. Food Chem., 21, 83 (1973).

72. K. Tsuji and J. H. Robertson, Anal. Chem., 45, 2136 (1973).

73. K. Tsuji, J. H. Robertson, and W. F. Beyer, Anal. Chem., 46, 539 (1974).

74. D. G. Kaiser, R. G. Carlson, and K. T. Kirton, J. Pharm. Sci., 63, 420 (1974).

75. H. P. Burchfield, E. E. Storrs, R. J. Wheeler, V. K. Bhat, and L. L. Green, Anal. Chem., 45, 916 (1973).

76. R. J. Daun, J. Assoc. Off. Anal. Chem., 54, 1277 (1971).

77. A. J. F. Wickramasinghe and S. R. Shaw, Biochem. J., 141, 179 (1974).

78. B. J. Gudzinowicz, Gas Chromatographic Analysis of Drugs and Pesticides, Marcel Dekker, New York, 1967.

79. A. B. Littlewood, Gas Chromatography, 2nd ed., Academic Press, New York, 1970.

80. B. B. Brodie, J. Axelrod, R. Soberman, and B. B. Levy, J. Biol. Chem., 179, 25 (1949).

81. H. M. Berman and M. A. Spirtes, Biochem. Pharmacol., 20, 2275 (1971).

82. P. Bertagni, F. Marcucci, E. Mussini, and S. Garattini, J. Pharm. Sci., 61, 965 (1972).

83. A. De Leenheer and A. Heyndrickx, J. Pharm. Sci., 62, 31 (1973).

84. A. J. Williams, T. W. G. Jones, and J. D. H. Cooper, Clin. Chim. Acta, 43, 327 (1973).

85. J. E. Lovelock, Nature, 189, 729 (1961).

86. E. D. Pellizzari, J. Chromatogr., 98, 323 (1974).

87. J. W. Blake, R. Huffman, J. Noonan, and R. Ray, Am. Lab., 5, 63 (1973).

88. R. B. Bruce, J. E. Pitts, and F. M. Pinchbeck, Anal. Chem., 40, 1246 (1968).

89. A. C. Moffat and E. C. Horning, Biochim. Biophys. Acta, 222, 248 (1970).

90. M. G. Horning, A. M. Moss, E. A. Boucher, and E. C. Horning, Anal. Lett., 1, 311 (1968).

91. I. A. Zingales, J. Chromatogr., 61, 237 (1971).

92. D. F. S. Natusch and T. M. Thorpe, Anal. Chem., 45, 1184A (1973).

93. P. J. Meffin, G. Moore, and J. Thomas, Anal. Chem., 45, 1964 (1973).

94. J. H. Goudie and D. Burnett, Clin. Chem. Acta, 43, 423 (1973).

95. D. D. Breimer and J. M. VanRossum, J. Chromatogr., 88, 235 (1974).

96. B. Caddy, F. Fish, and D. Scott, Chromatographia, 6, 335 (1973).

97. S. P. James and R. H. Waring, J. Chromatogr., 78, 417 (1973).

98. A. H. Beckett, J. F. Taylor, and P. Kouronakis, J. Pharm. Pharmacol., 22, 123 (1970).

99. H. Brötell, H. Ehrsson, and O. Gyllenhaal, J. Chromatogr., 78, 293 (1973).

100. L. T. Sennello and F. E. Kohn, Anal. Chem., 46, 752 (1974).

101. M. Riedmann, J. Chromatogr., 88, 376 (1974).

102. M. Riedmann, J. Chromatogr., 92, 55 (1974).

103. N. K. McCallum, J. Chromatogr. Sci., 11, 509 (1973).

104. D. C. Fenimore, R. R. Freeman, and P. R. Loy, Anal. Chem., 45, 2331 (1973).

105. C. Jackson, Jr., and P. J. Reynolds, J. Agric. Food Chem., 20, 972 (1972).

106. R. E. Schirmer and R. J. Pierson, J. Pharm. Sci., 62, 2052 (1973).

107. F. W. McLafferty, Interpretation of Mass Spectra, Benjamin, New York, 1966.

108. S. R. Schrader, Introductory Mass Spectrometry, Allyn & Bacon, Boston, 1971.

109. D. E. Wolf, W. J. A. VandenHeuvel, F. R. Koniuszy, T. R. Tyler, T. A. Jacob, and F. J. Wolf, J. Agric. Food Chem., 20, 1252 (1972).

110. G. R. Nakamura, T. T. Noguchi, D. Jackson, and D. Banks, Anal. Chem., 44, 408 (1972).

111. L. J. Fischer and J. J. Ambre, J. Chromatogr., 87, 379 (1973).

112. P. C. Cala, N. R. Trenner, R. P. Buhs, G. V. Downing, Jr., J. L. Smith, and W. J. A. VandenHeuvel, J. Agric. Food Chem., 20, 337 (1972).

113. L. Baczynskyj, D. J. Duchamp, J. F. Ziererl, Jr., and U. Axen, Anal. Chem., 45, 479 (1973).

114. C. J. Mirocha, C. M. Christensen, G. Davis, and G. H. Nelson, J. Agric. Food Chem., 21, 135 (1973).

115. C. C. Sweeley, W. H. Elliott, I. Fries, and R. Ryhage, Anal. Chem., 38, 1549 (1966).

116. C. J. W. Brooks and B. S. Middleditch, Clin. Chim. Acta, 34, 145 (1971).

117. C.-G. Hammar, B. Holmstedt, and R. Ryhage, Anal. Biochem., 25, 532 (1968).

118. B. Holmstedt and L. Palmer, in Advances in Biochemical Psychopharmacology, Vol. 7, Raven Press, New York, 1973.

119. D. J. Jenden and A. K. Cho, Annu. Rev. Pharmacol., 13, 371 (1973).

120. J. T. Watson, Annu. Rev. Pharmacol., 13, 391 (1973).

121. R. Fanelli and A. Frigerio, J. Chromatogr., 93, 441 (1974).

122. L. Palmer and B. Kolmodin-Hedman, J. Chromatogr., 74, 21 (1972).

123. R. N. Morris, G. A. Gunderson, S. W. Babcock, and J. F. Zaroslinski, Clin. Pharmacol. Therm., 13, 719 (1972).

124. M. Mitchard and M. E. Williams, J. Chromatogr., 72, 29 (1972).

125. G. Alvan, J.-E. Lindgren, C. Bogentoft, and O. Ericsson, Eur. J. Clin. Pharmacol., 6, 187 (1973).

126. A. K. Cho, B. J. Hodshon, B. Lindeke, and G. T. Miwa, J. Pharm. Sci., 62, 1491 (1973).

127. R. W. Walker, H. S. Ahn, G. Albers-Schonberg, L. R. Mandel, and W. J. A. VandenHeuvel, Biochem. Med., 8, 105 (1973).

128. A. Rane, M. Garle, O. Borga, and F. Sjoqvist, Clin. Pharmacol. Ther., 15, 39 (1974).

129. S. Agurell, B. Gustafsson, B. Holmstedt, K. Leander, J.-E. Lindgren, I. Nilsson, F. Sandberg, and M. Asberg, J. Pharm. Pharmacol., 25, 554 (1973).

130. J. F. Johnson, Guide to Modern Methods of Instrumental Analysis, Wiley (Interscience), New York, 1972.

131. H. M. McNair and E. J. Bonelli, Basic Gas Chromatography, 5th ed., Varian Aerograph, Walnut Creek, California, 1969.

132. E. Kovats, Helv. Chim. Acta, 41, 1915 (1958).

133. G. Schomburg, Adv. Chromatogr., 6, 211 (1968).

134. W. J. A. VandenHeuvel, W. L. Gardiner, and E. C. Horning, Anal. Chem., 36, 1550 (1964).

135. A. G. Zacchei and L. Weidner, J. Pharm. Sci., 64, 814 (1975).

136. D. L. Ball, W. E. Harris, and H. W. Habgood, Sep. Sci., 2, 81 (1967).

137. D. L. Ball, W. E. Harris, and H. W. Habgood, J. Gas Chromatogr., 5, 613 (1967).

138. D. L. Ball, W. E. Harris, and H. W. Habgood, Anal. Chem., 40, 129 (1968).

139. F. W. Karasek, Anal. Chem., 44, 32A (1972).

140. R. E. Dessy and J. A. Titus, Anal. Chem., 45, 124A (1973).

141. D. R. Deans, J. Chromatogr. Sci., 9, 729 (1971).

142. D. A. Craven, E. S. Everett, and M. Rubel, J. Chromatogr. Sci., 9, 541 (1971).

143. J. R. Watson, P. Crescuolo, and F. Matsui, J. Pharm. Sci., 60, 454 (1971).

144. J. R. Watson, F. Matsui, P. M. J. McConnell, and R. C. Lawrence, J. Pharm. Sci., 61, 929 (1972).

145. S. Patel, J. H. Perrin, and J. J. Windheuser, J. Pharm. Sci., 61, 1794 (1972).

146. E. C. Horning and M. G. Horning, J. Chromatogr. Sci., 9, 129 (1971).

147. A. Zlatkis, W. Bertsch, H. A. Lichtenstein, A. Tishbee, F. Shunbo, H. M. Liebich, A. M. Coscia, and N. Fleischer, Anal. Chem., 45, 763 (1973).

148. D. K. Fetter, M. F. Jacobs, and H. W. Rawlings, J. Pharm. Sci., 60, 913 (1971).

149. K. M. McErlane, J. Chromatogr. Sci., 12, 97 (1974).

150. I. Schroeder, J. C. Medina-Acevedo, and G. Lopez-Sanchez, J. Chromatogr. Sci., 10, 183 (1973).

151. R. Roman, C. H. Yates, J. F. Millar, and W. J. A. VandenHeuvel, Can. J. Pharm. Sci., 10, 12 (1975).

152. P. P. Karkhanis, D. O. Edlund, and J. R. Anfinsen, J. Pharm. Sci., 62, 804 (1973).

153. M. P. Gruber, R. W. Klein, M. E. Foxx, and J. Campisi, J. Pharm. Sci., 61, 1147 (1972).

154. C. K. Wong, J. R. Urbigkit, N. Conca, D. M. Cohen, and K. P. Munnelly, J. Pharm. Sci., 62, 1340 (1973).

155. L. T. Sennello, J. Pharm. Sci., 60, 595 (1971).

156. J. R. Watson, F. Matsui, R. C. Lawrence, and P. M. J. McConnell, J. Chromatogr., 76, 141 (1973).

157. B. L. Chang, B. F. Grabowski, and W. G. Haney, Jr., J. Pharm. Sci., 62, 1337 (1973).

158. P. F. Helgren, M. A. Thomas, and J. G. Theivagt, J. Pharm. Sci., 61, 103 (1972).

159. G. Manius, F. P. Mahn, V. S. Venturella, and B. Z. Senkowski, J. Pharm. Sci., 61, 1831 (1972).

160. H. W. Schultz and C. Paveenbampen, J. Pharm. Sci., 62, 1995 (1973).

161. H. A. Lloyd, H. M. Fales, P. F. Highet, W. J. A. VandenHeuvel, and W. C. Wildman, J. Am. Chem. Soc., 81, 3791 (1960).

162. D. L. Sondack and W. L. Koch, J. Pharm. Sci., 62, 101 (1973).

163. D. L. Sondack, J. Pharm. Sci., 63, 584 (1974).

164. L. T. Grady and R. O. Zimmerman, Jr., J. Pharm. Sci., 59, 1324 (1970).

165. J. J. Windheuser, J. L. Sutter, and A. Sarrif, J. Pharm. Sci., 61, 1311 (1972).

166. R. S. Santoro, P. P. Progner, E. A. Ambush, and D. E. Guttman, J. Pharm. Sci., 62, 1346 (1973).

167. B. F. Grabowski, B. J. Softly, B. L. Chang, and W. G. Haney, Jr., J. Pharm. Sci., 62, 806 (1973).

168. N. D. Greenwood and I. W. Guppy, Analyst, 99, 313 (1974).

169. J. M. Moore and F. E. Bena, Anal. Chem., 44, 385 (1972).

170. A. R. Sperling, J. Chromatogr. Sci., 12, 265 (1974).

171. R. E. Stoner and C. Parker, Clin. Chem., 20, 309 (1974).

172. A. B. Segelman and R. D. Sofia, J. Pharm. Sci., 62, 2044 (1973).

173. P. S. Fetterman, E. S. Keith, C. W. Waller, O. Guerrero, N. J. Doorenbos, and M. W. Quimby, J. Pharm. Sci., 60, 1246 (1971).

174. D. W. Cornish, D. M. Grossman, A. L. Jacobs, A. F. Michaelis, and B. Salsitz, Anal. Chem., 45, 221R (1973).

175. N. Gochman and D. S. Young, Anal. Chem., 45, 11R (1973).

176. E. S. Vesell and G. T. Passananti, Clin. Chem., 17, 851 (1971).

177. R. W. Reid, R. Katzen, and J. M. Clinger, Am. J. Clin. Pathol., 53, 462 (1970).

178. N. C. Jain and P. L. Kirk, Microbiol. J., 12, 229 (1967).

179. N. C. Jain and P. L. Kirk, Microbiol. J., 12, 242 (1967).

180. C. Cardini, V. Quercia, and A. Calo, J. Chromatogr., 37, 190 (1968).

181. J. E. Wallace and E. V. Dane, J. Forensic Sci., 12, 484 (1967).

182. H. B. Hucker and J. K. Miller, J. Chromatogr., 32, 408 (1968).

183. T. P. Michaels, W. W. Holl, and V. J. Greely, Ann. N.Y. Acad. Sci., 153, 493 (1968).

184. P. O. Lagerstrom, K. O. Borg, and D. Westerlund, Acta Pharm. Suec., 9, 47 (1972).

185. R. R. Braithwaite and B. Widdop, Clin. Chim. Acta, 35, 461 (1971).

186. K. Sabih and K. Sabih, Anal. Chem., 41, 1452 (1969).

187. R. J. Perchalski and B. J. Wilder, J. Pharm. Sci., 63, 806 (1974).

188. M. J. Kupperberg, Clin. Chim. Acta, 29, 283 (1970).

189. A. Berlin, S. Agurell, O. Borga, L. Lund, and F. Sjoquist, Scand. J. Clin. Lab. Invest., 29, 281 (1972).

190. D. Chin, E. Fastlich, and B. Davidow, J. Chromatogr., 71, 545 (1972).

191. W. J. A. VandenHeuvel, V. F. Gruber, R. W. Walker, and F. J. Wolf, J. Pharm. Sci., 64, 1309 (1975).

192. P. Friel and J. R. Green, Clin. Chim. Acta, 43, 69 (1973).

193. J. W. Meijer, Epilepsia, 12, 341 (1971).

194. H. V. Street, in Advances in Clinical Chemistry, Vol. 12 (O. Bodansky and C. P. Stewart, eds.), Academic Press, New York, 1969.

195. J. M. VanRossum, D. D. Breimer, G. A. M. van Ginneken, J. M. G. van Kordelaar, and T. B. Vree, Clin. Chim. Acta, 34, 311 (1971).

196. D. G. Kaiser and G. J. Van Giessen, J. Pharm. Sci., 63, 219 (1974).

197. D. G. Kaiser and E. M. Glenn, J. Pharm. Sci., 63, 785 (1974).

198. T. Yashiki, Y. Uda, T. Kondo, and H. Mima, Chem. Pharm. Bull., 19, 487 (1971).

199. T. Yashiki, T. Kondo, Y. Uda, and H. Mima, Chem. Pharm. Bull., 19, 478 (1971).

200. F. Marcucci, E. Mussini, L. Airoldi, R. Fanelli, A. Frigerio, F. DeNadai, A. Bizzi, M. Rizzo, P. Morselli, and S. Garattini, Clin. Chim. Acta, 34, 321 (1971).

201. C. P. Talley, N. R. Trenner, G. V. Downing, Jr., and W. J. A. VandenHeuvel, Anal. Chem., 43, 1379 (1971).

202. W. G. Smith and E. E. McNeil, Anal. Chem., 44, 1084 (1972).

203. D. E. Coffin and J.-C. Pilon, J. Assoc. Off. Anal. Chem., 56, 352 (1973).

204. A. L. Donoho, W. S. Johnson, R. F. Sheck, and W. L. Sullivan, J. Assoc. Off. Anal. Chem., 56, 785 (1973).

205. H. B. Hucker, A. Hochberg, and E. A. Hoffman, J. Pharm. Sci., 60, 1053 (1971).

206. H. B. Hucker and E. A. Hoffman, J. Pharm. Sci., 60, 1049 (1971).

207. E. C. Dinovo, L. A. Gothschalk, E. P. Noble, and R. Biener, Res. Commun. Chem. Pathol. Pharmacol., 7, 489 (1974).

208. R. C. Thomas and R. W. Judy, J. Med. Chem., 15, 964 (1972).

209. A. De Leenheer and A. Heyndrickx, J. Pharm. Sci., 61, 914 (1972).

210. S. S. Walkenstein, A. P. Intoccia, T. L. Flanagan, B. Hwang, D. Flint, J. Weinstock, A. J. Villani, D. Blackburn, and H. Green, J. Pharm. Sci., 62, 580 (1973).

211. H. B. Hucker, B. M. Michniewicz, and R. E. Rhodes, Biochem. Pharmacol., 20, 2123 (1971).

212. A. H. Beckett, R. T. Coutts, and F. A. Ugunbona, J. Pharm. Pharmacol., 25, 190 (1973).

213. N. R. Trenner, O. C. Speth, V. B. Gruber, and W. J. A. Vanden-Heuvel, J. Chromatogr., 71, 415 (1972).

214. E. C. Horning and W. J. A. VandenHeuvel, in Advances in Chromatography, Vol. 1 (J. C. Giddings and R. A. Keller, eds.), Marcel Dekker, New York, 1965.

215. D. Dell, D. R. Boreham, and B. K. Martin, J. Pharm. Sci., 60, 1368 (1971).

216. P. R. Vagelos, W. J. A. VandenHeuvel, and M. G. Horning, Anal. Biochem., 2, 50 (1961).

217. P. W. Feit, K. Roholt, and H. Sorensen, J. Pharm. Sci., 62, 375 (1973).

218. J. A. F. Wickramasinghe and S. R. Shaw, J. Pharm. Sci., 60, 1669 (1971).

219. H. V. Street, J. Chromatogr., 73, 73 (1972).

220. B. W. Barry and G. M. Saunders, J. Pharm. Sci., 60, 645 (1971).

221. H. R. Sullivan, S. L. Due, and R. E. McMahon, J. Pharm. Pharmacol., 25, 1009 (1973).

222. J. C. Craig, N. Y. Mary, and S. K. Roy, Anal. Chem., 36, 1142 (1964).

223. S. Y. Yeh, J. Pharm. Sci., 62, 1827 (1973).

224. W. J. A. VandenHeuvel, J. L. Smith, G. Albers-Schonberg, B. Plazonnet, and P. Belanger, in Modern Methods of Steroid Analysis (E. Heftmann, ed.), Academic Press, New York, 1973.

225. G. Bleau and W. J. A. VandenHeuvel, Steroids, 24, 549 (1974).

226. W. Sadee and E. Van der Kleijn, J. Pharm. Sci., 60, 135 (1971).

227. A. Frigerio, K. M. Baker, and G. Belvedere, Anal. Chem., 45, 1846 (1973).

228. J. W. Westley, R. H. Evans, Jr., and A. Stempel, Anal. Biochem., 59, 574 (1974).

229. K. M. Baker, A. Frigerio, P. L. Morselli, and G. Pifferi, J. Pharm. Sci., 62, 475 (1973).

230. W. J. A. VandenHeuvel, R. P. Buhs, J. R. Carlin, T. A. Jacob, F. R. Koniuszy, J. L. Smith, N. R. Trenner, R. W. Walker, D. E. Wolf, and F. J. Wolf, Anal. Chem., 44, 14 (1972).

231. L. Martis and R. H. Levy, J. Pharm. Sci., 61, 1343 (1972).

232. L. Martis and R. H. Levy, J. Pharm. Sci., 63, 834 (1974).

233. M. P. Rabinowitz, P. Reisberg, and J. I. Bodin, J. Pharm. Sci., 61, 1974 (1972).

234. W. J. A. VandenHeuvel, V. B. Gruber, and R. W. Walker, J. Chromatogr., 87, 341 (1973).

235. C. Rosenblum, N. R. Trenner, R. P. Buhs, C. B. Hiremath, F. R. Koniuszy, and D. E. Wolf, J. Agric. Food Chem., 20, 360 (1972).

236. E. M. Chambaz, G. Maume, B. Maume, and E. C. Horning, Anal. Lett., 1, 749 (1968).

237. R. M. Thompson and E. C. Horning, Steroids Lipids Res., 4, 135 (1973).

238. G. A. F. M. Rutten and J. A. Luyten, J. Chromatogr., 74, 177 (1972).

239. M. Novotny, R. Segura, and A. Zlatkis, Anal. Chem., 44, 9 (1972).

240. A. L. German and E. C. Horning, J. Chromatogr. Sci., 11, 76 (1973).

241. A. L. German, C. D. Pfaffenberger, J.-P. Thenot, M. G. Horning, and E. C. Horning, Anal. Chem., 45, 930 (1973).

242. B. F. Maume and J. A. Luyten, J. Chromatogr. Sci., 11, 607 (1973).

243. D. I. Carroll, I. Dzidic, R. N. Stillwell, M. G. Horning, and E. C. Horning, Anal. Chem., 46, 706 (1974).

244. M. G. Horning, J. Nowlin, K. Lertratanangkoon, R. N. Stillwell, W. G. Stillwell, and R. M. Hill, Clin. Chem., 19, 845 (1973).

245. C. E. Costello, H. S. Hertz, T. Sakai, and K. Biemann, Clin. Chem., 20, 255 (1974).

246. H. Nau and K. Biemann, Anal. Chem., 46, 426 (1974).

247. C. C. Sweeley, N. D. Young, J. F. Holland, and S. C. Gates, J. Chromatogr., 99, 507 (1974).

Chapter 6

THE INVESTIGATION OF COMPLEX ASSOCIATION BY GAS CHROMATOGRAPHY AND RELATED CHROMATOGRAPHIC AND ELECTROPHORETIC METHODS

C. L. de Ligny

Laboratory for Analytical Chemistry
University of Utrecht
The Netherlands

I. INTRODUCTION

Gas chromatography has now come to the age of maturity. This results in a gradual shift of the mainstream of research from the field of methodological improvements to the field of applications. A fascinating application of gas chromatography is the investigation of solution thermodynamics from data on retention volumes.

The borders of this field of research are as follows. On one side there are the apolar systems, the behavior of which is governed by the number of possible configurations, the degree of expansion of the solution components, and dispersion forces. Since the chemical nature of the solution components is alike, there must be a large size difference between the molecules of the volatile solute and the involatile stationary phase, the latter being typically a liquid polymer. This type of solutions can be described by the related theories of Flory et al. [1, 2] and Patterson et al. [3]. The unique features of gas chromatography as a tool for the investigation of polymer solutions are that data can be obtained at a polymer fraction ≈ 1 and over a large temperature range, extending from about 110° C below to 60° C above the boiling point of the solute. This offers the possibility of a very sensitive test of the aforementioned polymer solution theories [4-11].

On the other side there are the systems whose behavior is dominated by the reversible formation of weak complexes, often as a result of charge transfer or hydrogen bonding. They are the subject of this chapter. The remarkable success of gas chromatography in the determination of complex association constants is because gas chromatography can easily discriminate between the free solute and the complex, since these species show an appreciable difference in volatility. This discrimination may not always be feasible by liquid chromatography, and this may be one of the reasons why complex association has never been investigated using retention volumes obtained in liquid chromatography.* Another reason is possibly that liquid chromatography is still in its juvenile stage and that the research in this field is thus mainly directed to improvements in method.

Not only can complex association constants be determined by means of gas chromatography in an elegant way but there are also theoretical frameworks at hand for the interpretation of these data. For example, the influence of substituents in the solute molecule on the value of the association constant can be interpreted in the light of the work of Hammett [12] and Taft [13].

Between these borders there are the systems, the behavior of which is governed by inductive and dipole-dipole interactions, whereas the solution components may also show appreciable size differences. Though gas chromatography is perfectly suited for the investigation of the thermodynamics of these systems, until recently this was a rather unfruitful area of research since an adequate theoretical description of systems of this type could not be given. However, it seems that recent developments of perturbation theory are very promising [14]. An early and simple version by Pierotti was able to give a satisfactory description of the solubility of gases and vapors in apolar and polar [15-17] solvents and even in the solvent

---

*Affinity chromatography is an exception.

water [18, 19], but could cope less well with the enthalpy and entropy of solution separately [16]. Recent improvements yield excellent agreement between calculated and observed values of both the free enthalpy and the enthalpy of mixing in systems such as hexane-acetone [20]. However, the behavior of systems of which both components are polar still seems to be beyond the reach of any theory.

II. THEORY

A. Gas Chromatography

The involved equilibria are often formulated in terms of concentrations rather than in terms of activities, e.g., by Purnell in his classical survey of the potentialities of the gas-chromatographic method of determining association constants [21]. Other workers use concentrations for some of the reactants and activities for other reactants [22, 23]. Only a few papers have appeared in which activities are used throughout [24, 25]. Here, we follow the line of thought of de Ligny [25].

We can consider the formation of a complex AB from a volatile compound A and an involatile compound B. Generally, a mixture of the latter and an inert compound S is used as the stationary phase. The association equilibrium is governed by the equation

$$K_x = \frac{x_{AB}}{x_A x_B} \frac{f_{AB}}{f_A f_B} \tag{1}$$

where x is the mole fraction and f is the activity coefficient. The reference state to which the activity coefficients are related is specified later.

When a stationary phase consisting of S and one consisting of S + B are in equilibrium with a vapor phase containing A, it holds that

$$x_{A(S)} f_{A(S)} = x_{A(S+B)} f_{A(S+B)} \tag{2}$$

Further, it can be shown that the midpeak retention volume V at the mean column pressure and column temperature, corrected for the gas holdup in the apparatus, is given by

$$V = \frac{V_{St} c_{St}}{c_{Mob}} = \frac{N_{St} x_{St}}{c_{Mob}} \tag{3}$$

where

V_{St} = volume of the stationary phase in the column (liter),

c_{St} = solute concentration in the stationary phase (moles/liter),

c_{Mob} = solute concentration in the mobile phase, in equilibrium with c_{St} (moles/liter),

N_{St} = moles of stationary phase in the column,

x_{St} = solute mole fraction in the stationary phase.

The application of Eq. (3) to a solute A in a stationary phase consisting of S and one consisting of S + B gives, respectively,

$$V_{A(S)} = \frac{N_S x_{A(S)}}{c_{A(M)}} \tag{4}$$

and

$$V_{A(S+B)} = \frac{N_{S+B}(x_A + x_{AB})_{S+B}}{c_{A(M)}} \tag{5}$$

In Eqs. (4) and (5), x_A, x_{AB}, and c_A do not have specified values; they range from zero to a maximum midpeak value. However, the ratios $x_{A(S)}/c_{A(M)}$ and $(x_A + x_{AB})_{S+B}/c_{A(M)}$ are independent of c_A when the latter is small. Therefore, the value of c_A may be chosen arbitrarily and, in particular, the values of c_A in Eqs. (4) and (5) may be chosen equal.

Combination of Eqs. (1), (2), (4), and (5) gives

$$\frac{V_{A(S+B)}}{V_{A(S)}} = \frac{N_{S+B}}{N_S}\,\frac{(x_A + x_{AB})_{S+B}}{x_{A(S)}}$$

$$= \frac{N_{S+B}}{N_S}\,\frac{f_{A(S)}}{f_{A(S+B)}}\,\frac{(x_A + x_{AB})_{S+B}}{x_{A(S+B)}}$$

$$= \frac{N_{S+B}}{N_S}\,\frac{f_{A(S)}}{f_{A(S+B)}}\left(1 + K_x x_B \frac{f_A f_B}{f_{AB}}\right)_{S+B} \tag{6}$$

It is obvious that the analogs of Eq. (6) on the molal and molar concentration scales are, respectively,

$$\frac{V_{A(S+B)}}{V_{A(S)}} = \frac{G_{S+B}}{G_S}\,\frac{\gamma_{A(S)}}{\gamma_{A(S+B)}}\left(1 + K_m m_B \frac{\gamma_A \gamma_B}{\gamma_{AB}}\right)_{S+B} \tag{6'}$$

and

$$\frac{V_{A(S+B)}}{V_{A(S)}} = \frac{V_{St(S+B)}}{V_{St(S)}}\,\frac{y_{A(S)}}{y_{A(S+B)}}\left(1 + K_c c_B \frac{y_A y_B}{y_{AB}}\right)_{S+B} \tag{6''}$$

where

G	= weight of the stationary phase in the column (kg),
V_{St}	= volume of the stationary phase in the column (liter),
γ, y	= activity coefficients of the molal and molar concentration scales, respectively,
K_m, K_c	= association constants on the molal and molar concentration scales (kg/mole and liters/mole), respectively,
m	= molality (moles/kg).

If, instead of the single complex AB, a series of complexes A_pB_q is formed, the analog of Eq. (6) is

$$\frac{V_{A(S+B)}}{V_{A(S)}} = \frac{N_{S+B}}{N_S}\,\frac{f_{A(S)}}{f_{A(S+B)}}\left(1 + \sum_p\sum_q pK_x x_A^{p-1} x_B^q \frac{f_A^p f_B^q}{f_{A_pB_q}}\right)_{S+B} \tag{7}$$

Since in elution chromatography x_A is not constant during the process and is very small, it follows from Eq. (7) that complex association constants can only be determined from chromatographic retention data for $p = 1$. For that case Eq. (7) simplifies to

$$\frac{V_{A(S+B)}}{V_{A(S)}} = \frac{N_{S+B}}{N_S}\,\frac{f_{A(S)}}{f_{A(S+B)}}\left(1 + \sum_q K_x x_B^q \frac{f_A f_B^q}{f_{AB_q}}\right)_{S+B} \tag{8}$$

In the following, the determination of complex association constants from gas-chromatographic data is elucidated using Eq. (6). The application of Eqs. (6'), (6''), or (7) instead of (6) is straightforward.

In principle, the method is based on the determination of the left-hand side of Eq. (6) for a range of mixtures S + B, and extrapolation to some suitably chosen reference state in which the activity coefficients are by definition unity. When K_x is small, the change of $V_{A(S+B)}N_S/V_{A(S)}N_{S+B}$ with varying composition of the mixtures S + B may be caused mainly by the simultaneous change of $f_{A(S+B)}/f_{A(S)}$. Even when K_x is large, small variations of K_x, e.g., caused by the introduction of substituents in A or B, may be overshadowed by the effect of substituents on the change of $f_{A(S+B)}/f_{A(S)}$ with solvent composition. Therefore, this ratio should be estimated or eliminated before the determination of K_x is attempted.

The estimation of $f_{A(S+B)}/f_{A(S)}$ is a difficult problem. Generally, both A and B will have a dipole moment and, as stated in Sect. I, the calculation of activity coefficients in such systems is still beyond the reach of theory. Only in the case where A or B has zero dipole moment can the activity coefficient ratio be calculated in principle, preferably using perturbation theory.

Eon and co-workers [24] have suggested that the activity coefficient ratio be calculated by the Flory-Huggins equation. This equation takes into account only the effect of the differences in size of A, B, and S on the (configurational part of the) activity coefficients, not the effect of the differences in chemical nature. Probably, the latter effect is often much larger than the former, so that the value of this procedure is questionable.

Experimental values of activity coefficients at infinite dilution in some mixtures of components of widely differing polarities are compared with values calculated by perturbation theory, regular solution theory, and the Flory-Huggins equation in Table 1. It follows that perturbation theory gives the best results, and the Flory-Huggins equation the poorest ones.

In view of the difficulties associated with the <u>calculation</u> of the activity coefficient ratio, it was proposed to <u>eliminate</u> it by the use of a reference solute A* that is closely related to A but does not yield a complex with B [22, 23, 25]. For this solute the analog of Eq. (6) is

$$\frac{V_{A*(S+B)}}{V_{A*(S)}} = \frac{N_{S+B}}{N_S} \frac{f_{A*(S)}}{f_{A*(S+B)}} \tag{9}$$

The combination of Eqs. (6) and (9) yields

$$\frac{V_{A(S+B)}}{V_{A(S)}} \frac{V_{A*(S)}}{V_{A*(S+B)}} = \frac{f_{A(S)}}{f_{A(S+B)}} \frac{f_{A*(S+B)}}{f_{A*(S)}} \left(1 + K_x x_B \frac{f_A f_B}{f_{AB}}\right)_{S+B} \tag{10}$$

or

$$\left(\frac{V_A}{V_{A*}}\right)_{S+B}\left(\frac{V_{A*}}{V_A}\right)_S \approx 1 + K_x x_B \left(\frac{f_A f_B}{f_{AB}}\right)_{S+B} \qquad (10')$$

In a study of the association of substituted alkenes (A) with $AgNO_3$ (B) in ethylene glycol (S) [27] the corresponding substituted alkanes were used as the reference solutes A*. The reference solvent in this case was not S, but a solution of $LiNO_3$ (C) in ethylene glycol of the same molality as the $AgNO_3$ solution. Table 2 shows that the ratio $f_{A*(S+C)}/f_{A*(S+B)}$ [which is used as an estimate for the ratio $f_{A(S+C)}/f_{A(S+B)}$; see Eqs. (10) and (10')] deviates markedly from unity for polar solutes at high salt molalities. Thus, even in this rather ideal case it appears to be incorrect to set the factor $f_{A(S)}/f_{A(S+B)}$ in Eq. (6) simply equal to unity, and the use of Eq. (10') should be preferred.

In a study [28] of the association of substituted alcohols A with monofunctional hexadecyl derivatives the corresponding substituted alkyl chlorides were used as the reference solutes A*. The reference solvent was hexadecane.

An empirical estimate of $f_{A(S)}f_{A*(S+B)}/f_{A(S+B)}f_{A*(S)}$ was made as follows. Littlewood [29] has shown that for alkanes (alk) and polar solutes (pol), dissolved in hexadecane S or a monofunctional hexadecyl derivative B, the following equation holds:

$$\frac{f_{alk(B)}}{f_{alk(S)}}\,\frac{f_{pol(S)}}{f_{pol(B)}} \approx \exp(r\mu_B\mu_{pol}^s) \qquad (11)$$

where r and s are constants and μ is the dipole moment.

From a statistical analysis of Littlewood's data it follows that the most probable value of the term within parentheses is (at 60° C) $0.029\mu_B\mu_{pol}^{1.47}$. When Eq. (11) is applied successively to the polar solutes A and A*, it follows that

$$\frac{f_{A(S)}}{f_{A(B)}}\,\frac{f_{A*(B)}}{f_{A*(S)}} = \exp[0.029\mu_B(\mu_A^{1.47} - \mu_{A*}^{1.47})] \qquad (12)$$

It was assumed that Eq. (11) could also be applied to mixtures S + B by multiplying the exponent by x_B, and to bifunctional polar solutes, with $\mu_{pol}^{1.47} = \mu_1^{1.47} + \mu_2^{1.47}$, where μ_1 and μ_2 are the individual group dipole moments. In the present case, where A is a substituted alcohol and A* is the corresponding substituted alkyl chloride, Eq. (12) reads

TABLE 1

Activity Coefficients at Infinite Dilution at 25° C in Some Mixtures of Components of Widely Differing Polarities

		f_2			
Solvent (1)	Solute (2)	Experiment[a]	Perturbation theory[a]	Regular solution theory[a]	Flory-Huggins[b]
Methanol	Pentane	27.0	28.8	151	0.5
Methanol	Hexane	27.0	24.6	242	0.4
Aniline	Heptane	23.1	26.8	29.3	0.9
Aniline	Benzene	2.2	1.4	3.8	1.0
Nitrobenzene	Hexane	7.0	6.7	10.2	1.0
Nitrobenzene	Benzene	1.2	1.2	2.2	1.0
Ethylene glycol	Cyclohexane	316	432	451	0.8
Ethylene glycol	Hexane	625	618	2346	0.8
Dimethyl sulfoxide	Pentane	61.3	55.2	49.1	
Dimethyl sulfoxide	Cyclohexane	40.6	54.5	28.2	
Acetophenone	Benzene	1.2	1.2	1.5	0.9
Acetophenone	Pentane	5.4	5.4	4.8	1.0
Acetophenone	Isoprene	2.1	3.8	2.7	1.0
Methyl cellosolve	Pentane	14.5	12.7		1.0

Methyl cellosolve	Hexane	16.0	14.1		0.9
Methyl cellosolve	1-Pentene	8.6	13.6		1.0
Acetone	Pentane	5.3	5.7	4.3	0.9
Acetone	Hexane	6.5	6.9	4.7	0.8
Acetone	1-Pentene	3.2	5.8	4.0	0.9

[a]Reprinted from Tiepel and Gubbins [20], by courtesy of the authors.
[b]Calculated from van der Waals volumes [26].

$$\frac{f_{A(S)}}{f_{A(S+B)}} \frac{f_{A^*(S+B)}}{f_{A^*(S)}} = \exp[0.029\mu_{S+B}(\mu^{1.47}_{-OH} - \mu^{1.47}_{-Cl})] \tag{13}$$

Some values of the right-hand side of Eq. (13) are given in Table 3. In this case it appears to be even incorrect to use Eq. (10'), and Eq. (10) should be preferred in combination with an estimate of the factor $f_{A(S)}f_{A^*(S+B)}/f_{A(S+B)}f_{A^*(S)}$.

The use of the reference compound A* has three other, practical advantages [see Eq. (10)]:

1. Instead of absolute retention volumes, only relative retention volumes have to be determined. This means that no knowledge of inlet and outlet pressures, gas velocity, and recorder chart speed is required.
2. The amount of stationary phase in the columns need not be known.
3. Any systematic errors due to adsorption of A on the surface of the solid support for the stationary liquid or at the surface of the stationary liquid, or due to a nonlinear solution isotherm, cancel partially.

From the foregoing, we see that it is possible to calculate $K_x(f_A f_B/f_{AB})_{S+B}$ in a number of ways, which can be arranged in the following order of preference:

1. From Eq. (10), if a reasonable estimate of the activity coefficient ratio can be made [28].
2. From Eq. (10') [22, 23, 27, 30-32].
3. From Eq. (6). When A or B has zero dipole moment, the activity coefficient ratio can be calculated by means of perturbation theory or, less accurately, by regular solution theory [33].
4. From Eq. (6), setting the activity coefficient ratio equal to unity (all other investigations).

The next step is the choice of a reference state in which the activity coefficients are by definition equal to unity, and the extrapolation of $K_x(f_A f_B/f_{AB})_{S+B}$, determined for a range of mixtures S + B, to this reference state. In this way, the thermodynamic association constant in the chosen reference solvent is found. The following conventions are possible.

1. Gil-Av and Herling [34], in a study of the association of olefins (A) with $AgNO_3$ (B), proposed the hypothetical ideal solution in pure liquid A as the reference state. This seems to be an

TABLE 2

Values of $f_{A^*(S+C)}/f_{A^*(S+B)}$ [a,b]

Solute	Molality				
	0.25	1	2	3	Mean
Hexane	0.97	0.96	0.99	1.02	0.98
Heptane	0.98	1.01	0.96	1.04	1.00
Octane	0.97	1.05	0.95	1.06	1.01
2-Methylhexane	1.01	0.97	1.03	1.04	1.01
3-Methylhexane	0.98	1.00	0.98	1.06	1.00
Ethyl propyl ether	0.95	1.04	0.97	1.06	1.01
Propyl acetate	0.93	1.01	0.88	0.84	
Propyl chloride	1.02	1.10	1.14	1.27	

[a] A* is a (substituted) alkane, and S + C and S + B are $LiNO_3$ and $AgNO_3$ solutions of equal molality in ethylene glycol.

[b] Reprinted from de Ligny et al. [27], by courtesy of Elsevier Scientific Publishing Company, Amsterdam.

TABLE 3

Values of $f_{A(S)}f_{A^*(B)}/f_{A(B)}f_{A^*(S)}$ [a]

Hexadecyl derivative		Hexadecyl derivative	
Hexadecyl fluoride	0.92	Dioctyl ether	0.95
Hexadecyl chloride	0.91	Dioctyl ketone	0.90
Hexadecyl bromide	0.92	Dioctyl sulfide	0.93
Hexadecyl iodide	0.92	Dioctyl disulfide	0.92
Hexadecyl cyanide	0.83	Dioctylmethylamine	0.97
2-Pentadecanone	0.88		

[a] A is a (substituted) alcohol, A* is the corresponding (substituted) alkyl chloride, S is hexadecane, and B is a monofunctional hexadecyl derivative. Data at 60° C [calculated by Eq. (13)].

inappropriate choice. First, for each olefin A a separate reference state is introduced. Consequently, the effect of variations in the chemical nature of the olefin on the value of K cannot be interpreted, since not only the reactant but also the reference state is changed. Second, an extrapolation of the data on $K_x(f_A f_B/f_{AB})_{S+B}$ to this reference state is clearly impossible. Consequently, the activity coefficients must be determined or calculated. This is always difficult and in the present case even impossible for the ratio f_B/f_{AB}. Thus, only values of $K_x(f_B/f_{AB})_{S+B}$ could be tabulated.

2. A better choice of reference state seems to be the hypothetical ideal solution in pure liquid B, or in one particular mixture S + B [35, 36]. It must be admitted that in this case for each involatile reactant B a separate reference state is introduced, so that now the effect of variations in the chemical nature of B on the value of K cannot be interpreted. However, this will seldom be the goal of a gas-chromatographic investigation of complex association, since it is much easier to investigate a series of volatile solutes (A) than a series of involatile reactants (B). A practical advantage of this choice of reference state is that one may limit oneself to measurements on S and B (or the reference state mixture S + B) instead of making measurements on a range of mixtures S + B, since, by definition,

$$K_x\left(\frac{f_A f_B}{f_{AB}}\right)_{B \text{ or } S+B} \equiv K_x \tag{14}$$

This procedure reduces both the experimental effort and the accuracy of the results.

3. The best choice of reference state, used by the majority of investigators, is the hypothetical ideal solution in pure liquid S. In this case the effect of variations in the chemical nature of both reactants A and B on the value of K can be interpreted. Further, one need not bother about the possible formation of complexes AB_q: This convention involves an extrapolation to $x_B = 0$ where the complex AB dominates [see Eq. (8)].

Table 4 shows the effect of choosing different conventions (2 and 3) for the reference state and of using the approximate equation (10') instead of (10). Especially with convention 2, use of Eq. (10') introduces considerable errors. The same conclusion applies a fortiori to Eq. (6).

TABLE 4

Association Constants K_X of Substituted Alcohols with Hexadecyl Cyanide[a,b]

Alcohol	In hexadecyl cyanide		In hexadecane	
	Eq. (10)	Eq. (10')	Eq. (10)	Eq. (10')
Ethanol	2.82	2.18	2.42	2.22
Propanol	2.71	2.07	2.27	2.07
Butanol	2.72	2.08	2.36	2.17
2-Fluoroethanol	2.55	1.94	2.72	2.53
2-Chloroethanol	3.48	2.71	3.98	3.76
2-Bromoethanol	3.88	3.06	4.36	4.15
2-Methoxyethanol	1.33	0.94	1.29	1.11
1-Butene-4-ol	2.67	2.06	2.33	2.15
2-Hydroxy ethyl methyl sulfide	2.3	1.7	2.1	1.7

[a]In the solvents hexadecyl cyanide and hexadecane, and calculated by Eqs. (10) and (10').

[b]From de Ligny et al. [28].

B. Liquid Chromatography and Electrophoresis

Equations (1) to (10') also hold for liquid chromatography, if the complex is insoluble in the mobile phase. However, in contrast to the situation in gas chromatography, both B and AB will often have an appreciable solubility in the mobile phase in liquid chromatography. It can be shown that in this case the analog of Eq. (6) contains a factor:

$$F = \frac{1 + K_X x_B (f_A f_B / f_{AB})_{S+B}}{1 + K'_X x'_B (f_A f_B / f_{AB})_{M+B}} \tag{15}$$

where the primed symbols refer to the mobile phase. It is clear that from this factor neither K_X nor K'_X can be found easily.

The situation is more favorable in gel chromatography, where S and S' and therefore K_X and K'_X are identical. x'_B is known and the ratio x_B/x'_B

can be found from the retention volume of B. Thus, when the activity coefficients are set equal to unity as a first approximation, F contains only one unknown, K_x. One may thus determine K_x for a range of values of x'_B and extrapolate to $x'_B = 0$, thus finding K_x in the pure solvent S. Since the ratio x_B/x'_B should be different from unity, B should be at least partially excluded from the interior of the gel beads. A restriction of the method is therefore that B must be a rather large molecule or ion. If B is totally excluded from the interior of the gel beads, the numerator is equal to one and the situation is fully analogous to that in gas chromatography. Equations (6), (10), and (10') hold when the symbol S is changed into M and the inverse of the factor $1 + K_x x_B f_A f_B / f_{AB}$ is taken.

However, to the author's knowledge complex association constants have never been determined from chromatographic retention data in the fields of liquid or gel chromatography.

It is interesting to observe that the situation in electrophoresis is closely analogous to that in gas chromatography, when only one complex AB is formed. In the pure solvent S the mobility of the sample ν_S is of course equal to that of the solute ν_A:

$$\nu_S = \nu_A \tag{16}$$

In a mixture of the solvent S with the complex-forming compound B the following relationship holds for the mobilities of the sample and the species A and AB:

$$\nu_{S+B} = \nu_A \left(\frac{x_A}{x_A + x_{AB}} \right)_{S+B} + \nu_{AB} \left(\frac{x_{AB}}{x_A + x_{AB}} \right)_{S+B} \tag{17}$$

When we refer all mobilities to that of the species AB, the last term of Eq. (17) vanishes. Combination of Eqs. (16) and (17) then gives

$$\frac{\nu_S}{\nu_{S+B}} = \left(\frac{x_A + x_{AB}}{x_A} \right)_{S+B} = 1 + K_x x_B \left(\frac{f_A f_B}{f_{AB}} \right)_{S+B} \tag{18}$$

which is analogous to Eq. (10'). More complex equations, in which the analogy to the chromatographic case is less apparent, hold when a series of complexes AB_q is formed.

A limited amount of work has been done along these lines. Alberty and King [37] determined the association constants of the complexes CdI^+, CdI_2, CdI_3^-, and CdI_4^{2-} by a moving boundary method. Russian workers, using quartz powder as a stabilizing medium, investigated the complex association of La, Ce, Pm, Eu, and Y with ethylenediaminotetraacetic acid, EDTA [38]; of La, Ce, Nd, Pm, and Eu with citric acid [39]; of Pu with EDTA [40]; of Am with oxalic acid [41]; of Ce with hydroxyl ions [42]; of Am and Cm with EDTA [43]; and of Am, Cm, Ce, and Pm with α-hydroxyisobutyric acid [44]. Jokl investigated the association of Cu, Co(II), Ni, Cd, Man, and Zn with glycine, alanine, leucine, methionine, glutamic acid, and nitrilotriacetic acid [45] and of rare earth ions with ethylenediaminotetraacetic acid by means of paper electrophoresis [46]. Chabard et al. [47] used thin-layer electrophoresis in a study of the dissociation of weak acids. Paper electrophoresis was also applied by Ohyoshi and co-workers [48] for the investigation of ruthenium chloride complexes, whereas Sakanoue and Nakatani [49] used electrophoresis on cellulose acetate for measuring the association constants of the complexes of Pm, Eu, Am, and Cm with lactic acid.

It must be realized that there are two difficulties in the use of electrophoresis as a method for studying complex association. The first one is that the velocity of the complex(es) must be estimated, generally by extrapolation of observations on equilibrium mixtures. The second one is that high experimental skill is required to achieve the necessary accuracy in techniques such as paper and thin-layer electrophoresis.

An interesting procedure to obtain complex association constants from quantitative gel-chromatographic data, instead of from retention data, was first described by Hummel and Dreyer [50]. As will be clear from the following description, this method can also be applied in the field of electrophoresis. However, this has not yet been done actually.

The principles of the method are well illustrated by a study by Colman [51] of the binding of the manganous ion to nucleotides:

$$Mn^{2+} + ADP \rightleftharpoons ADP - Mn^{2+}$$

Although nucleotides are relatively small molecules, choice of the appropriate type of gel enabled the requirements of the technique to be fulfilled with ease. A column containing Sephadex G-10 was equilibrated with eluent containing $MnSO_4$ at the concentration and pH at which complex formation was to be studied. A sample containing a known quantity of the nucleotide dissolved in the same eluent was applied, and the concentrations of Mn^{2+} and nucleotide in the effluent were determined (Fig. 1).

The sample when applied contains nucleotide, free Mn^{2+} ions, and nucleotide-Mn^{2+} complex. As the sample zone moves through the column, the nucleotide continues to bind free Mn^{2+} ions until equilibrium is reached. The first peak to emerge from the column thus contains a mixture of nucleotide, Mn^{2+}, and complex in equilibrium with the concentration of Mn^{2+} ions

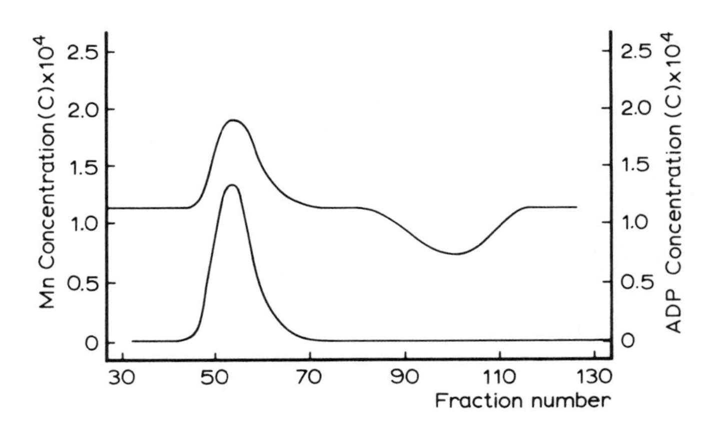

FIG. 1. Determination of the association constant of ADP with Mn^{2+} ions: 1.40 μmoles ADP was applied to a Sephadex G-10 column equilibrated with 1.14×10^{-4} M $MnSO_4$ in 0.108 M triethanolamine chloride buffer. The association constant was calculated to be 1.68×10^4 liters/mole. (Reprinted from Colman [51] by courtesy of Academic Press, New York.)

in the eluent. The trough which follows represents the amount of Mn^{2+} in the complex. The stoichiometry of the complex formation can be deduced directly from the composition of the equilibrium mixture, and the association constant can be derived either from these data or from Scatchard plots. The basic requirements of the method are the following:

1. The peak containing the complex must of course be separated from the trough corresponding to the bound ligand.
2. The complex and the added binding molecule, ADP in Fig. 1, must migrate as a single peak. This is achieved if the two substances have the same elution volume or if equilibrium between them is so rapid that their separation is prevented under the conditions of elution.

Hummel and Dreyer [50] demonstrated the virtues of the method by a study of the binding of 2'-cytidylic acid by ribonuclease, and Pfleiderer [52] used it for investigating the binding of reduced diphosphopyridine nucleotide by lactic dehydrogenase.

The interaction of serum proteins with sulfonamides was studied by Clausen [53]. Fairclough and Fruton [54] investigated the association of bovine serum albumin with tryptophan and tryptophan derivatives, including peptides. Evidence was offered for the utility of the method in the quantitative measurement of the competition of two small molecules for a protein binding site. As mentioned above, Colman [51] extended the method to the study of the association of rather small molecules ($M \approx 500$) with metal ions. The association of even smaller molecules with metal ions might be investigated by means of electrophoresis.

C. Comparison of Gas-Chromatographic, Electrophoretic, and Spectroscopic Methods for the Determination of Complex Association Constants

Up to the present, the most frequently applied methods for the determination of association constants of organic complexes are spectroscopic ones: infrared, UV, visible, and nuclear magnetic resonance spectroscopy.

Since there is a close analogy between the spectroscopic, gas-chromatographic, and electrophoretic methods for the determination of complex association constants from the theoretical point of view, a comparison of these methods is interesting.

In principle, all these methods start from the equation

$$R = R_A \frac{x_A}{x_A + x_{AB}} + R_{AB} \frac{x_{AB}}{x_A + x_{AB}} \tag{19}$$

where R is an observed quantity which is additively composed of contributions of the unassociated molecules A and of the complexes AB. In gas chromatography,

$$R = \left(\frac{V_{A*}}{V_A}\right)_{S+B} \left(\frac{V_A}{V_{A*}}\right)_S, \qquad R_A = 1$$

and the last term in Eq. (19) vanishes [see Eqs. (6) and (10')]. In electrophoresis, R, R_A, and R_{AB} are the mobilities of the sample, species A, and species AB, respectively [see Eq. (17)]. In optical spectroscopy, R is the molar absorbance at the chosen wavelength, and in NMR R is the chemical shift.

The methods also have in common that generally one of the reactants, B, is used in large excess, so that its concentration does not appreciably change upon complex formation.

Substitution of Eq. (1) into Eq. (19) and introduction of the symbol K_x'' for $K_x f_A f_B / f_{AB}$ gives

$$R = R_A \frac{1}{1 + K''_x x_B} + R_{AB} \frac{K''_x x_B}{1 + K''_x x_B} \tag{20}$$

To establish K''_x from the observations R by Eq. (20), R_A and R_{AB} must be known. In the case of gas chromatography, $R_{AB} = 0$ and $R_A = 1$, as mentioned above.

In the cases of infrared spectroscopy and of investigations of charge transfer complexes by means of visible or UV spectroscopy, the spectra of A and AB are often so different that a wavelength can be selected for which $R_{AB} \approx 0$. R_A is the molar absorbance at that wavelength in the solvent S. In the case of electrophoresis, R_{AB} is generally nonzero. The same holds for investigations of hydrogen bonding by means of visible or UV spectroscopy, since the resulting spectral changes are usually so small that the spectra of A and AB overlap considerably.

The poorest technique seems to be NMR. Not only is R_{AB} always nonzero, but when A shows strong self-association, R_A cannot be measured since the limited sensitivity of the technique precludes measurements at such low x_A values in the solvent S that A is essentially monomeric.

It follows that, in principle, gas chromatography is among the best techniques, if not the best one. (The same conclusion applies to gel chromatography, if the compound B is totally excluded from the interior of the gel beads.)

Of course, each technique has its own special limitations. The principal limitation of gas chromatography is that A must be far more volatile than B. Up to now, investigations have only been performed on rather volatile solutes. However, by using high-pressure gas chromatography [55], measurements on less volatile substances might be easily made.

III. EXPERIMENTAL PRECAUTIONS WITH THE GAS-CHROMATOGRAPHIC METHOD

With polar solutes in apolar stationary phases a rather strong dependence of the retention volume on the injected amount of solute is often observed [23]. A linear relationship between retention volume and amount of solute often exists over a limited range (~0.2-2 μl for 2-m-long columns of 4 mm ID) which can be extrapolated to zero sample size. (At very small sample size, deviations from this linear relationship are often observed when siliceous solid supports are used, because of adsorption [23].)

In that case it is wise to use porous Teflon as the support material. At room temperature this material is difficult to handle but at 8° C it loses its stickiness (probably as a result of a change to another modification, as judged by the large heat effect occurring at that temperature in differential thermal analysis). Below 8° C, efficient columns having a plate height of 1.5 mm at 10% liquid loading can be packed using this support [28].

Use of a large amount of stationary phase also reduces the effect of adsorption, both at the solid support and at the surface of the stationary liquid, since the ratio of support surface area to liquid volume and the liquid surface area both decrease with increasing amount of stationary phase. The occurrence of adsorption at the surface of the stationary liquid has been amply demonstrated [56-63], the only case in which it can be safely assumed to be negligible being that where both the solvent and the solute are apolar [56, 58, 64].

Eventually, it may be necessary to obtain retention data for a series of liquid loadings and to extrapolate to $1/V_{St} \longrightarrow 0$. Two different methods have been devised for this extrapolation [32, 65]; a comparative study was made by Liao and Martire [66].

Fueno et al. [67] observed a slight dependence of retention volumes on gas flow rate.

IV. RESULTS FROM THE GAS-CHROMATOGRAPHIC METHOD

A. Comparison with Results Obtained by Other Methods

Purnell and Srivastava [68] have started a broad program designed to establish the reliability of association constants derived from gas-chromatographic retention data. In the course of this work it became necessary to formulate viewpoints regarding alternative techniques also. Thus, complexing between benzene, toluene, ethylbenzene, and the three xylenes, respectively, with 2,4,7-trinitro-9-fluorenone in each of the solvents di-n-butyl succinate, di-n-butyl adipate, and di-n-butyl sebacate, over the temperature range 40° to 60° C, was studied by gas chromatography, optical spectroscopy, and NMR.

A substantial discrepancy between the data stemming from the two spectroscopic techniques was evident and, more significantly, a substantial number of negative association constants was found. It was concluded that all the spectroscopic results obtained for the investigated systems (where the association constants are quite small) are totally unreliable. In contrast, the association constants derived from gas chromatography were of reasonable magnitude and had a high degree of consistency and coherence from solute to solute and from solvent to solvent. The authors conclude that gas chromatography provides a more reliable technique for evaluation of small stability constants than do the spectroscopic methods. In a more favorable case, where the association constants are larger, good agreement between gas-chromatographic and infrared spectroscopic data was found [69].

B. Complexes with Coordinate Covalent Bonds

Gas chromatography was applied for the determination of association constants in inorganic fused-salt systems by Juvet et al. [70]. From measurements of the retention volume of $SbCl_3$ as a function of the amount of alkali metal chloride added to tetrachloroaluminate and tetrachloroferrate liquid phases, the equilibrium constant of the equilibrium

$$SbCl_3 + Cl^- \rightleftharpoons SbCl_4^-$$

was established.

At 290° C K_c is equal to 40 ± 10 in the solvent potassium tetrachloroaluminate as well as in potassium tetrachloroferrate; at 315° C K_c is equal to 0.8 ± 0.2 in thallium tetrachloroaluminate.

From the curvature of plots of the retention volume of $SbCl_3$ versus the molarity of the added alkali metal chloride at low values of the latter, it was concluded that the solvents dissociate according to the equilibrium

$$2MCl_4^- \rightleftharpoons M_2Cl_7^- + Cl^-$$

The dissociation constant was estimated to be equal to 3×10^{-4} at 290° C for potassium tetrachloroaluminate, 4×10^{-4} at 290° C for potassium tetrachloroferrate, and 6.5×10^{-3} at 315° C for thallium tetrachloroaluminate.

C. Charge Transfer Complexes

1. Complexes with Silver Ions

Complexes of Ag^+ ions and olefins were investigated by means of gas chromatography as early as 1962.

Hartley [71] has pointed out that data obtained for metal-olefin interactions can be used to answer two questions: "Which olefins form the most stable complexes" and "Why does one olefin form a more stable complex than another?" The first question requires only association constants.

However, to answer the second question, it is necessary to analyze the free enthalpy change on complex formation into its enthalpy and entropy components. There are many examples in the literature where authors, having obtained stability constants at a single temperature, have attempted to interpret these in terms of both steric and electronic influences on the silver ion-olefin bond. While a number of such explanations have been shown by subsequent enthalpy studies to be valid, there are other cases where modification of the structure of the olefin has very little influence on the enthalpy of complex formation, but a considerable effect on the entropy of complex formation and hence on the stability constant.*

This important remark holds also for the study of other types of complexes.

The pioneer in the field of gas-chromatographic investigations of Ag^+ ion-olefin complexes was Gil-Av [34], who determined the association constants of a number of cycloalkenes in 1.77 N $AgNO_3$ in ethylene glycol at 30° C. As mentioned in Sect. II A, he used an inappropriate reference state (the pure liquid olefin A). From these data (K_0' in the last column of his Table I) no conclusions can be drawn about the influence of the chemical nature of the olefin on the value of the association constant. Fortunately, he also gives a set of data (K in the fifth column of his Table I) which represents the association constants according to convention 2 (the $AgNO_3$ solution as a reference state). Still more fortunately, from a consideration of these data the same conclusions follow as the author drew from his K_0'. data.

A very elaborate investigation was made in the same year by Muhs and Weiss [72]. They determined the association constants of more than 100 complexes of alkenes, alkynes, cycloalkenes, polyenes, and aromatics with Ag^+ ions in ethylene glycol at 40° C, and drew interesting conclusions about the influence of the chemical nature of the unsaturated compound on the value of the association constant. These conclusions must be considered with a touch of reserve, however, as it was shown later [27] that their work suffers from some slight experimental errors.

Genkin and Boguslavskaya [73] measured the association constants of some alkenes and dienes in the temperature range 15° to 40° C in the solvents ethylene glycol and dihydroxyethyl ether.

Very careful measurements were made by Cvetanovic et al. [74]. Though the authors pointed out a possible error of -15% in their association constants because of the slow decomposition of $AgNO_3$ columns (a fact that was not mentioned by the older investigators), their values are in good

---

*Reprinted with permission from Hartley [71]. Copyright by the American Chemical Society.

agreement with those of de Ligny et al. [27] who accounted for the decomposition of $AgNO_3$ (see Table 7). Thus, the accuracy is probably much better than estimated by Cvetanovic et al., the systematic error caused by the decomposition of $AgNO_3$ being negligible. Thus, we can have confidence in the values given by these authors for the enthalpy and standard entropy of association of alkenes and deuterated alkenes with Ag^+ ions in ethylene glycol.

The bonding in these complexes can be described in the following way [75]: A σ-type bond, which is the result of donation of π electrons from the occupied 2p bonding orbital of the unsaturated compound into the vacant 5s orbital of the Ag^+ ion, and a π-type bond, which is the result of back-donation of d electrons from occupied 4d orbitals of the Ag^+ ion into the unoccupied π^*-2p antibonding orbitals of the unsaturated compound. It has been argued [76] that the σ-type and π-type contributions to the stability may be equally important.

Consequently, as Cvetanovic et al. point out, it is difficult to interpret the observed influence of the chemical nature of the alkene on complex stability. The decrease in stability with increasing alkyl substitution suggests steric hindrance to complex formation by the bulky substituents. However, the mechanism might also be a decrease in back-donation of d electrons because of the inductive electron release by the alkyl substituents to the double bond. Nevertheless, the authors give the following tentative interpretation of their data (Table 5).

TABLE 5

Enthalpies ΔH and Standard Entropies ΔS^0 of Alkene-Ag^+ Ion Complex Formation in Ethylene Glycol[a]

Alkene	$-\Delta H$ (kcal/mole)	$-\Delta S^0$ (cal/mole · deg)
Ethene	3.5	6.0
Propene	3.5	7.5
Butene-1	3.7	8.0
Pentene-1	3.6	8.1
3-Methylbutene-1	3.9	8.8
2-Methylbutene-1	3.5	8.6
Trimethylethene	2.4	7.8
Tetramethylethene	1.9	8.5

[a]Reprinted with permission from Cvetanovic et al. [74]. Copyright by the American Chemical Society.

The entropy of complex formation tends to decrease somewhat with increasing size of the alkyl substituents. This signifies a somewhat greater physical restraint in the complexes of the olefins containing bulkier substituent groups. The increase in the enthalpy of complex formation (and the decrease in the corresponding entropy) with increasing degree of substitution is also caused by steric hindrance. The greater strength of complexes formed by deuterated alkenes (which is apparent both from the enthalpy and the entropy of complex formation) must be caused by the greater inductive electron release from C-D bonds than from C-H bonds. This implies that an increasing electron release to the double bond increases the complex stability, which, in turn, suggests that σ-type bonding predominates in the complexes.

Wasik and Tsang [77] determined association constants in aqueous $AgNO_3$ solutions from gas-chromatographic retention data for 0.2 to 3.5 N $AgNO_3$ and the known solubility of the alkenes in water. They also found a deuterium isotope effect of the same magnitude as that observed by Cvetanovic et al. in the solvent ethylene glycol.

Association constants of substituted alkenes were measured by Schnecko (nitriles) [78], Fueno et al. (ethers) [67], and de Ligny et al. (various substituents) [27]. Fueno et al. drew the following conclusions from their data (Table 6):

1. Cis isomers have K values several times greater than those of the corresponding trans isomers because of the greater enthalpy loss of the former upon complexation. This is ascribable to the relief of strain of the cis alkyl and alkoxy groups, when the complex is formed.

   The authors point out that an alternative explanation cannot be excluded. Steric hindrance to complex formation by the two bulky groups may be greater when they are trans to each other. Thus, cis isomers may tend to give more stable complexes. This interpretation assumes that the difference in the (free) enthalpies of geometrical isomers is a less important factor than that of the resulting complexes.

2. The K values largely decrease with increasing degree of β substitution, because of the corresponding decrease in the degree of freedom of the coordinated Ag^+ ion.

3. When the substituent on the β position varies in the order $-CH_3$, $-C_2H_5$, and $i-C_3H_7$, the enthalpy of complex formation decreases. The increase of the favorable inductive effect probably outweighs the increase of the unfavorable steric effect.
   In the entropy of complex formation the steric effect dominates. In the cis isomers the favorable effect on the enthalpy outweighs the unfavorable effect on the entropy, so that the association

TABLE 6

Equilibrium Constants at 20° C, Enthalpies, and Standard Entropies of Complex Formation of Ag^+ Ions with Alkenyl Alkyl Ethers

$$(R_1)(R_2)C{=}C(H)(OC_2H_5)$$

in Ethylene Glycol[a]

R_1	R_2	K_c	$-\Delta H$ (kcal/mole)	$-\Delta S^0$ (cal/mole · deg)
H	H	6.72	4.4	11.0
H	CH_3	3.13	3.9	11.0
CH_3	H	0.54	3.2	12.1
H	C_2H_5	4.24	4.5	12.6
C_2H_5	H	0.73	3.3	12.0
H	i-C_3H_7	4.46	4.7	12.9
i-C_3H_7	H	0.59	3.5	12.9
CH_3	CH_3	0.21	3.5	15

[a]Reprinted from Fueno et al. [67] by courtesy of the Chemical Society of Japan.

constants increase in the order $-CH_3$, $-C_2H_5$, and i$-C_3H_7$. In the trans isomers the effects nearly cancel. The same is true for the 1-alkenes, investigated by Cvetanovic et al. [74] (Table 5).

De Ligny et al. [27] used the corresponding subsituted alkanes as reference solutes and calculated the association constants from Eq. (10'). The advantages of this procedure are stated in Sect. II A.

Owing to the rather large variation in the chemical nature of the substituents, their electronic effects can be investigated. In the concerned allyl compounds, resonance between the substituents and the reaction center is impossible since they are isolated by a CH_2 group. Therefore, the following functional relationship can be expected:

$$\log\left(\frac{K_m}{K_m^0}\right) = \rho_I \sigma_I \tag{21}$$

where K_m^0 is the association constant of the unsubstituted alkene (propene), ρ_I is the reaction constant, and σ_I is the inductive substituent constant.

Figure 2 shows that this relationship is very well obeyed by the substituents $-CH_3$, $-OC_2H_5$, $-OCOCH_3$, and -Cl, with $\rho = -2.3$. This means that the association constants are much larger as the substituents are more electron donating, which is another argument in favor of the view that σ-type bonding predominates in alkene-Ag^+ ion complexes. Once the electronic effects of substituents had been established, it became possible to investigate the steric effect of substituents attached to position 3 (Table 7). Whereas the steric effect of even three methyl substituents at this position is balanced by their inductive effect, the data on heptene, octene, and 3-methylhexene show that n-butyl and larger n-alkyl groups, or the combination of a methyl and an n-propyl group, attached to position 3, have a steric effect that outweighs their inductive effect.

The data on hexene, the methylhexenes, and heptene show that the steric effect of a methyl group depends on its position as follows: $2 > 4 > 3 \geq 5 \geq 6$. The same sequence was found earlier by Muhs and Weiss [72].

TABLE 7

Association Constants of Alkene-Ag^+ Ion Complexes in Ethylene Glycol at 40° C[a]

Alkene	K_m [27]	K_c [27]	K_c [74]
Propene	6.8 ± 0.4	6.2	5.9
Butene-1	7.0 ± 0.4	6.4	6.8
3-Methylbutene-1	6.5 ± 0.7	5.9	6.1
3,3-Dimethylbutene-1	6.6 ± 0.3		
Heptene-1	6.0 ± 0.2		
Octene-1	5.9 ± 0.3		
2-Methylhexene-1	3.0 ± 0.2		
3-Methylhexene-1	5.5 ± 0.1		
4-Methylhexene-1	4.0 ± 0.2		
5-Methylhexene-1	5.8 ± 0.4		

[a]Reprinted from de Ligny et al. [27], by courtesy of Elsevier Scientific Publishing Company, Amsterdam.

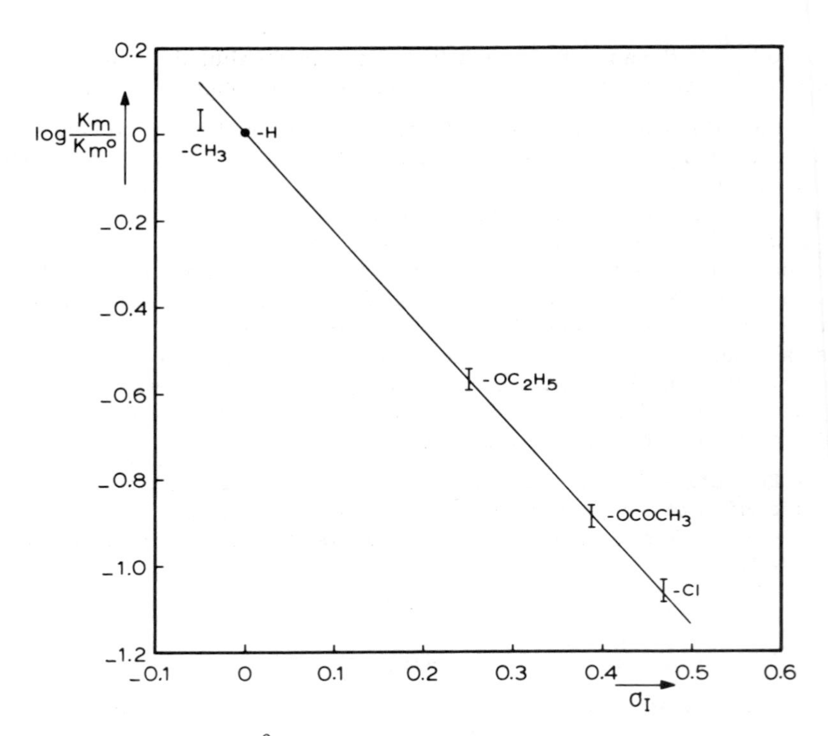

FIG. 2. Log (K_m/K_m^0) of alkene-Ag^+ ion complexes in ethylene glycol at 40° C as a function of σ_I. (Reprinted from de Ligny et al. [27], by courtesy of Elsevier Scientific Publishing Company, Amsterdam.)

2. Complexes with Other Metals

In a search for new complex-forming reagents suitable for gas chromatography, a series of $Rh(I)(CO)_2$ β-diketonates in squalane solution was found to interact rapidly and reversibly with olefins by Gil-Av and Schurig [35]. The compound derived from 3-trifluoroacetylcamphor was examined in detail and its complexation with 27 C_2-C_6 n-alkenes and 12 cycloalkenes with four- to seven-membered rings was determined.

The association with the Rh compound (in squalane at 50° C) is in general far stronger than that with the Ag^+ ion (in ethylene glycol at 40° C). Furthermore, the effect of substituents is usually more pronounced, and in some cases has a different sense than with the Ag^+ ion.

Kraitr et al. [36] investigated the thermodynamics of interaction of hexenes with Pd^{2+} (and Ag^+) ions in N-methylacetamide at 30° C. The association constants of the Pd^{2+} complexes are nearly a factor of 2 larger than those of the Ag^+ complexes, except when the alkene has a substituent adjacent to the double bond. Apparently, steric hindrance to complex formation is much greater with the Pd^{2+} ion than with the Ag^+ ion in this case.

3. Organic Complexes

Cadogan and Purnell [79] investigated the thermodynamics of complex formation of benzene, toluene, and the xylenes with di-n-propyl tetrachloroterephthalate in the solvent squalane. The phthalate ester appeared to be a rather weak acceptor compared to acceptors such as s-trinitrobenzene, tetracyanoethylene, and picric acid. This is mainly caused by the low value of the entropy of complex formation (about -8 cal/mole · deg for benzene, toluene, and o-xylene, and -9.5 cal/mole · deg for m- and p-xylene). This indicates considerable steric hindrance in the complexes, consistent with a configuration in which the aromatic rings attempt to lie together but are impeded by the out-of-plane ester groups of the phthalate. Further, the m- and p-xylenes would then be expected to experience hindrance extra to that of benzene, toluene, and o-xylene.

Eon and co-workers [80, 81] studied the thermodynamics of complex formation of derivatives of furan, thiophene, and pyrrole, and of some other solutes with dibutyl tetrachlorophthalate in the solvent squalane. Some of their results are rather remarkable, e.g., for diethyl ether values of -28.4 kcal/mole and -90.8 cal/mole · deg are given for the enthalpy and entropy of complex formation, respectively.

Complex formation of the methyl benzenes with another aromatic ester, tetra-n-butyl pyromellitate, was investigated in the solvent squalane by Sheridan et al. [82]. It was apparent from the trend in the association constants (roughly, benzene > toluene > xylenes > mesitylene) that steric hindrance plays an important role in these systems also. The out-of-plane ester groups of the pyromellitate prevent closeness of approach of the methyl benzenes, thus impeding the formation of coplanar complexes. In the absence of steric factors, electronic effects alone should produce the reverse trend. Unlike the case with di-n-propylterephthalate [79], the values of neither the enthalpy nor the entropy of complex formation show a regular trend with increasing methyl substitution.

Whereas solutions of tetramethyl pyromellitate in aromatic solvents do show charge transfer bands, the authors could not detect any charge transfer bands in solutions of tetra-n-butyl pyromellitate in aromatic solvents. Therefore, they are inclined to the view that classical electrostatic forces are primarily stabilizing the investigated complexes.

Martire et al. investigated the interaction between haloalkanes and the electron donors di-n-octyl ether and di-n-octyl thioether [30] and di-n-octylmethylamine and tri-n-hexylamine [31], using alkanes as reference solutes A*.

Their K values are based on two different reference states for the activity coefficients, namely, the hypothetical ideal solution in the electron donor B for y_A and y_{AB} and the hypothetical ideal solution in the reference solvent S (n-octadecane) for y_B. This seems to be a rather disturbing convention. Fortunately, it is possible to calculate K_x and K_c with the hypothetical ideal solution in B as the reference state for all involved activity coefficients from Martire's values of K, the activity coefficient of B, γ_D, and the concentration of B, C_D: $K_x = K' = K\gamma_D c_D$ and $K_c = K\gamma_D$ (see Martire et al. [30, Eq. (2)]).

The initial purpose of this work was to carry out a study of the hydrogen-bonding properties of the haloforms and dihalomethanes. However, the results indicated that there are two possible types of interactions present within these systems: (a) hydrogen bonding of the hydrogen atoms of the haloalkane to the electron donor and (b) charge transfer between the halogen atoms and the electron donor. (The authors point out the possibility that the interactions involving the halogen atoms and the ether or thioether do not arise from charge transfer but rather may be described by electrostatic considerations alone, as no charge transfer bands have been observed in these systems.)

The haloalkanes can be divided into four groups, as shown in Table 8. The hydrogen-bonding power increases in the order: group 1 = group 4 < group 2 < group 3, whereas the ability to form charge transfer complexes increases in the order: group 2 < group 3 < group 1 < group 4.

The stability of the complexes formed with the two amines, as judged from the values of ΔH as well as of ΔS^0, increases in the order: group 1 < group 2 < group 3. Moreover, within groups 2 and 3 the values of ΔH and ΔS^0 roughly increase when -Cl is substituted by -Br, i.e., with decreasing acidity of the H atoms. This shows that with the amine donors primarily hydrogen-bonded complexes are formed. However, carbon tetrachloride and 1,1,1-trichloroethane do form weak complexes, and bromotrichloromethane rather strong ones which must be due to charge transfer.

The stabilities of the complexes formed with the ether end thioether show a different pattern. As judged from the values of ΔH, the stability increases in the order: group 1 = group 2 < group 3 = group 4. From the values of ΔS^0 the following order results: group 2 < group 3 < group 1 < group 4. This is exactly the order of increasing ability to form charge transfer complexes. Within groups 2 and 3 the values of ΔH and ΔS^0 are insensitive to substitution of -Cl by -Br. Taken together, this evidence shows that with the ether and thioether both hydrogen-bonded and charge transfer complexes are formed.

TABLE 8

Enthalpies ΔH (kcal/mole) and Standard Entropies ΔS^0 (cal/mole · deg) of Complex Formation of Haloalkanes with Some Electron Donors[a]

	Di-n-octyl thioether		Di-n-octyl ether		Tri-n-hexyl amine		Di-n-octyl methylamine	
	$-\Delta H$	$-\Delta S^0$	$-\Delta H$	$-\Delta S^0$	$-\Delta H$	$-\Delta S^0$	$-\Delta H$	$-\Delta S^0$
Group 1								
Carbon tetrachloride	1.4	8.4	1.6	10.7	0.8	9.7	0.7	7.9
1,1,1-Trichloroethane	1.6	9.0	1.8	10.4	0.8	9.4	1.6	10.6
Group 2								
Dichloromethane	1.3	6.4	1.5	7.5	1.8	10.7	2.4	11.1
Bromochloromethane	1.5	6.7	1.9	8.7	2.0	11.0	2.3	10.5
Dibromomethane	1.4	6.1	1.8	8.3	1.3	8.7	1.9	9.0
Group 3								
Chloroform	1.9	8.1	2.4	9.7	3.4	14.4	3.5	13.1
Dichlorobromomethane	2.0	7.8	2.3	9.4	3.2	13.5	3.4	12.7
Dibromochloromethane	2.0	7.5	2.8	10.8	2.6	11.7	3.0	10.9
Bromoform	2.3	8.1	2.2	8.9	3.0	12.6	3.2	11.4
Group 4								
Bromotrichloromethane	1.6	7.7	2.5	12.6	2.4	13.1	2.8	12.2
Carbon tetrabromide	3.0	10.1	2.5	12.5				

[a]Reprinted with permission from Martire et al. [30, 31]. Copyright by the American Chemical Society.

A comparison between the electron donors should be made with some reserve, in view of the different reference states involved. However, we may conclude from both the values of ΔH and those of ΔS^0 that the ether is a stronger electron donor than the thioether. It is interesting to realize that, if we had based the comparison on the values of K, i.e., on $\Delta H - T\,\Delta S^0$, we would have erroneously concluded that the thioether is the stronger electron donor.

As regards the two amines, the values of ΔH are lower for di-n-octylmethylamine, and the values of ΔS^0 are slightly higher, resulting in far larger association constants with this amine. The cause is probably a larger amount of steric hindrance to complex formation with tri-n-hexylamine. Further, the amines appear to be much stronger electron donors than the ether.

Complex formation between aromatic hydrocarbons and 1,3,5-trinitrobenzene dissolved in dinonylphthalate was investigated by Castells [33]. He demonstrated clearly the inadequacy of the approach of Eon et al. [24], of estimating the activity coefficient ratio $f_{A(S+B)}/f_{A(S)}$ in Eq. (6) from the Flory-Huggins equation alone and calculated it by using the Flory-Huggins equation and regular solution theory.

Meen and associates [83] determined formation constants for alkyl benzene complexes with 2,4,7-trinitro-9-fluorenone in six alkyl ester solvents and in β,β'-thiodipropionitrile. The activity coefficient ratio $f_{A(S+B)}/f_{A(S)}$ in Eq. (6) was supposed to be equal to unity. Their results were recalculated by Eon and Karger [84], who estimated the activity coefficient ratio by the Flory-Huggins equation. They observed deviations between the results of both calculation procedures ranging up to 50% for small (≈ 0.5) K_x values. While their work shows clearly that it is not justified to set the activity coefficient ratio simply equal to unity, in view of Castells' results [33] it is doubtful that their K_x values are correct.

D. Hydrogen Bond Complexes

Iogansen et al. [85] tried to measure hydrogen bond energies by means of gas chromatography as early as 1966. Their work was criticized by Novak [86].

Martire and Riedl [23] investigated the hydrogen bonding of aliphatic alcohols with di-n-octyl ether and di-n-octyl ketone. Because of the suspicion [66] that these measurements might contain small systematic errors caused by solute adsorption at the gas-liquid interface, the measurements on the ether were repeated very carefully and extended to di-n-octyl thioether and di-n-octylmethylamine [32]. It appeared that the association constants found earlier were low by 10 to 20% because of adsorption at the surface of the reference solvent S, n-heptadecane.

In these investigations, alkanes were used as reference solutes A*. The K values are based on two different reference states, namely, the hypothetical ideal solution in the electron donor B for y_A and y_{AB} and the hypothetical ideal solution in the reference solvent S for y_B. (If desired, values of K_c and K_x can be calculated from the K values given by Liao and Martire; see Sect. IV C 3.)

As in the case of complex formation with haloalkanes [30, 31] it follows from the order of the values of ΔH, ΔS^0, and K that the amine is the strongest electron donor, and the thioether the weakest one. The average values of ΔH for hydrogen bonding with the amine and ether are in perfect agreement with spectroscopic values. It is interesting that the ΔH values of all (3×6) systems can be described within experimental error by the two-parameter equation

$$\Delta H_{ij} = Q_i Q_j \qquad (22)$$

where Q_i depends only on the alcohol and Q_j only on the electron donor.

Queignec and Wojtkowiak [69] determined the equilibrium constants for the association of alkynes with diethylacetamide in the solvent squalane by means of gas chromatography and infrared spectroscopy. For $(CH_3)_3C{-}C{\equiv}CH$, $(CH_3)_3Si{-}C{\equiv}CH$, $(CH_3)_2ClC{-}C{\equiv}CH$, and n-heptyne-1, the association constants obtained by the two techniques agreed within experimental error. However, for $ClCH_2C{\equiv}CH$ and $BrCH_2C{\equiv}CH$, larger values were found by gas chromatography ($K_c = 1.73 \pm 0.07$ and 1.67 ± 0.07, respectively) than by infrared spectroscopy ($K_c = 1.54$ and 1.50, respectively). The cause of this discrepancy is probably that charge transfer complexes are also formed, by interaction of the halide and nitrogen atoms (see Martire et al. [31]). These complexes are not observed in the infrared method, where the absorption of the $\equiv C{-}H$ vibration is measured.

A detailed study of complex formation of bicyclic alcohols and some ethers with tris-(p-t-butylphenyl) phosphate in the solvent dotriacontane was made by Vivilecchia and Karger [87]. Their results, which were shown to be free of adsorption effects, are summarized in Table 9. Several interesting conclusions can be drawn from them. The saturated ethers, 16, 17, and 18, show negligibly small K_x values. The differences from zero are probably caused by inadequately accounting for the activity coefficient ratio $f_{A(S+B)}/f_{A(S)}$ in Eq. (6), which was estimated by the Flory-Huggins equation. The unsaturated ethers, 5 and 6, and anisole, 24, show small but genuine K_x values. Charge transfer complexes are probably formed with the phenyl groups of the phosphate. Spectroscopic evidence exists for this type of complex. Consequently, the unsaturated alcohols are able to form charge transfer as well as hydrogen bond complexes, and their K_x values should be larger than those of their saturated counterparts: compare 2 with 7 and 4 with 9. Some of the unsaturated alcohols can form an intramolecular hydrogen bond with the double bond. This reduces the tendency to form an intermolecular hydrogen bond: compare 1 with 8, 1 with 2, 3 with 4, and 19 with 20. From the ΔH values of these compounds it follows that the enthalpy of formation of the intramolecular hydrogen bond is about -0.5 kcal/mole, in excellent agreement with spectroscopic results. From the

TABLE 9

Equilibrium Constants at 115° C, Enthalpies, and Standard Entropies of Complex Formation of Bicyclic Alcohols and Ethers with Tris-(p-t-butylphenyl) Phosphate in the Solvent Dotriacontane[a]

A. / B. (structures)						K_x	$-\Delta H$ (kcal/mole)	$-\Delta S^0$ (cal/mole · deg)
A. Bicyclo[2.2.1]heptanols								
No.	R_1	R_2	R_3	R_4				
1	H	OH	H	H		2.04	3.0	6.3
2	OH	H	H	H		2.84	3.5	6.9
3	H	H	OH	H		1.16	3.0	7.4
4	H	H	H	OH		3.53	3.6	6.8
5	H	OCH_3	H	H		0.52		
6	H	H	H	OCH_3		0.37		
B. Bicyclo[2.2.1]heptanols								
No.	R_1	R_2	R_3	R_4	R_5			
7	H	OH	H	H	H	2.37	3.5	7.3
8	H	H	OH	H	H	2.47	3.6	7.5
9	H	H	H	H	OH	2.70	3.7	7.6
10	H	CH_3	OH	H	H	1.72	2.7	5.9
11	H	OH	CH_3	H	H	1.83	2.6	5.5
12	CH_3	OH	H	H	H	1.93	3.3	7
13	CH_3	H	OH	H	H	2.16	3.5	7
14	CH_3	OH	H	CH_3	CH_3	2.21	3.1	6.4
15	CH_3	H	OH	CH_3	CH_3	2.39	3.3	6.8
16	CH_3	OCH_3	H	CH_3	CH_3	0.03		
17	CH_3	H	OCH_3	CH_3	CH_3	0		
18	H	CH_3	OCH_3	H	H	0.24		

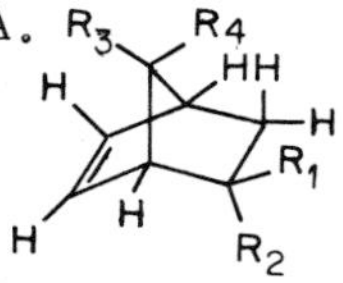

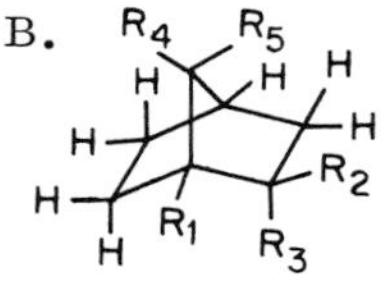

TABLE 9 (Continued)

		K_X	$-\Delta H$ (kcal/mole)	$-\Delta S^0$ (cal/mole · deg)
C. Other Alcohols and Ethers				
No.	Structure			
19	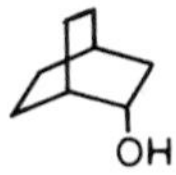 [bicyclo[2.2.2]octan-2-ol] OH	1.99		
20	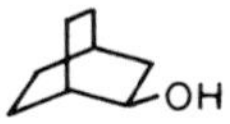 [bicyclic alcohol] OH	2.78		
21	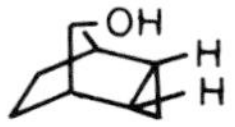 [tricyclic alcohol with cyclopropyl] OH, H, H	2.35		
22	 [tricyclic alcohol with cyclopropyl] HO, H, H	3.33		
23	C_6H_5OH	21.4	6	9
24	$C_6H_5OCH_3$	0.90		

[a]Reprinted with permission from Vivillecchia and Karger [87]. Copyright by the American Chemical Society.

K_X values of 21 and 22 it follows that the cyclopropyl group is also able to form an intramolecular hydrogen bond. The K_X values of the pairs 8-10, 7-11, and to a lesser degree those of the pairs 7-12, 8-13, 7-14, and 8-15 show the effect of steric hindrance to complex formation by methyl substituents.

Littlewood and Willmott [22] succeeded in determining the association constants of alkanols with 1-dodecanol in the solvent squalane. A complication arises in this case because of the strong self-association of 1-dodecanol. Since the association constants of the alkanols with 1-dodecanol did not depend on chain length, the same value was assumed to hold for the self-association constant of 1-dodecanol. Taking into account the self-association of 1-dodecanol, the authors also established the association constants of some ethers, acetates, and alkyl halides with 1-dodecanol.

E. Dipole Complexes

In 1964, Littlewood [29] obtained retention volumes of polar and apolar solutes in hexadecane and a number of polar hexadecyl derivatives.

The enhanced retention volumes of the polar solutes in the polar hexadecyl derivatives were attributed to classical electrostatic interactions, primarily dipole-dipole interaction. Complex association resulting from these interactions was not invoked to explain the data (see Sect. II A).

In 1966, Littlewood and Willmott [22] obtained analogous data for the solvents squalane and lauronitrile, and mixtures of the two. This time they interpreted their data in terms of complex formation of the polar solutes A with lauronitrile, due to dipole-dipole attraction. As reference solutes A* they used alkanes (Table 10). This raises the interesting philosophical point as to which of the two alternative explanations: a change of activity coefficients due to electrostatic interactions or a change in activity coefficients due to size and dispersion interaction effects only, combined with the invocation of an associated species, is to be preferred. Certainly, the physical model underlying the former explanation is simpler. However, it must be admitted that it is difficult to describe the effect of dipole-dipole interaction on activity coefficients in a theoretically satisfying way (see Sect. I). On one hand, it is difficult to imagine that activity coefficients can be influenced by dispersion interaction, but not by induction and orientation interactions. On the other hand, the enthalpy of dissociation of the complex $-\Delta H$ is so large that it will be stable for a number of intermolecular vibrations (Table 10).

Lambert et al. [88] also invoked the concept of dipole-dipole complexes to interpret the values of the second virial coefficients of the vapors of ethyl chloride, acetone, and acetonitrile. They showed that the experimental values of the enthalpy of association are in good agreement with the calculated values of the dipole-dipole interaction energy. In another branch of chemistry, the analogous concept of electrostatically bound ion pairs has been generally accepted to explain the electrical conductivity of "strong" electrolytes in nonaqueous solutions. Taking all arguments together, it seems not unreasonable to interpret the effects of dipole attraction in terms of the formation of definite complexes.

The data in Table 10 show that association constants in polar systems (e.g., in hydrogen-bonding systems) will contain an appreciable contribution from dipole-dipole interaction, and thus be composite association constants, unless in the calculation of the activity coefficient ratio $f_{A(S+B)}/f_{A(S)}$ dipole-dipole interactions are taken into account. (An analogous situation has been observed in the case of complex formation of haloalkanes with di-n-octyl ether, di-n-octyl thioether, di-n-octylmethylamine, and tri-n-hexylamine [30, 31], and of unsaturated alcohols with tris-(p-t-butylphenyl) phosphate [87], where both charge transfer and hydrogen bond complexes are formed.) It must be realized that spectroscopic techniques "see" only hydrogen bond complexes (or charge transfer complexes, as the case may be) but not dipole-dipole associates. Therefore, when the results of a gas-chromatographic investigation are to be compared with spectroscopic

TABLE 10

Equilibrium Constants at 40° C and Enthalpies of Complex Formation of Dipolar Solutes with Lauronitrile in the Solvent Squalane[a]

Solute	K_x	$-\Delta H$ (kcal/mole)	μ(D)[b]
1-Hexene	1.1		0.4
1-Heptene	1.1		0.4
Diethyl ether	3.2		1.15
Di-n-propyl ether	2.3		1.30
n-Propyl chloride	4.2		2.05
n-Propyl bromide	4.0		2.18
Ethyl acetate	7.9		1.78
n-Propyl acetate	6.7		
n-Butyl acetate	6.2		
Nitromethane	14.2	3.9	3.46
Nitroethane	9.9	2.2	3.65
Nitropropane	8.3	2.1	3.66
Ethyl cyanide	9.6	2.1	4.02
n-Propyl cyanide	8.2	2.0	4.07

[a]Reprinted with permission from Littlewood and Willmott [22]. Copyright by the American Chemical Society.

[b]Solute dipole moments [89].

results, the influence of the orientation effect on the activity coefficient ratio $f_{A(S+B)}/f_{A(S)}$ should be accounted for.

At the beginning of Sect. IV we observed that gas chromatography is superior to spectroscopic techniques for the determination of very small complex association constants. Here, at the end of this section, we have become acquainted with a type of complex that is not observed at all by spectroscopic techniques (and, indeed, by very few other techniques) but that can be investigated easily by means of gas chromatography. It is justifiable to conclude that gas chromatography is among the most versatile tools for the study of intermolecular association.

REFERENCES

1. P. J. Flory, R. A. Orwoll, and A. Vrij, J. Am. Chem. Soc., 86, 3507 (1964).
2. P. J. Flory, R. A. Orwoll, and A. Vrij, J. Am. Chem. Soc., 86, 3515 (1964).
3. G. Delmas, D. Patterson, and T. Somcynsky, J. Polymer Sci., 57, 79 (1962).
4. W. E. Hammers and C. L. de Ligny, Rec. Trav. Chim., 90, 819 (1971).
5. W. E. Hammers and C. L. de Ligny, Rec. Trav. Chim., 90, 912 (1971).
6. W. E. Hammers and C. L. de Ligny, J. Polymer Sci. (C), 39, 273 (1972).
7. W. R. Summers, Y. B. Tewari, and H. P. Schreiber, Macromolecules, 5, 12 (1972).
8. D. Patterson, Y. B. Tewari, and H. P. Schreiber, J. Chem. Soc. Farady II, 68, 885 (1972).
9. H. P. Schreiber, Y. B. Tewari, and D. Patterson, J. Polymer Sci. (A-2), 11, 15 (1973).
10. R. D. Newman and J. M. Prausnitz, J. Phys. Chem., 76, 1492 (1972).
11. R. D. Newman and J. M. Prausnitz, AIChE J., 19, 704 (1973).
12. L. P. Hammett, Physical Organic Chemistry, McGraw-Hill, New York, 1940, p. 184.
13. R. W. Taft, J. Am. Chem. Soc., 79, 1045 (1957).
14. K. E. Gubbins, AIChE J., 19, 684 (1973).
15. R. A. Pierotti, J. Phys. Chem., 67, 1840 (1963).
16. C. L. de Ligny and N. G. van der Veen, Chem. Eng. Sci., 27, 391 (1972).
17. R. Battino, F. D. Evans, W. F. Danforth, and E. Wilhelm, J. Chem. Thermodynamics, 3, 743 (1971).
18. R. A. Pierotti, J. Phys. Chem., 69, 281 (1965).
19. H. D. Nelson and C. L. de Ligny, Rec. Trav. Chim., 87, 623 (1968).
20. E. W. Tiepel and K. E. Gubbins, Proc. Int. Solvent Extr. Conf., Soc. Chem. Industry, The Hague, 1971, p. 25.

21. J. H. Purnell, in Gas Chromatography 1966 (A. B. Littlewood, ed.), Elsevier, Amsterdam, 1967, p. 3.

22. A. B. Littlewood and F. W. Willmott, Anal. Chem., 38, 1031 (1966).

23. D. E. Martire and P. Riedl, J. Phys. Chem., 72, 3478 (1968).

24. C. Eon, C. Pommier, and G. Guiochon, Chromatographia, 4, 235 (1971).

25. C. L. de Ligny, J. Chromatogr., 69, 243 (1972).

26. A. Bondi, J. Phys. Chem., 68, 441 (1964).

27. C. L. de Ligny, T. van't Verlaat, and F. Karthaus, J. Chromatogr., 76, 115 (1973).

28. C. L. de Ligny, N. J. Koole, H. D. Nelson, and G. H. E. Nieuwdorp, J. Chromatogr., 1161, 63 (1975).

29. A. B. Littlewood, Anal. Chem., 36, 1441 (1964).

30. J. P. Sheridan, D. E. Martire, and Y. B. Tewari, J. Am. Chem. Soc., 94, 3294 (1972).

31. J. P. Sheridan, D. E. Martire, and F. P. Banda, J. Am. Chem. Soc., 95, 4788 (1973).

32. H. L. Liao and D. E. Martire, J. Am. Chem. Soc., 96, 2058 (1974).

33. R. C. Castells, Chromatographia, 6, 57 (1973).

34. E. Gil-Av and J. Herling, J. Phys. Chem., 66, 1208 (1962).

35. E. Gil-Av and V. Schurig, Anal. Chem., 43, 2030 (1971).

36. M. Kraitr, R. Komers, and F. Cuta, J. Chromatogr., 86, 1 (1973).

37. R. A. Alberty and E. L. King, J. Am. Chem. Soc., 73, 517 (1951).

38. V. P. Shvedov and A. V. Stepanov, Radiokhimiya, 1, 162 (1959).

39. A. V. Stepanov and V. P. Shvedov, Radiokhimiya, 1, 668 (1959).

40. A. V. Stepanov and T. P. Makarova, Radiokhimiya, 7, 664 (1965).

41. A. V. Stepanov and T. P. Makarova, Radiokhimiya, 7, 670 (1965).

42. A. V. Stepanov and V. P. Shvedov, Zh. Neorgan. Khim., 10, 1000 (1965).

43. I. A. Lebedev, A. M. Maksimova, A. V. Stepanov, and A. B. Shalinets, Radiokhimiya, 9, 707 (1967).

44. I. A. Lebedev and A. B. Shalinets, Khim. Transuranovykh Oskolochnykh Elem., Akad. Nauk SSSR, Otd. Obshch. Tekh. Khim., 1967, 140.

45. V. Jokl, J. Chromatogr., 14, 71 (1964).

46. V. Jokl and I. Valaskova, J. Chromatogr., 72, 373 (1972).

47. J. L. Chabard, G. Besse, G. Voissiere, J. Petit, and J. A. Berger, Bull. Soc. Chim. Fr., 1970, 2425.

48. E. Ohyoshi, A. Ohyoshi, and M. Shinagawa, Radiochim. Acta, 13, 10 (1970).

49. M. Sakanoue and M. Nakatani, Bull. Chem. Soc. Jap., 45, 3429 (1972).

50. J. P. Hummel and W. J. Dreyer, Biochim. Biophys. Acta, 63, 530 (1962).

51. R. F. Colman, Anal. Biochem., 46, 358 (1972).

52. G. Pfleiderer, Mechanismen enzymatischer Reaktionen, Springer-Verlag, Berlin and New York, 1964, p. 300.

53. J. Clausen, J. Pharmacol. Exp. Ther., 153, 167 (1966).

54. G. F. Fairclough and J. S. Fruton, Biochemistry, 5, 673 (1966).

55. S. T. Sie and G. W. A. Rijnders, Anal. Chim. Acta, 38, 31 (1967).

56. R. L. Martin, Anal. Chem., 33, 347 (1961).

57. R. L. Martin, Anal. Chem., 35, 116 (1963).

58. R. L. Pecsok, A. de Yllana, and A. Abdul-Karim, Anal. Chem., 36, 452 (1964).

59. D. E. Martire, R. L. Pecsok, and J. H. Purnell, Nature, 203, 1279 (1964).

60. D. E. Martire, R. L. Pecsok, and J. H. Purnell, Trans. Faraday Soc., 61, 2495 (1965).

61. D. E. Martire, Anal. Chem., 38, 244 (1966).

62. R. L. Pecsok and B. H. Gump, J. Phys. Chem., 71, 2202 (1967).

63. D. F. Cadogan, J. R. Conder, D. C. Locke, and J. H. Purnell, J. Phys. Chem., 73, 708 (1969).

64. A. J. Ashworth and D. H. Everett, Trans. Faraday Soc., 56, 1609 (1960).

65. D. F. Cadogan and J. H. Purnell, J. Phys. Chem., 73, 3849 (1969).

66. H. L. Liao and D. E. Martire, Anal. Chem., 44, 498 (1972).

67. T. Fueno, O. Kajimoto, T. Okuyama, and J. Furukawa, Bull. Chem. Soc. Jap., 41, 785 (1968).

68. J. H. Purnell and O. P. Srivastava, Anal. Chem., 45, 1111 (1973).

69. R. Queignec and B. Wojtkowiak, Bull. Soc. Chim. Fr., 1970, 860.

70. R. S. Juvet, V. R. Shaw, and M. A. Khan, J. Am. Chem. Soc., 91, 3788 (1969).

71. F. R. Hartley, Chem. Rev., 73, 163 (1973).

72. M. A. Muhs and F. T. Weiss, J. Am. Chem. Soc., 84, 4697 (1962).

73. A. N. Genkin and B. I. Boguslavskaya, Neftekhimiya, 5, 897 (1965).

74. R. J. Cvetanovic, F. J. Duncan, W. E. Falconer, and R. S. Irwin, J. Am. Chem. Soc., 87, 1827 (1965).

75. M. J. S. Dewar, Bull. Soc. Chim. Fr., 1951, C71.

76. H. Hosoya and S. Nagakura, Bull. Chem. Soc. Jap., 37, 249 (1964).

77. S. P. Wasik and W. Tsang, J. Phys. Chem., 15, 2970 (1970).

78. H. Schnecko, Anal. Chem., 40, 1391 (1968).

79. D. F. Cadogan and J. H. Purnell, J. Chem. Soc. (A), 1968, 2133.

80. C. Eon, C. Pommier, and G. Guiochon, J. Phys. Chem., 75, 2632 (1971).

81. C. Eon, C. Pommier, and G. Guiochon, Chromatographia, 4, 241 (1971).

82. J. P. Sheridan, M. A. Capeless, and D. E. Martire, J. Am. Chem. Soc., 94, 3298 (1972).

83. D. L. Meen, F. Morris, and J. H. Purnell, J. Chromatogr. Sci., 9, 281 (1971).

84. C. Eon and B. Karger, J. Chromatogr. Sci., 10, 140 (1972).

85. A. G. Ioganssen, G. A. Kurkchi, and O. V. Levina, in Gas Chromatography 1966 (A. B. Littlewood, ed.), Elsevier, Amsterdam, 1967, p. 35.

86. J. Novak, J. Chromatogr., 28, 391 (1967).

87. R. Vivilecchia and B. L. Karger, J. Am. Chem. Soc., 93, 6598 (1971).

88. J. D. Lambert, G. A. H. Roberts, J. S. Rowlinson, and V. J. Wilkinson, Proc. R. Soc. (London) A, 196, 113 (1949).

89. Handbook of Chemistry and Physics, Chem. Rubber Pub. Co., Cleveland, Ohio, 1968-1969, p. E66.

Some interesting papers that came to the author's attention in the time interval between writing and proof-reading are given below:

L. Mathiasson, J. Chromatogr., 114, 47 (1975).

A. W. Girotti and E. Breslow, J. Biol. Chem., 243, 216 (1968).

V. M. Chernajenko and S. E. Bresler, Biophys. Chem., 1, 227 (1974).

P. Andrews, B. J. Kitchen, and D. J. Winzor, Biochem. J., 135, 897 (1973).

L. W. Nichol, A. G. Ogston, D. J. Winzor, and W. H. Sawyer, Biochem. J., 143, 435 (1974).

R. I. Brinkworth, C. J. Masters, and D. J. Winzor, Biochem. J., 151 631 (1975).

B. M. Dunn and I. M. Chaiken, Biochem., 14, 2343 (1975).

K. Kasai and S. Ishii, J. Biochem., 77, 261 (1975).

D. E. Martire, Anal. Chem., 48, 398 (1976).

R. J. Laub and J. H. Purnell, J. Am. Chem. Soc., 98, 35 (1976).

M. Saleem, M. Aslam Khan, M. Shahid, and K. Iqbal, Chromatographia, 8, 699 (1975).

V. Schurig, R. C. Chang, A. Zlatkis, and B. Feibush, J. Chromatography, 99, 147 (1974).

C. E. Doering, R. Geyer, and G. Burkhardt, Z. Chem., 15, 319 (1975).

A. N. Genkin and N. A. Petrova, J. Chromatogr., 105, 25 (1975).

R. J. Laub and R. L. Pecsok, Anal. Chem., 46, 1214 (1974).

R. J. Laub and R. L. Pecsok, Anal. Chem., 46, 1659 (1974).

L. Mathiasson and R. Jonsson, J. Chromatogr., 101, 339 (1974).

L. Mathiasson, J. Chromatogr., 114, 39 (1975).

A. V. Bratchikov, G. L. Ryzhova, and L. G. Kostina, Russ. J. Phys. Chem., 49, 95 (1975).

R. J. Laub and R. L. Pecsok, J. Chromatog., 113, 47 (1975).

A. Kratochwill, J. U. Weidner, and H. Zimmermann, Ber. Bunsenges, 77, 408 (1973).

G. Fini and P. Mirone, J. Chem. Soc. Faraday Transact. II, 70, 1776 (1974).

Chapter 7

GAS-LIQUID-SOLID CHROMATOGRAPHY

Antonio Di Corcia and Arnaldo Liberti

Istituto di Chimica Analitica
Universita di Roma
Rome, Italy

I. INTRODUCTION

Since the development of gas chromatography, adsorption (gas-solid) chromatography (GSC) was mainly confined to the separation of permanent gases and light hydrocarbon mixtures. The limiting factors to the use of GSC were the lack of adsorbing media with geometrically and chemically homogeneous surfaces and the poor reproducibility of products obtained in various batches. Conversely, gas-liquid chromatography (GLC) proved to be much more versatile for the great variety of liquid phases that could be used; it also offered the advantage of virtually linear solubility isotherms within a wide range of concentrations.

Today, GSC is used on a much larger scale and has been applied to the separation of various mixtures from hydrogen isotopes to polycyclic hydrocarbons. This development has been made possible because of the introduction of a greater control of the homogeneity and specificity of the molecular adsorbents obtained by specific synthesis of adsorbents and chemical modification of their surfaces. In particular, the introduction of porous polymers [1] and graphitized carbon blacks (GCB) [2-4] has made possible the enjoyment of the obvious advantages that GSC holds over GLC [5]. Porous polymers are adsorbing materials with a microporous surface. These nonpolar polymers may be produced by blocking polymerization of styrene with divinylbenzene as cross-linking agent. By using polar monomers as starting materials, polar adsorbing surfaces may also be obtained. The latter are adsorbing materials having essentially a nonporous and nonpolar surface. As will be shown later, GCBs especially lend themselves for use in gas-liquid-solid chromatography (GLSC). For this reason, we feel the necessity to summarize briefly their preparation and adsorptive characteristics.

GCBs are prepared by heating ordinary carbon blacks to about 3000° C in an inert gas. Volatile tarry substances are removed below 1000° C; above this temperature crystallites grow, the various functional groups originally present on the carbon black particle surface [6] being destroyed. Experimental evidence by x-ray and electron diffraction studies [7, 8] has shown that particles become polyhedral with faces formed by graphite crystals growing from within the particles. The extent of the graphitization, which leads to surface homogeneity, is a function of the type of carbon black initially used. It has been shown [7] that the growth of graphitic crystallites is more extensive for particles of larger diameter. As a matter of fact, Graphon, Sterling FT-G, and Sterling MT-G, which constitute a series of GCBs having particle diameters of approximately 310, 2250, and 5600 Å [9], respectively, possess an increasing degree of surface homogeneity in the order listed. These three carbons have specific surface areas of about 85, 14, and 7.5 m^2/g.

A graphitized carbon black surface is almost completely free of unsaturated bonds, lone electron pairs, free radicals, and ions. The adsorption is due mainly to attraction via London dispersion forces. Therefore, molecules having functional groups or π bonds, or atoms with lone electron pairs capable of specific interactions are also adsorbed on the surface of GCBs nonspecifically and practically in the same way as noble gas and saturated hydrocarbons.

The energy of nonspecific interaction depends greatly on the distance between the adsorbent surface and the force centers of individual linkages of the adsorbed molecule. Hence, the flat surfaces of GCBs are particularly able to separate molecules on the basis of their differences in geometrical structures and in the polarizability of their linkages.

Despite the great progress made in GSC some difficulties limiting an extensive use of such techniques still remain. One is that retention times are generally very high as compared to those obtainable with GLC. This is because the adsorption heat is generally higher than the solution heat for a given compound. This drawback has been partially overcome by the expansion of the working temperature range of the gas-chromatographic columns, up to 500° C. Owing to the leveling effect of the temperature, however, working at high temperatures can decrease the separating power of an adsorbing medium to such an extent that its use becomes meaningless.

The second limitation arises from the fact that only a limited number of relatively high homogeneous adsorbing surfaces is available. Generally, the selectivity of a gas-solid column is very high. Nevertheless, in some separation problems there may be the need either for enhancing the selectivity power of a gas-solid column or for directly modifying the type of selectivity. In GLC, the problem of varying the chemical composition of the stationary phase is far easier since a wide range of liquids carrying the necessary bonds and functional groups is available.

The third limitation is that a so-called uniform surface is always contaminated by small residual surface heterogeneities which can be geometric or chemical in nature, or both. The adsorption energy may not be constant over the entire surface of an adsorbent. If this is the case, the molecules will be adsorbed preferentially on those sites where a maximum in adsorption heat will be displayed; we can call these places the most active sites of the surface.

To the extent that adsorption on surface heterogeneities is responsible for a larger adsorption heat than on the rest of the surface, we can identify surface heterogeneities as the most active sites of the surface. When few in number, surface heterogeneities may yield nonlinear adsorption isotherms at low surface coverages. From a chromatographic point of view, the effect of residual heterogeneities on the Gaussian concentration distribution of the eluate in the gas phase is to retain for the longest period of time those portions of the peak that represent the lowest partial pressures. Therefore, as the slug of sample moves down the column it gradually skews to the rear. The extent of this unwelcome effect becomes more pronounced as the sample size of a given eluate is made smaller.

II. GEOMETRIC AND CHEMICAL RESIDUAL HETEROGENEITIES OF AN ADSORBING MEDIUM

The surface of a solid is rarely smooth, but is usually inundated by cracks, crevices, cavities, corners, and edges. Even on a molecular scale, roughness is frequently introduced by lattice disorder or spiral dislocation.

Steps and recesses, which may be present in GCBs, can produce very relevant effects. Graham [10] reported that the adsorption of nitrogen on these sites gave an adsorption heat nearly twice that for the adsorption on the predominant, weaker sites. Isirikyan and Kiselev [11] showed that hexane adsorbed on Graphon exhibited, at very low surface coverages, an abnormally high initial adsorption heat of about 14 kcal/mole. Di Corcia and Samperi [12] gave further evidence of the presence of active sites on GCB upon adsorption of nonpolar molecules. By using the chromatographic technique, they measured an adsorption heat equal to 12.2 kcal/mole for butane at zero surface coverage. After hydrogen treatment at 1000° C, which was seen to be effective in erasing topographical irregularities, the adsorption heat of butane decreased to 7.35 kcal/mole.

In chromatography, geometric irregularities can affect the symmetry of the chromatographic peak if their surface concentration is not negligible. In our experience, we find that, when injected in small amounts, even hydrocarbons are eluted as tailed peaks on Graphon, which contains about 1% of geometric active sites. The same is not true when Sterling FT, which has no more than 0.1% of geometric active sites [10], is used.

The second kind of surface heterogeneity is chemical in nature. The presence of chemical impurities can be due to several factors. The utmost effort to obtain completely homogeneous substrates has not yet succeeded in avoiding residual chemical inhomogeneities. On GCB, chemical impurities composed of oxygen surface complexes are presumably a burnt-off residue left over from the heating of carbon blacks in producing graphitic carbons [13]. On porous polymers, chemical impurities can be produced by oxidation of the surface catalyzed by heat, light, etc. The presence of traces of metal salts used in the production of polymers [14] might also be responsible for chemical heterogeneities. On polytetrafluoroethylene, impurities can be introduced by surfactant molecules used in emulsion polymerization.

Residual chemical inhomogeneity on an adsorbing surface has a profound influence on the adsorption of polar compounds. Let us take, for example, an otherwise uniform, nonpolar surface that is contaminated by a small amount of a nonvolatile, polar impurity of much higher adsorptive potential. In van der Waals adsorption of nonpolar molecules, such sites would be no stronger than the rest of the surface and, to the extent that they represent -OH groups, they would even be much weaker. Their effect would therefore either be very small or would appear only in the upper part of the isotherm, which is usually obscured by the beginning of second-layer adsorption. From a chromatographic point of view, this means that the elution of nonpolar molecules is not affected by the presence of chemical impurities, providing they are few in number.

On the contrary, the adsorption isotherms of polar vapors on this substrate show a small but marked "knee" at the low-pressure end of the isotherm, thereby indicating the presence of the impurity. The shape of the isotherm would be the same whether the impurity is considered as a coating on a part of the surface or as a mechanically separate ingredient of the mixture. Although not described quite in this way, a number of adsorbents have been previously reported that are, effectively, mixtures of two surfaces, each with its own distribution of adsorptive energies. The most clear-cut example is a slightly oxidized graphitic surface for which the adsorption isotherm of water vapor shows a pronounced knee whose height varies with the degree of oxidation of the surface; reducing the surface with hydrogen at 1000° C eliminates the knee [15, 16].

Further experimental evidence of the effect of chemical impurities is given by the fact that the lower alcohols are eluted on untreated GCBs as very tailed peaks. Treating these adsorbents with hydrogen results in perfectly symmetrical peaks, even for a few nanograms of methanol [17].

Moreover, adsorption on GCBs of molecules carrying strongly basic functional groups can directly result in irreversible adsorption at low surface coverages. This effect can be accounted for by the presence of surface oxygen complexes that are acidic in nature. Conversely, the measurement of propylamine adsorption heat at very low surface coverages on H_2-treated GCB gave a value in good agreement with that theoretically obtainable considering the interaction of a propylamine molecule with a pure basal plane of graphite [18].

It is apparent that such behavior can only be explained on the basis of two types of adsorption sites on GCB. The vast majority of the surface sites are nonpolar and must correspond to the graphitelike array of carbon atoms. These sites show no tendency to interact preferentially with molecules carrying functional groups. The polar sites are few in number but they can establish specific, strong interactions with polar molecules. These sites are probably of unequal energies but as a group are still vastly different from the remainder of the surface.

III. METHODS FOR ELIMINATING SURFACE HETEROGENEITIES

Surface heterogeneities, especially chemical ones, on a solid substrate can give rise to various undesirable effects in gas chromatography, such as loss of sample and ghosting phenomena, in addition to badly tailed elution peaks. These effects are commonly encountered when the analysis of strong hydrogen-bonding compounds is attempted. The source of these unwelcome effects is always traced to abnormally strong solid-gas interactions which take place by using both GLC and GSC. Therefore, efforts for improving the gas-chromatographic technique for the analysis of polar compounds have been generally centered on the search for means of eliminating the effect of heterogeneous patches of the surface.

In GSC, several methods have been adopted for obtaining symmetrical peaks by linearizing adsorption isotherms. One method is the blocking of chemical surface impurities by mixing in a carrier gas various strongly adsorbed vapors, such as water [19, 20], ammonia [21], and formic acid [22, 23]. However, this procedure is not always convenient and sometimes can be impracticable, e.g., when high-sensitivity ionization detectors are used. Another method which applies specifically to GCB surfaces is treating these adsorbents at 1000° C in a stream of hydrogen. Hydrogen-treated GCBs have proved to be very effective in yielding symmetrical peaks for the elution of very polar compounds, such as aliphatic amines and acids [17], amphetamines [24], and inorganic acids [25].

Although these procedures are effective in eliminating surface heterogeneities, they do not contribute to overcoming the other two limitations of GBC, e.g., high retention times and no flexibility in varying the selectivity characteristics of the packing material. All three objectives can be reached by the deposition on the solid surface of small quantities of liquids that are themselves capable of establishing strong interactions with the anomalous active surface sites. The use of such compounds, commonly referred to as "tailing reducers," is a well-known practice in obtaining symmetrical peaks. It was first suggested by Eggertsen et al. [26] who added up to 1.5% by weight of squalane to Pelletex carbon. They observed an enhanced peak symmetry that vastly improved the resolution of a hydrocarbon mixture.

Moreover, in the submonolayer region of liquid concentration they observed a sharp decrease of the retention times as the liquid solid ratio was increased [27]. Other authors [19, 28] emphasize that the addition of a liquid to an adsorbing surface also has the effect of profoundly modifying the characteristics of the packing even though the liquids are added in tiny amounts. This modification in the chromatographic process arises from the combined effects of the force centers of both the liquid and the solid phase. The elution of a certain compound along a column where these effects are acting together was defined as gas-liquid-solid chromatography (GLSC) by Purnell [29].

According to Halasz and Heine [30], the term GLSC is not adequate to describe the chromatographic features of a liquid-modified solid. In fact, whereas for the solid it is reasonable to assume that its external force field remains unperturbed in spite of the presence of molecules deposited on its surface, this consideration cannot be made for the absorbed molecules of the modifying liquid. Even where molecules coming from a liquid are added to an adsorbing medium to form several layers, their partitioning properties are distinctly different from those of the bulk liquid [31-35]. From a theoretical point of view, then, the chromatographic process on a liquid-modified solid should be discussed in terms of eluate-solid surface interaction plus interactions occurring on the solid surface between eluate-adsorbed molecules and individual macromolecules preadsorbed from a liquid. In this respect, it should be more appropriate to speak not of GLSC but of gas-mixed adsorption chromatography. Nevertheless, the conventional term of GLSC is still maintained to indicate roughly that this chromatographic technique is lying somewhere between gas-liquid and gas-solid chromatography.

A further method for eliminating peak asymmetry is that of depositing monomolecular layers of strongly adsorbed substances, especially polymers [36-39], on a nonporous or macroporous adsorbing surface. In this way, adsorption on the solid is replaced by adsorption on the monolayer. Similarly to GLSC, in this case also the GS method uses a property of the GL method, i.e., the capacity of varying the selectivity characteristics of the chromatographic column. Though considered apart, gas chromatography on monolayers can be regarded as a particular case of GLSC since the solid surface is still acting in the chromatographic process. In fact, it can be safely assumed that the adsorbing properties of the monolayer are strictly dependent on the surface activity of the solid used as supporting material, so that modifications in the chromatographic process can be induced by changing either the liquid or the solid substrate. Therefore, in this chromatographic method, too, there is a synergetic action of both the liquid and the solid substrate.

IV. MULTILAYER ADSORPTION ON A QUASI-HOMOGENEOUS SURFACE

Fo understand fully the adsorptive modifications brought in by preadsorbing molecules coming from an involatile liquid it is useful first to consider adsorption of a single compound on a bare, homogeneous surface that contains residual chemical and/or geometrical inhomogeneity. Upon increasing the surface coverage θ of the adsorbate, which is the ratio between the number of adsorbed molecules and that needed to form one complete monolayer, throughout the range of coverage from zero up to two or three layers, the process of adsorption can be suitably followed by plotting adsorption heat Q_{ads} as a function of θ. The typical shape of an adsorption heat-coverage curve is shown in Fig. 1.

Starting from zero coverage there is initially a decrease of adsorption heat. This effect is accounted for by considering that molecules will be adsorbed first at those spots of the surface where the attraction is greatest and consequently the heat of adsorption assumes the greatest values; we may call these sites the most active of the surface. After these sites have been progressively filled by increasing the surface coverage, one should expect the heat of adsorption caused by the mutual attraction of admolecules and surface to be practically constant over a long range. The relative value of the adsorption heat should correspond to that given by adsorption on the homogeneous patches of the surface.

Conversely, the heat of adsorption does not fall to a constant value, but passes through a minimum and increases again. When the degree of occupation becomes greater, the adsorbed molecules will influence each other not only because of the space they occupy but also because of the mutual attractions rising among them. Hence, there will be an additional

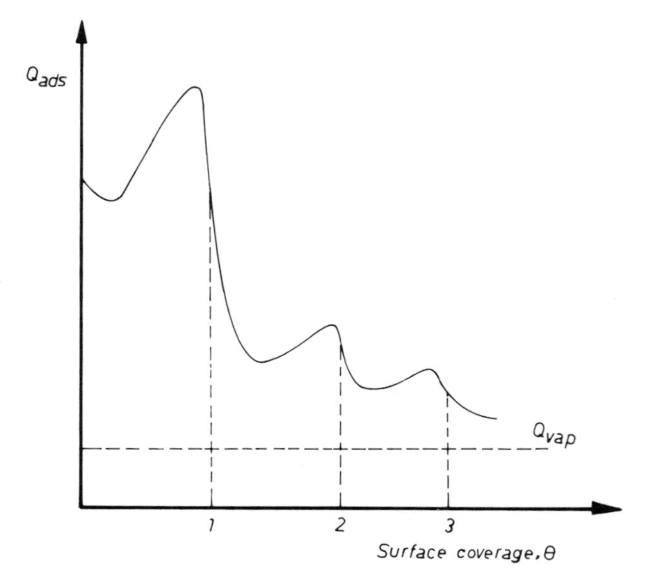

FIG. 1. Typical shape of a heat-surface coverage curve upon adsorption from the gas phase on a quasi-homogeneous surface.

contribution to the adsorption heat by virtue of the mutual attraction of molecules already adsorbed, and this effect becomes greater as the surface population increases.

By increasing the surface coverage further, a maximum in the adsorption heat is observed followed immediately by a sharp decrease. This indicates that the first monolayer, which is built up by more or less closely two-dimensionally packed molecules, is completed.

Molecules striking on this already adsorbed layer of molecules may be attracted by sufficiently great forces. If this event occurs, conditions are fulfilled for adsorption in more than one layer, i.e., for multimolecular adsorption. The filling of the second and third layers will proceed in a way similar to that of the first one. Therefore, there is the appearance of a second and a third maximum in adsorption heat, indicating that a second layer of molecules followed by a third one is formed on the solid medium. As the multilayer adsorption proceeds, the heat of adsorption gradually approaches the heat of condensation, thus indicating that a slab of liquid starts to form on the solid medium.

The requirements for the occurrence of this process of adsorption are, in addition to the need for a homogeneous surface, as follows: (a) there must be sufficiently strong lateral interactions between adsorbed molecules, (b) the affinity of the first adsorbed molecules for the residual chemical surface inhomogeneities must not be too great, and (c) the temperature must be such that thermal agitation of the adsorbed molecules does not erase the discontinuities between the filling of adsorbed layers.

The general shape of the adsorption heat-surface coverage curves discussed above may be naturally modified to a greater or lesser extent depending on the particular adsorbent-adsorbate system being considered.

The degree of surface homogeneity of a certain material plays an important role in determining the quality of the adsorption process. In Fig. 2 are shown heat-coverage curves for adsorption of hexane on two GCBs (Sterling FT-G and Graphon) differing in their degree of surface homogeneity [11]. Owing to the greater inhomogeneity of Graphon, the initial heats of adsorption are found to be higher and adsorbate-adsorbate interactions are partially hindered. In addition, the drop in heat after the maximum is smoothed out as compared to adsorption on the higher homogeneous GCB, and there is no evidence of a second heat maximum. This indicates that surface heterogeneity hampers to some extent a regular deposition of hexane molecules. Where the adsorbate is capable of establishing strong interactions (e.g., hydrogen bonding) with the residual chemical heterogeneities of the GCB, the profile of the heat-coverage curve may be profoundly altered, as shown in Figs. 3 and 4 [40, 41]. In this instance, the nearly constant value of the adsorption heat for ammonia following the initial drop can be explained by assuming that adsorption is initiated on the

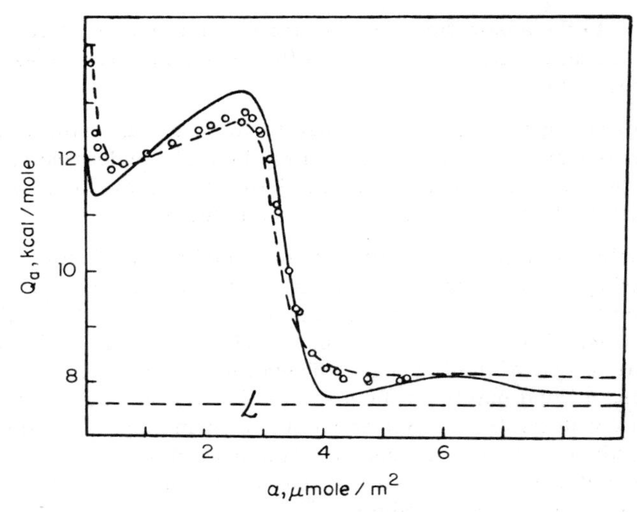

FIG. 2. The differential heats of adsorption of n-hexanol at 20° C on Graphon (dashed line) and on Sterling MT-1 (3100° C) (solid line); L = heat of vaporization of hexane [11]. (Reproduced by permission of the Journal of Physical Chemistry.

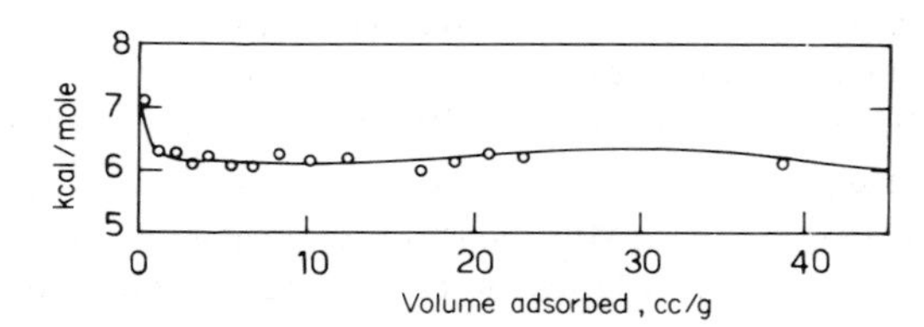

FIG. 3. Calorimetric heat of adsorption of ammonia on Graphon at -78° C [40]. (Reproduced by permission of the Journal of Physical Chemistry.)

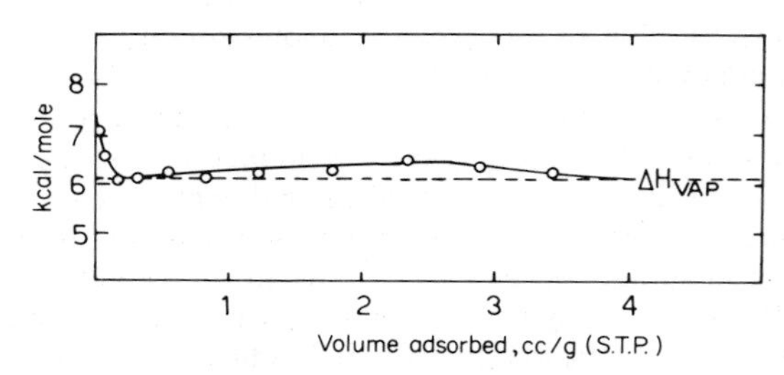

FIG. 4. Calorimetric heats of adsorption of ammonia on Sterling MT (3100° C) at -79° C [41]. (Reproduced by permission of the Journal of Physical Chemistry.)

oxygen complexes present in traces on the GCB surfaces. Thereafter, the adsorption proceeds with the formation of patches around these centers.

V. ANALOGY BETWEEN MIXED ADSORPTION AND ADSORPTION ON A LIQUID-MODIFIED SOLID

The model of adsorption on a liquid-modified solid is analogous to that of adsorption of a particular binary gas mixture. Let one of the two gases of the mixture, gas 1, be not only more strongly adsorbed by far than gas 2 but also have such a great affinity for the solid that its equilibrium concentration in the gas phase may be neglected. Also, let us gradually increase the amount of gas 1 keeping the amount of gas 2 constant. Under these conditions, as the surface concentration of gas 1 molecules is increased, the adsorption of gas 2 molecules is modified more and more.

Two-dimensional sketches of adsorption of gas 2 at increasing surface concentrations of gas 1 on an adsorbent containing some inhomogeneities are shown in Fig. 5 [42]. Inhomogeneities are represented as discontinuities of the surface. Figure 5A represents adsorption of gas 2 on the bare surface of the adsorbent. In Fig. 5B, since inhomogeneities are preferentially occupied by molecules of gas 1, adsorption of gas 2 is shown to occur on the homogeneous part of the surface. In Fig. 5C, owing to the increased degree of surface occupation of gas 1, lateral interactions between the two

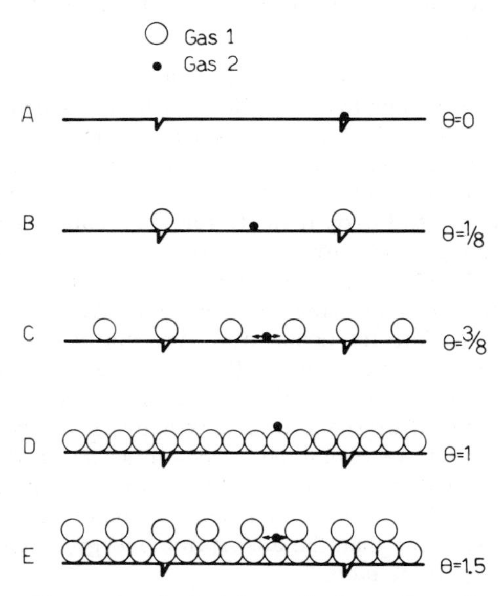

FIG. 5. Two-dimensional sketches of adsorption of gas (•) at varying degrees of surface coverage of gas 1 (○) [42]. (Courtesy of G. Crescentini.)

adsorbed gases come into play. In Fig. 5D, adsorption of gas 2 molecules is shown to occur on the top of the monomolecular layer of gas 1. From here and up to about three layers, the process of adsorption of gas 2 will be the same as above except that the solid surface is replaced by one layer of molecules of gas 1. Figure 5E illustrates the mode of adsorption of gas 1 when the surface coverage of gas 1 is intermediate between $\theta = 1$ and $\theta = 2$.

Measuring adsorption heats of gas 2 with increasing surface coverage of gas 1 can provide information on the mode of deposition of gas 1, and on modifications on adsorption of gas 2 caused by preferential adsorption of gas 1 molecules as well as on the magnitude of the interactions occurring on the solid between admolecules 1 and 2. In this case, obviously, as multilayer deposition of adsorbate 1 takes place, the adsorption heat of adsorbate 2 will gradually approach its heat of solution in liquid 1. By simply assimilating the molecules of gas 1 to molecules preadsorbed from a nonvolatile liquid we can discuss adsorption on liquid-modified solids in the same way as described in the preceding section.

In Fig. 6 are reported isosteric adsorption heat curves of some adsorbates on two GCBs, differing in their degree of surface homogeneity, both modified with increasing amounts of polyethylene glycol 1500 (PEG 1500) macromolecules [43]. The measurement of adsorption heats was carried out by using the gas-chromatographic technique. Heats of solution of the same eluates in PEG 1500 were obtained by using a conventional gas-liquid column.

Let us examine first the behavior of the adsorbing Sterling FT-G + PEG 1500 system. It can be assumed that the first molecules of PEG 1500 will be preferentially adsorbed on the residual chemical inhomogeneities, since these polymeric macromolecules contain terminal OH groups which can establish specific interactions with polar sites of the surface. There is no evidence of the initial drop in heats for the eluates considered. This means that the Sterling FT-G surface homogeneity is so high that only a few molecules of PEG 1500 suffice to block its residual inhomogeneity. The progressive rise in adsorption heat for both polar and nonpolar eluates is due to lateral, specific, and nonspecific interactions occurring on the solid between PEG 1500 molecules and those of eluates. Obviously, the increment in the adsorption heat will be greatest for the most polar eluates. In the case of pentane, after reaching a maximum, adsorption heat decreases sharply. This effect can be explained by assuming that a first layer of PEG 1500 macromolecules is correspondingly formed, and that adsorption energy for pentane on the top of a more or less closely packed monolayer of PEG 1500 is much smaller than the adsorption energy on the still unoccupied patches of the GCB surface.

On the top of this monolayer, other macromolecules of PEG 1500 can be adsorbed and a second layer starts. The completion of this layer is

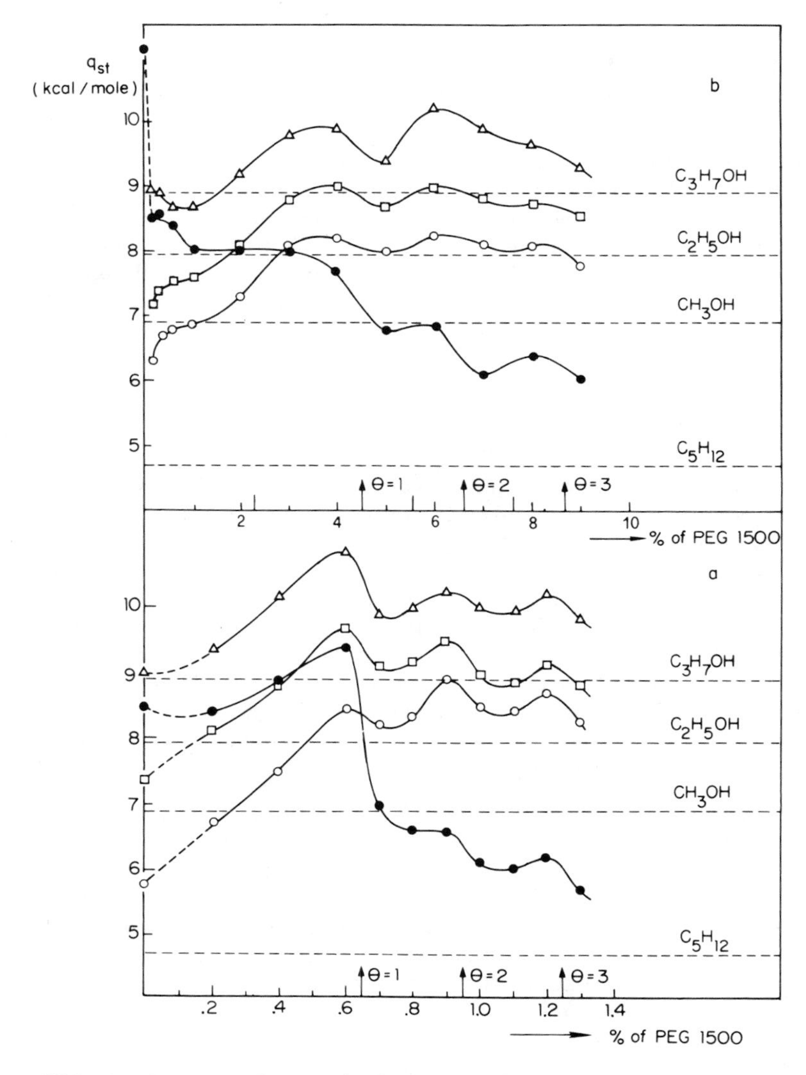

FIG. 6. Isosteric heats of adsorption of pentane (●), methanol (○), ethanol (□), and propanol (Δ) on (a) Sterling FT-G + PEG 1500; (b) Graphon + PEG 1500. Dashed lines indicate heats of solution in PEG 1500 [43]. (Reproduced by permission of Analytical Chemistry.)

again indicated by the presence of a discontinuity in the adsorption heat curves. The maxima in heat following the first one are less pronounced since the adsorption on the second and succeeding layers is not as well defined as on the first. As the deposition of layers proceeds, it appears that heats of adsorption of eluates approach heats of solution into the PEG 1500 liquid.

By varying the degree of polarity of eluates, it is interesting to evaluate differences in the adsorption process in terms of the surface coverage of a polar-modifying agent. At low surface concentrations, e.g., 0.2% PEG 1500, alcohols show a well-defined rise in heat, which is steeper passing from propanol to methanol. Conversely, adding small amounts of PEG 1500 does not cause an appreciable change in heat for pentane. The functional group of adsorbed alcohol molecules can establish long-range, specific, attractive forces with both the terminal OH groups and the ether oxygen atom of $-CH_2-O-CH_2$ links of polyethylene glycol macromolecules. In contrast, only weak, nonspecific, short-range London dispersion forces can be operative for pentane.

The heat-PEG 1500 coverage curve for methanol as opposed to that for pentane is particularly interesting in two respects: (a) the first discontinuity is scarcely pronounced and (b) the second maximum is higher than the first. These observations suggest that when a nonpolar surface is modified by a polar liquid, there are some differences in the mechanism of adsorption of eluates according to their degree of polarity. For a nonpolar molecule such as pentane, heat data throughout the range of PEG 1500 surface coverage from zero to just before unity are consistent with the model in which adsorption occurs unchangeably on the graphitic planes of the surface and the position of the adsorbate molecules are scarcely affected by the presence of PEG 1500 molecules.

Conversely, heat data for methanol suggest that highly polar molecules are preferentially adsorbed on those spots of the surface which are close to functional groups of PEG 1500, especially OH groups. Moreover, the relatively slight drop in heat corresponding to the completion of the first layer of PEG 1500 seems to indicate that adsorption of an eluate having good affinity with the modifying liquid may occur on top of the macromolecules even though the first layer is not yet completed. This trend will be more and more attenuated as the polarity of the eluate decreases. As a matter of fact, in passing from methanol to propanol the shape of the heat curves tends to become similar to the one of pentane.

Surface heterogeneities can hinder to a greater or lesser extent multilayer deposition of macromolecules coming from a liquid. As it has been pointed out, Graphon possesses a higher degree of both geometrical and chemical surface heterogeneity than Sterling FT-G. For this reason, the shapes of heat curves relative to the Graphon + PEG 1500 system differ to some extent from those just discussed. The initial sharp drop in heat from pentane indicates that adsorption of PEG 1500 is initiated not only on the chemical heterogeneities but also on the geometrical ones. As anomalous sites are progressively filled by PEG 1500, they become unavailable for adsorption of eluates. The appearance of maxima and minima in the heat-coverage curves is not as well defined as those for adsorption on the Sterling FT-G + PEG 1500 system. The smoothness in the heat curves makes it clear that the deposition of liquid molecules does not proceed in a regular way. That is, the adsorption of PEG 1500 macromolecules on top of one another can get underway before the first layer is virtually completed. On the other hand, even on a highly homogeneous adsorbing surface multilayer adsorption of liquid molecules may be hindered to some extent when binding among the liquid molecules is too strong.

Let us take, for example, a homogeneous hydrophobic surface. That a surface is hydrophobic means, in simplest terms, that water stands on it in the form of drops and a large contact angle can be measured. On a molecular scale, then, one does not expect intuitively to find that adsorption of water molecules will proceed to form regular adsorbed layers. Instead of this, a more reasonable model of adsorption is that in which patches of an adsorbed monolayer are initiated at some kind of "active" centers of the surface, that these patches grow, and then finally merge by virtue of lateral hydrogen bonding between adsorbed water molecules. Even if a surface were completely free of inhomogeneities, the very first molecules of water adsorbed could act as active centers for further adsorption of water.

The incidence of strong binding among liquid molecules in determining the mode of deposition of the modifying agent was stressed by Bruner et al. [44]. They reported adsorption heat curves for saturated hydrocarbons on both Sterling FT-G + squalane and Sterling FT-G + glycerol adsorbing systems (Fig. 7). From the shape of the heat curves, it is clear that adsorption of squalane, which is a C_{30} saturated hydrocarbon, takes place on GCB as multilayer adsorption. This is not the case when a very polar compound, such as glycerol, is added to a nonpolar surface, such as that of Sterling FT-G.

Incomplete stretching of long-chain macromolecules may also be the cause of a certain disorder in the deposition of the modifying agent. In Table 1, monolayer capacities of increasing molecular weight of polymers (polyethylene glycol) deposited on a graphitized carbon black experimentally obtained are compared with the theoretically calculated monolayer capacity [39]. It is apparent that the possibility of formation of a regularly ordered layer is greatest in the case of the lowest molecular weight PEG.

In summary, that the deposition of the modifying liquid agent occurs as a multilayer adsorption on the solid surface is proved by (a) the absence

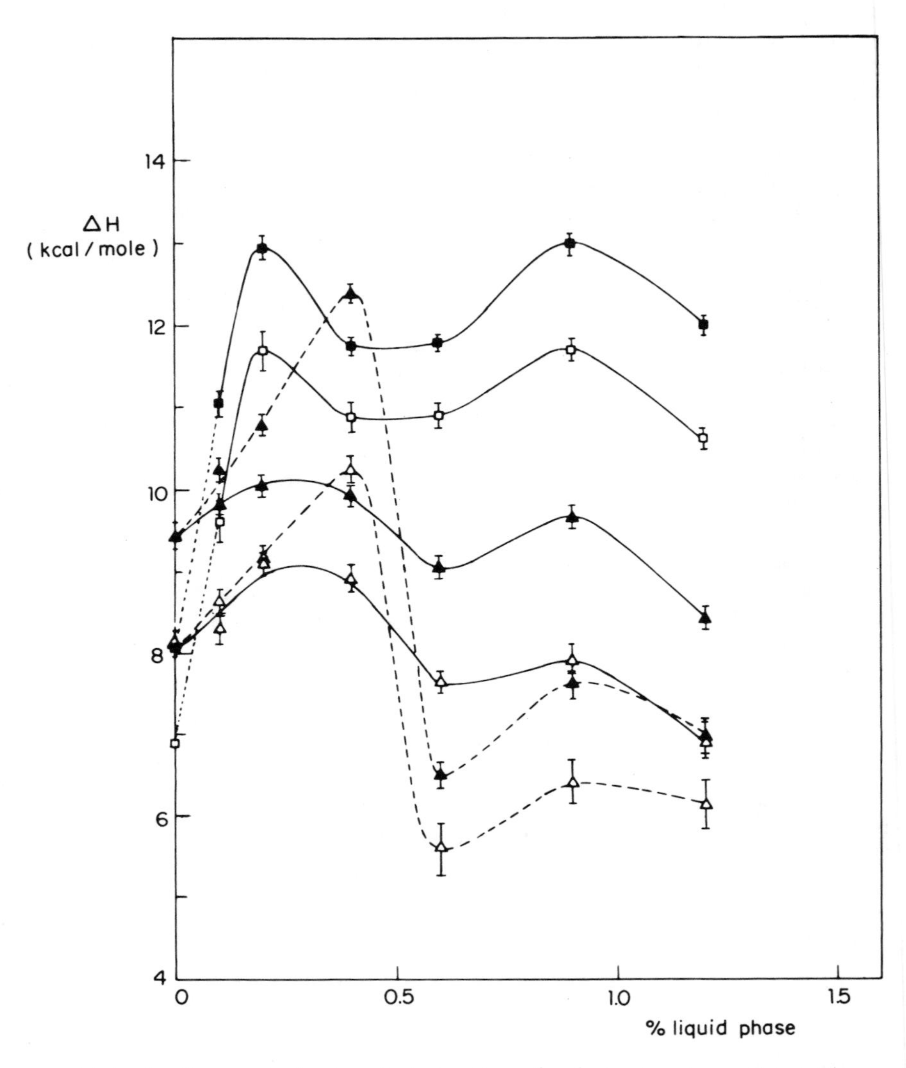

FIG. 7. Isosteric heats of adsorption of n-butane (△), pentane (▲), ethanol (□), and propanol (■) on Sterling FT-G modified with various percentages of squalane (dashed lines) and glycerol (solid lines) [44]. (Reproduced by permission of Analytical Chemistry.)

TABLE 1

Capacities of Dense Monolayers of PEG Deposited on Graphitized Channel Carbon Black

Modifying agent	Capacity of dense monolayer (mg/m^2)		
	Calculated from the van der Waals dimensions of the monomer unit of the macromolecule	Determined from adsorption isotherms from solution data	Determined from gas chromatograpic data
PEG 300	0.44	0.43	0.42
PEG 3000	0.44	0.60	0.57
PEG 15,000	0.44	0.74	0.73

of an initial steep drop in heat for eluates, (b) a well-defined maximum at $\theta \sim 1$, (c) an almost vertical fall in heat at $\theta = 1$, and (d) the appearance of small but real maxima in heat at $\theta \sim 2$ and $\theta \sim 3$.

In conclusion, it can be said that plotting adsorption heats of eluates as a function of the surface coverage of the modifying agent is an effective procedure for obtaining information on the mechanism of adsorption on liquid-modified solids. In addition, these heat curves can provide useful indications on the mode of deposition of very large molecules, providing a careful choice of the eluate is made. As a general rule, the eluate suitable to this purpose must display an affinity with the solid surface that is much higher than that of the liquid. In this way, even at high surface concentrations of liquid molecules, adsorption of the eluate can still occur on the remaining unoccupied patches of the solid.

Additional information can also be obtained by plotting differential standard entropy changes on adsorption of eluates versus the degree of surface occupation of the molecules of the liquid. An example of this is given in Fig. 8 [43], which refers to differential standard entropy changes upon adsorption of pentane and ethanol from the gas phase to two GCB surfaces (Sterling FT-G and Graphon) modified by various amounts of PEG 1500. In all cases, a more or less steady increase in the entropy loss of adsorption is observed as the surface concentration of PEG 1500 is increased and before a macromolecular monolayer is completed. This effect can be explained by considering that the addition of the liquid phase progressively reduces thermal motions of adsorbates and decreases the number of possible configurations of the eluate-adsorbing solid system. As far as the Sterling FT-G + PEG 1500 + pentane ternary system is concerned, a steady

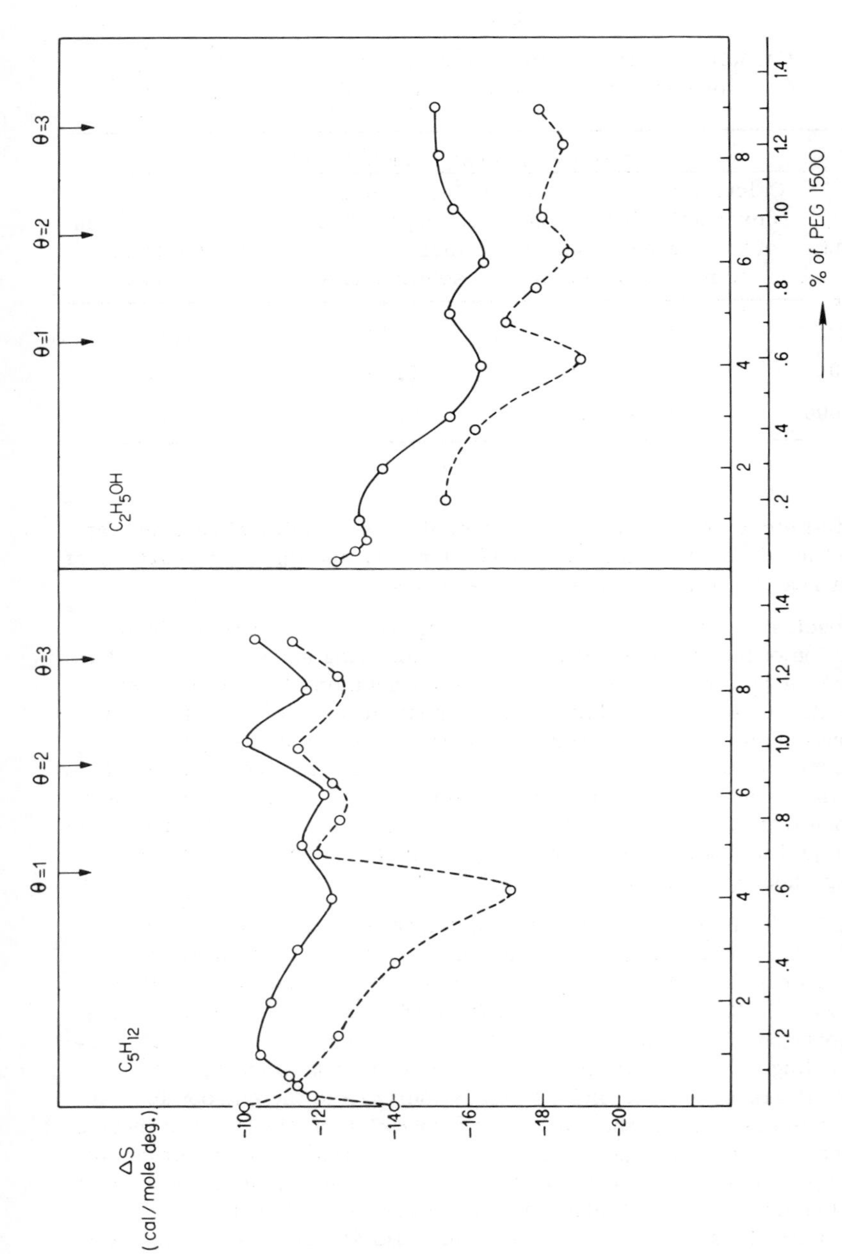

FIG. 8. Differential entropy changes at 89° C upon adsorption of pentane and ethanol versus the amount of PEG 1500 deposited on Graphon (upper scale abscissa) and Sterling FT-G (lower scale abscissa). Dashed lines refer to adsorption on Sterling FT-G + PEG 1500 and solid lines to adsorption on Graphon + PEG 1500 [43]. (Reproduced by permission of Analytical Chemistry.)

decrease in the adsorption entropy followed by a sharp increase can be noted. The relative, pronounced minimum may be interpreted in terms of a relatively small number of configurations of the system which are possible when a regularly packed monolayer of PEG 1500 nears completion. The sharp increase after the minimum indicates that the first layer of PEG 1500 is completed and adsorption of pentane takes place on top of this layer.

The adsorption of molecules with functional groups on the top of a polar monolayer is strongly localized. This behavior is accounted for by observing that, at the completion of the monolayer, the subsequent rise in adsorption entropy for ethanol is not high as compared to that for pentane.

As far as PEG 1500-modified Graphon is concerned, the initial increase in the adsorption entropy for pentane confirms the presence of some geometric irregularities on the surface of this adsorbing material, where the adsorption for both polar and nonpolar eluates is localized to some extent. Adding small amounts of a polar, nonvolatile liquid is sufficient to block both geometric and chemical irregularities. This initial, local concentration of the modifying agent on the most active sites is responsible for some irregularities in the deposition of the liquid phase, as it is shown by the smoothness of entropy curves and particularly by the fact that the first minimum is not very pronounced for both pentane and ethanol.

From a chromatographic point of view, some general considerations can be made on the role played by preadsorbed, nonvolatile molecules on a quasi-homogeneous adsorbing surface. At very low surface concentrations, the modifying agent acts mainly as a "tailing reducer"; i.e., it deactivates anomalous sites of the surface. By increasing the degree of surface occupation, mutual attractions between eluate molecules and those of the liquid merge via lateral interactions. At this stage, the liquid can act as a "selectivity modifier," since the original mechanism of adsorption appears to be remarkably modified. At the same time, the liquid may also behave as a "retention volume reducer" because of the loss of solid surface area that invariably accompanies the addition of a liquid in the submonolayer region. At the surface concentration needed to build up one densely packed monomolecular layer of nonvolatile molecules, adsorption of eluates occurs on top of this layer whose selectivity characteristics depend to a large extent on the activity of the solid surface. Further addition of the modifying agent can give rise to a film many molecular diameters thick. Thus, retention of eluates along such a chromatographic column is due to the combined effects of sorption into the multimolecular film and adsorption on its outer layer. Obviously, partitioning properties of this film are different from those of the bulk liquid, since forces exerted by the solid surface and by the outside monomolecular layer [45-47] provoke orientation as well as electrical and magnetic modifications of the intermediate layers. The extent of such effects depends on the surface activity of the adsorbing medium, and the nature and relative amount of the liquid phase added to the solid.

VI. CHOICE OF THE LIQUID PHASE IN GAS-LIQUID-SOLID CHROMATOGRAPHY

When one is resolved to use GLSC in order to obtain or more simply to improve the separation of a certain mixture of compounds, given a solid surface the first step in finding the optimum column is the selection of the best liquid phase. In many cases, the need for blocking solid surface heterogeneities is the conditioning factor in the choice of the liquid phase. Basically, the aim of deactivating the solid surface can be reached by using

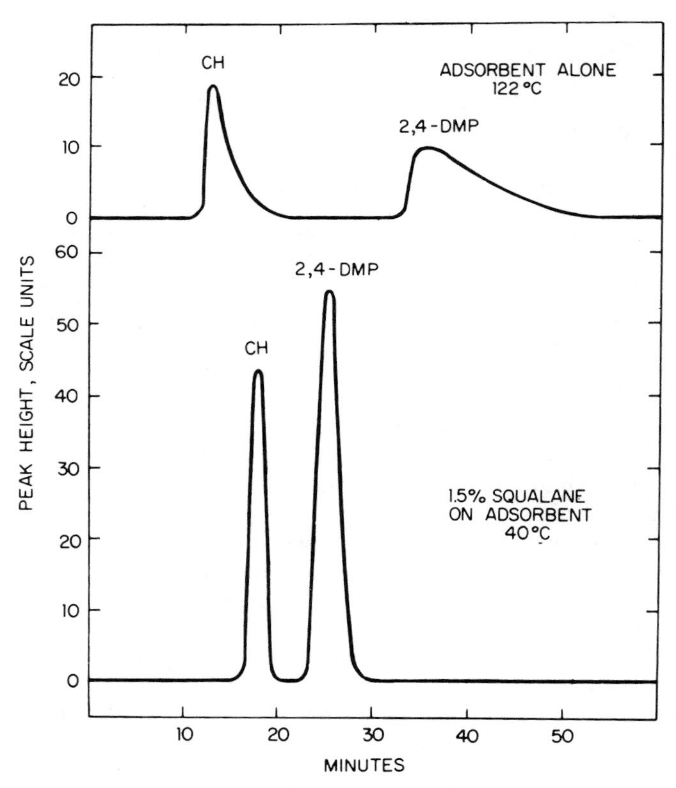

FIG. 9. Effect of adding squalane to Pelletex carbon in the elution of cyclohexane (CH) and dimethylpentane (DMP) [26]. (Reproduced by permission of Analytical Chemistry.)

a liquid phase that is itself capable of strong interactions with the "hot surface sites."

Where geometric, nonspecific heterogeneities are the only ones responsible for peak tailing, the addition of a nonpolar liquid suffices to rule out the influence of active centers in the chromatographic process. Two meaningful examples are given in Figs. 9 [26] and 10 [48].

Chemical heterogeneities are responsible for unwelcome effects during elution of molecules with a locally concentrated peripheral electron density. On a graphitic surface, chemical heterogeneities, which are composed primarily of oxygen complexes, vary in their chemical nature [6, 49]. Therefore, depending strictly on the chemical nature of the eluate,

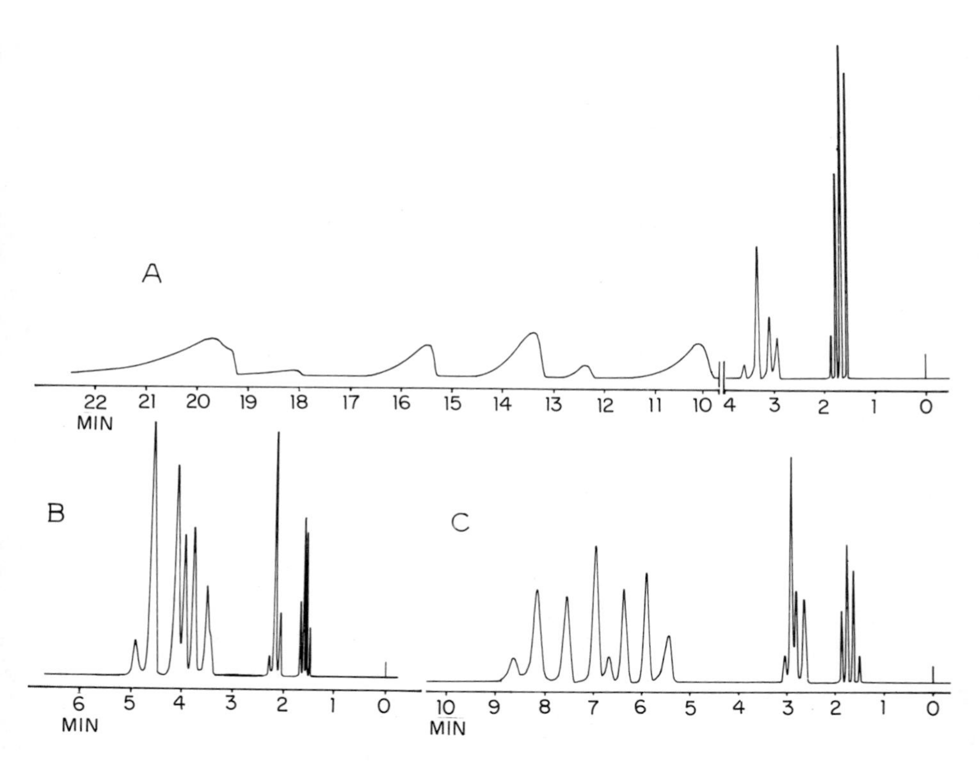

FIG. 10. Effect of adding nonpolar liquids to a graphitized carbon black (~70 m^2/g) in the elution of C_1-C_4 hydrocarbons. (A) Packed capillary column (PCC) with GCB alone; (B) PCC with GCB modified with n-decane; (C) PCC with GCB modified with 0.4% squalane [48]. (Reproduced by permission of Analytical Chemistry.)

various interactions may occur on these sites, ranging from a relatively weak dipole-induced dipole to a strong donor-acceptor coordination chemical bond. In this respect, the choice of the best deactivating liquid phase should be made on the basis of its own capability of being more strongly adsorbed on chemical heterogeneities than are given eluate molecules. A clarifying example is given by a surface of GCB shielded by a densely packed layer of polyethylene glycol molecules. Although there may be hydrogen bonding between terminal —OH groups of the preadsorbed macromolecules and surface oxygen complexes, aliphatic amines and fatty acids are eluted as

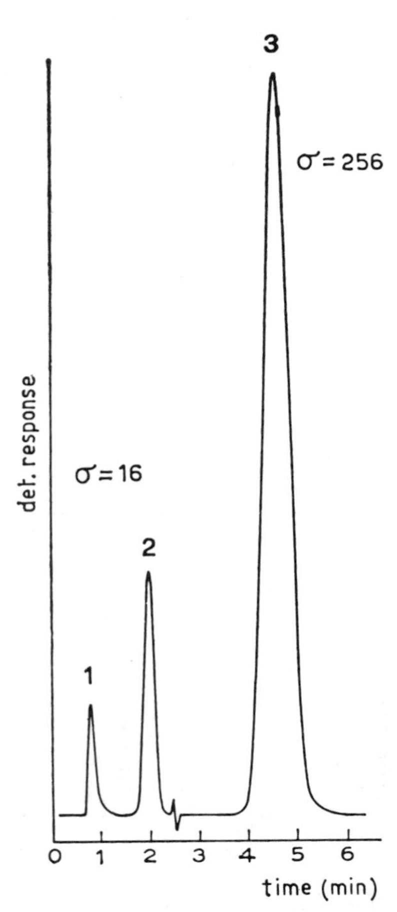

FIG. 11. Chromatogram showing the analysis of methylamine (1) and dimethylamine (2) in commercial trimethylamine (3). Column, Graphon (40-60 mesh) modified with 2% tetraethylenepentamine; sample size, 0.2 μl of a solution of 1% trimethylamine in water [50]. (Reproduced by permission of Analytical Chemistry.)

severely tailed peaks along such a column. This effect can be explained by assuming that only a segment of the macromolecule needs to desorb from the surface to accommodate a smaller molecule that is capable of giving rise to much stronger interactions with the acidic or basic fractions of the solid surface heterogeneities. Conversely, as shown in Figs. 11, 12, and 13

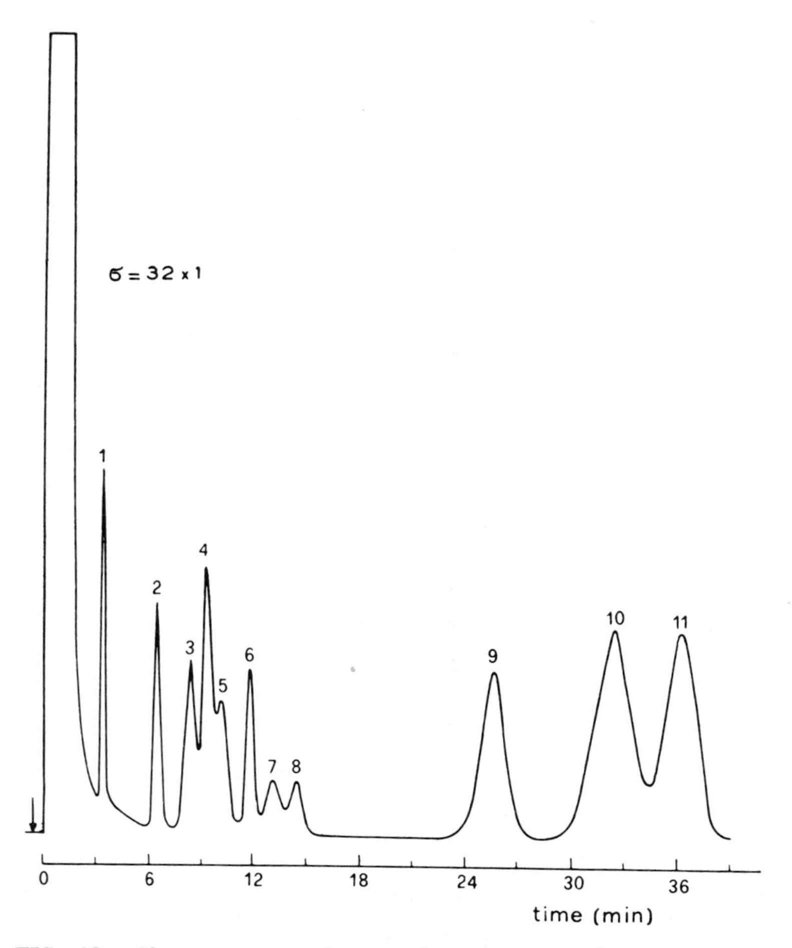

FIG. 12. Chromatogram showing the separation of some aromatic acids. Column, 1.3 m × 2 mm; 0.3% FFAP on Sterling FT-G (60-80 mesh); linear carrier gas velocity, 12 cm/sec; temperature, 230° C. 1, Benzoic; 2, o-toluic; 3, m-toluic; 4, salicylic; 5, p-toluic; 6, o-chlorobenzoic; 7, m-chlorobenzoic; 8, p-chlorobenzoic; 9, m-hydroxybenzoic; 10, p-hydroxybenzoic; 11, o-nitrobenzoic [51]. (Reproduced by permission of Analytical Chemistry.)

the use of liquid phases carrying suitable functional groups, such as tetraethylenepentamine and a high-molecular-weight acid (FFAP), results in linear elution of low-molecular-weight amines [50], acids [51], and diphenols [52], even at nanogram levels.

In order to obtain linear elution of the classes of compounds considered above, another useful way of blocking surface chemical inhomogeneities

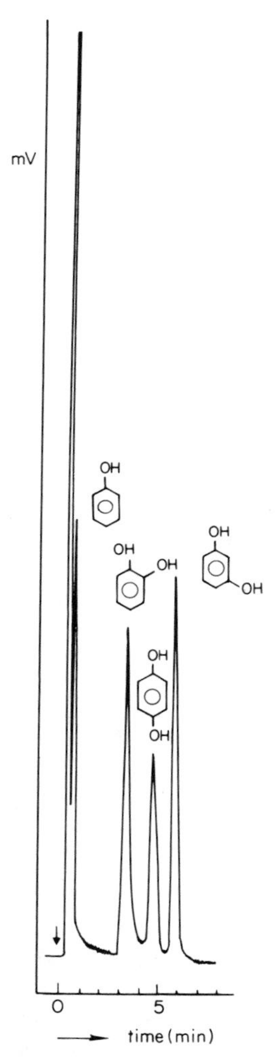

FIG. 13. Chromatogram showing the separation of diphenols. Column, Sterling FT-G (60-80 mesh) + 0.5% FFAP [52]. (Reproduced by permission of the Institute of Petroleum.)

can be that of neutralizing them by means of tiny amounts of strongly acidic or basic inorganic compounds, such as H_3PO_4 or KOH. Yet, the mere addition of such inorganic compounds to a bare surface of an adsorbing medium does not suffice to eliminate peak asymmetry completely. This effect demonstrates that although some of the active surface sites are basic or acidic in nature, there is another type of polar site capable of retarding polar eluates. Therefore, there is the need once again for choosing a liquid phase able to eliminate residual peak asymmetry. In Fig. 14 is shown the elution of C_2-C_5 fatty acids in an extremely dilute aqueous solution obtained by adding proper amounts of both H_3PO_4 and PEG 20 M to the graphitic surface [53]. As illustrated in Figs. 15, 16, and 17, by mixing tiny amounts of KOH to PEG 20 M [54] or to polyethyleneimine (PEI) 40 M [55], it is possible to take advantage of GLSC for the analysis of strong basic compounds. Also, because of the addition of GE-XE 60 to H_3PO_4-modified GCB [56], the determination of sulfur dioxide at the parts-per-billion levels can be performed in quite a short analysis time, as shown in Fig. 18.

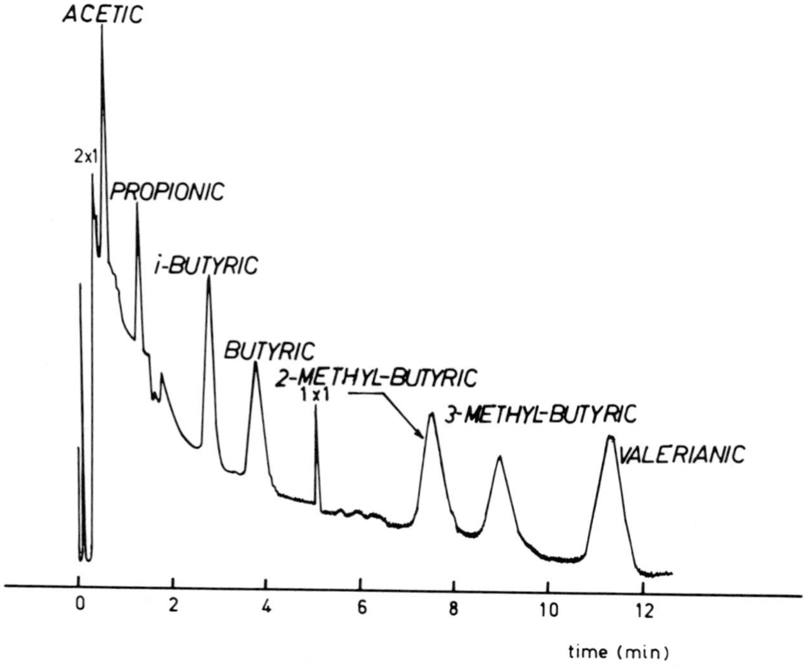

FIG. 14. Chromatogram showing the separation of free acids in aqueous solution. Column, 0.7 m × 4 mm; Graphon (60-80 mesh) + 0.5% H_3PO_4 + 3% PEG 20 M; linear carrier gas velocity, 8 cm/sec; concentration of each component, 0.3 ppm; temperature, 168° C [53]. (Reproduced by permission of Analytical Chemistry.)

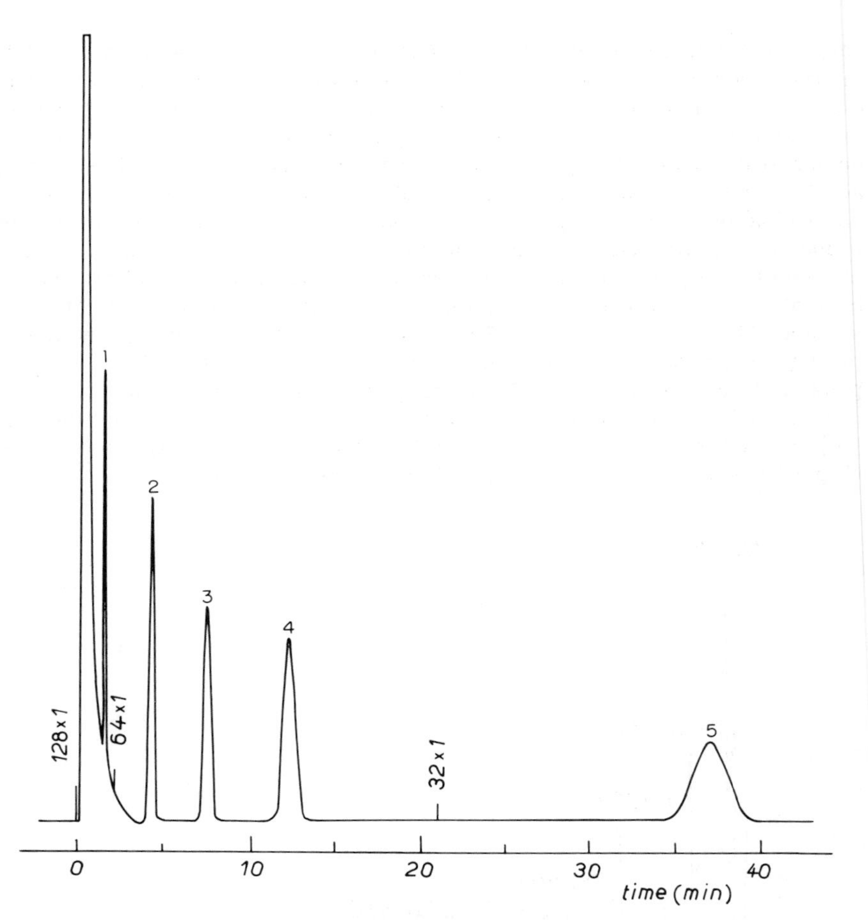

FIG. 15. Chromatogram showing the separation of some high-boiling, aliphatic amines. 1, Decylamine; 2, dodecylamine; 3, cyclododecylamine; 4, tetradecylamine; 5, hexadecylamine. Column, 1.4 m × 2 mm; Sterling FT-G (60-80 mesh) + 0.3% KOH + 1.3% PEG 20 M; temperature, 220° C; linear carrier gas velocity, 8.3 cm/sec; sample size, 1 μl containing about 80 ng of each component [54]. (Reproduced by permission of Analytical Chemistry.)

However, in some cases the need for modifying the selectivity characteristics of a GS column may be the decisive factor in the choice of the modifying agent. For example, a limiting factor in the use of a Graphon column for the fractionation of a mixture of C_1-C_4 hydrocarbons is the incomplete separation of cis-2-butene from butane in addition to some peak

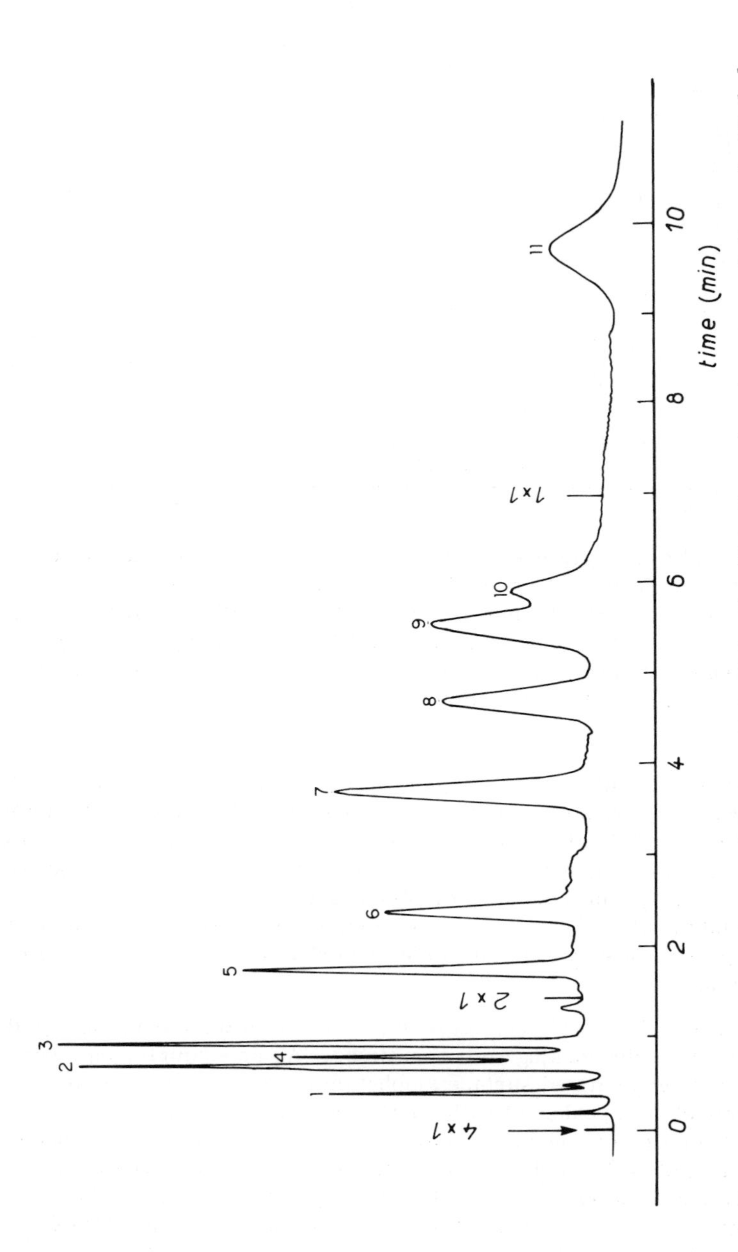

FIG. 16. Chromatogram showing the separation of C_1–C_4 aliphatic amines in water solution. 1, Methylamine; 2, dimethylamine; 3, trimethylamine; 4, ethylamine; 5, isopropylamine; 6, propylamine; 7, t-butylamine; 8, diethylamine; 9, s-butylamine; 10, isobutylamine; 11, butylamine. Column, 1.4 m × 2 mm; Sterling FT-G (60–80 mesh) + 0.2% KOH + 0.5% PEG 1500; temperature, 75° C; linear carrier gas velocity, 11 cm/sec; sample size, 1 μl containing about 2 ppm of each component [54]. (Reproduced by permission of Analytical Chemistry.)

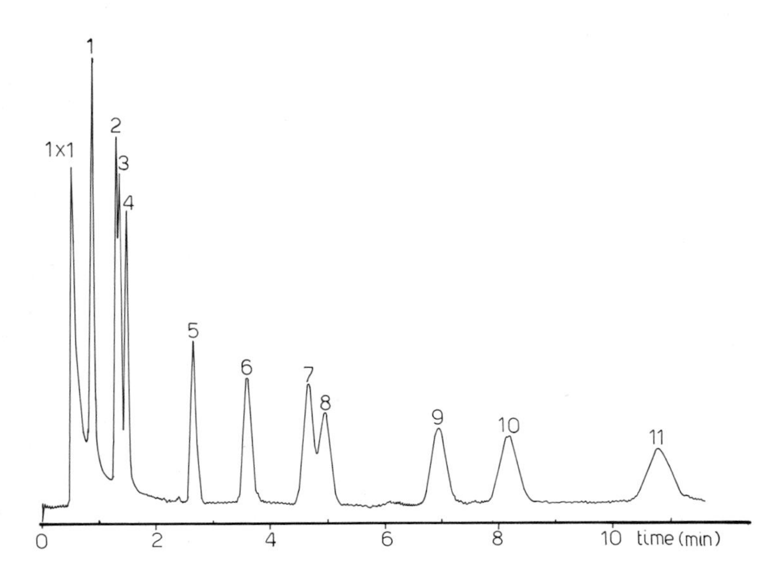

FIG. 17. Chromatogram showing the separation of C_1-C_4 aliphatic amines in water solution. Peak numbering as in Fig. 16. Column 1, 1.8 m × 2 mm; Carbopack C (80-100 mesh) + 0.3% KOH + 1% PEI 40 M; linear carrier gas velocity, 7 cm/sec; sample size, 1 μl containing about 1 ppm of each component; temperature, 41° C [55]. (Reproduced by permission of the Journal of Chromatographic Science.)

tailing. Although the addition of squalane as well as other modifying liquids eliminates peak tailing, no improvement is obtained in the separation factor for the cited pair. This aim can be achieved by partially shielding the carbon surface with polyethylene glycol molecules, as shown in the chromatogram of Fig. 19 [57]. In this case, dipole-induced dipole types of lateral interactions taking place between olefins and the modifying agent on the carbon surface provoke a slight but sufficient retardation of cis-2-butene with respect to butane.

In other cases, the decisive factor in the choice of the modifying liquid may be the need for decreasing retention times. As an example, on a bare hydrogen-treated graphitized surface amphetamine-type compounds are eluted as symmetrical peaks but with high retention times. Figure 20 illustrates the elution of some amphetamines performed within a reasonable elution time at quite a low column temperature by coating H_2-treated Sterling FT-G with 2% of a nonpolar liquid, such as Dexsil [24]. A further example is given in Fig. 21, which shows the separation of some complex deuterated and tritiated hydrocarbon mixtures, performed on 0.2% squalane-modified Sterling FT-G [58]. In this case, again, the addition of liquid has the main purpose of decreasing the analysis time since the selectivity power of these compounds during adsorption on GCB is sufficiently high [58-60].

The few examples just reported make it clear that in GLSC the choice of the best modifying liquid should not be made only on the basis of its ability to yield symmetrical peaks through the blocking of solid surface inhomogeneities but rather by taking into account any effect that is involved in the deposition of the liquid. When a certain liquid is selected, the proper amount of the liquid to be added to the adsorbing medium is the second step in obtaining the optimum GLS column for the resolution of a given mixture of substances.
given mixture of substances.

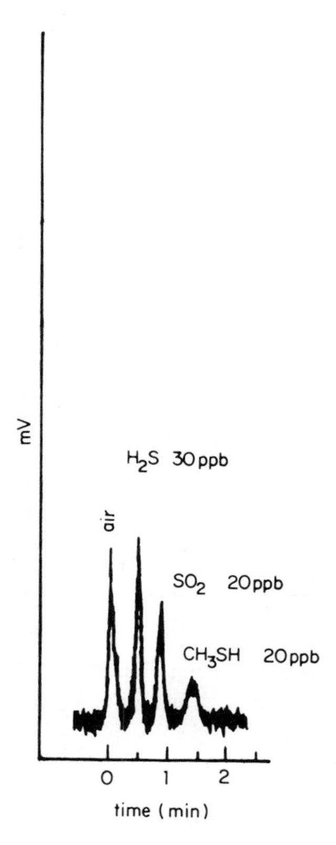

FIG. 18. Chromatogram showing the elution of sulfur compounds in air. Column, 0.8 m × 4 mm; Graphon (40-60 mesh) + 0.7% H_3PO_4 + 0.7% XE-60; room temperature; carrier gas N_2; flow rate, 100 ml/min [56]. (Reproduced by permission of Analytical Chemistry.)

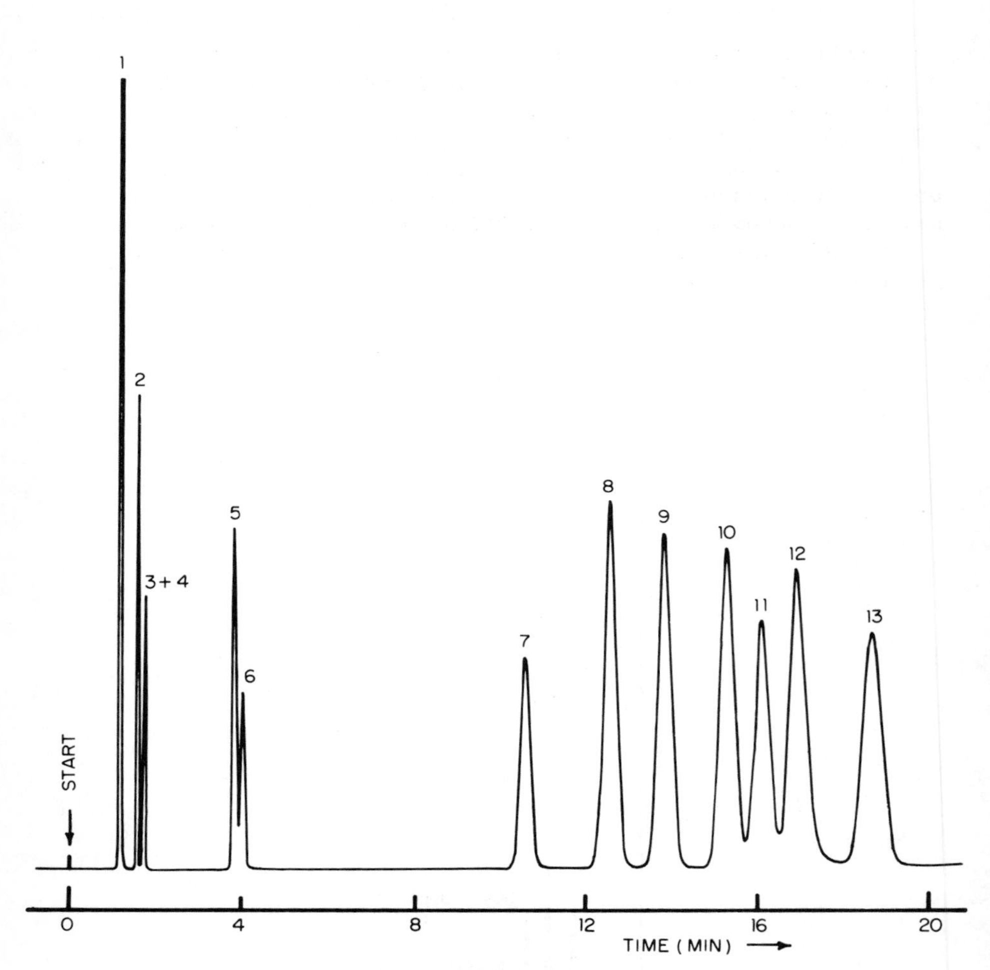

FIG. 19. Chromatogram showing the separation of C_1-C_4 hydrocarbons. 1, Methane; 2, ethylene; 3, acetylene; 4, ethane; 5, propene; 6, propane; 7, isobutane; 8, 1-butene; 9, isobutene; 10, butane; 11, cis-2-butene; 12, 1,3-butadiene; 13, trans-2-butene. Column, 3m × 2 mm; Carbopack B (100-120 mesh) + 2.8% PEG 1500; pressure drop, 4.9 kg/cm^2; temperature, 48° C [57]. (Reproduced by permission of the Journal of Chromatography.)

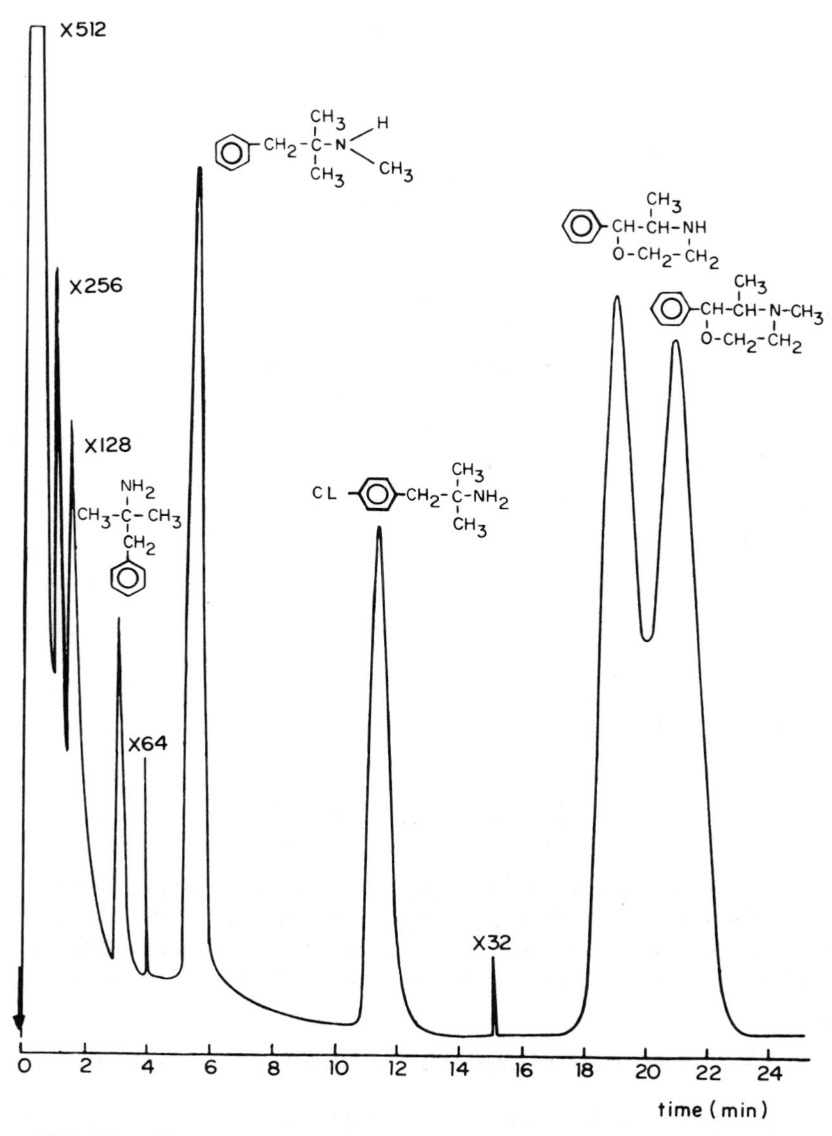

FIG. 20. Chromatogram showing the separation of some amphetamine-type compounds. Column: 0.8 m × 3 mm; hydrogen-treated Sterling FT-G + 2% Dexsil; temperature, 110° C [24]. (Reproduced by permission of the Journal of Chromatography.)

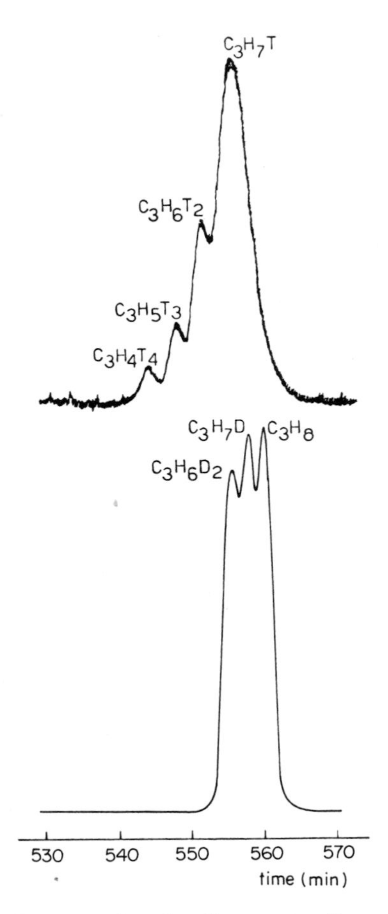

FIG. 21. Chromatogram showing the separation of tritiated and deuterated propanes. Column, 105 m × 2mm; Sterling FT-G + 0.2% squalane; pressure drop, 10.5 kg/cm^2; temperature, 16° C; carrier gas, hydrogen [60]. (Reproduced by permission of Analytical Chemistry.)

VII. EFFECT OF LIQUID LOAD ON RETENTION TIME

The experimentally determined retention time t_R for a given eluate is obtained by subtracting the time spent by the eluate in the mobile gas phase from the total time spent in the column. The retention time is equal to the average stay time of a molecule per collision with the sorbent, times the average number of such collisions as the molecules move along the column. Obviously, the number of collisions during the adsorption process is directly proportional to the surface development of the solid medium.

Let us assume that contributions to the average stay time of the eluate arising from attractive forces exerted by the preadsorbed, nonvolatile molecules are negligible. In this connection, then, the first effect of submonolayer concentrations of a nonvolatile liquid on a homogeneous adsorbing medium is to decrease the retention time of eluates, because of the reduction of the specific area available to a gaseous adsorbate. Under the same conditions, as the adsorbing medium is more porous, the rate of the decrease of retention time is higher. The submicroscopic roughness is responsible for this effect. The relatively large molecule of the preadsorbed liquid can bridge depressions and consequently covers more area of the adsorbing medium than would be accounted for merely by its molecular planar projection.

Moreover, the reduction of the solid surface involves a decrease of the adsorption configurational entropy owing to a decrease in the number of possible ways of arranging molecules among the surface sites. Hence, there is an additional contribution to the decrease of retention of eluate which is particularly remarkable when the first layer of macromolecules nears completion [61].

Two examples of uninterrupted decrease of retention time with increasing liquid/solid ratio are shown in Figs. 22 [27] and 23 [24]. It may be seen that in both cases retention times decrease sharply as small amounts of liquid are added to the carbon surface up to a certain percentage of liquid. In the case of the squalane + Pelletex system, this amount corresponds to the calculated quantity of squalane required to coat the adsorbent with a

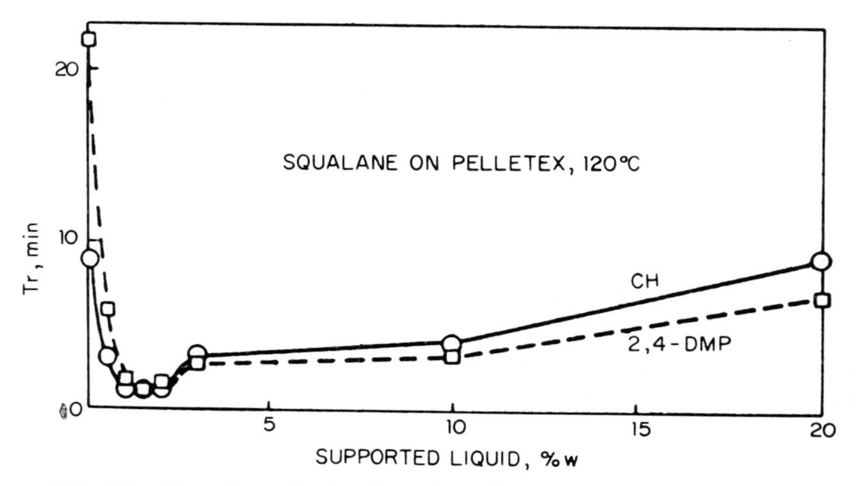

FIG. 22. Variation of retention time of cyclohexane and 2,4-dimethylpentane as a function of percentage of squalane on Pelletex at 120° C [27]. (Reproduced by permission of Analytical Chemistry.)

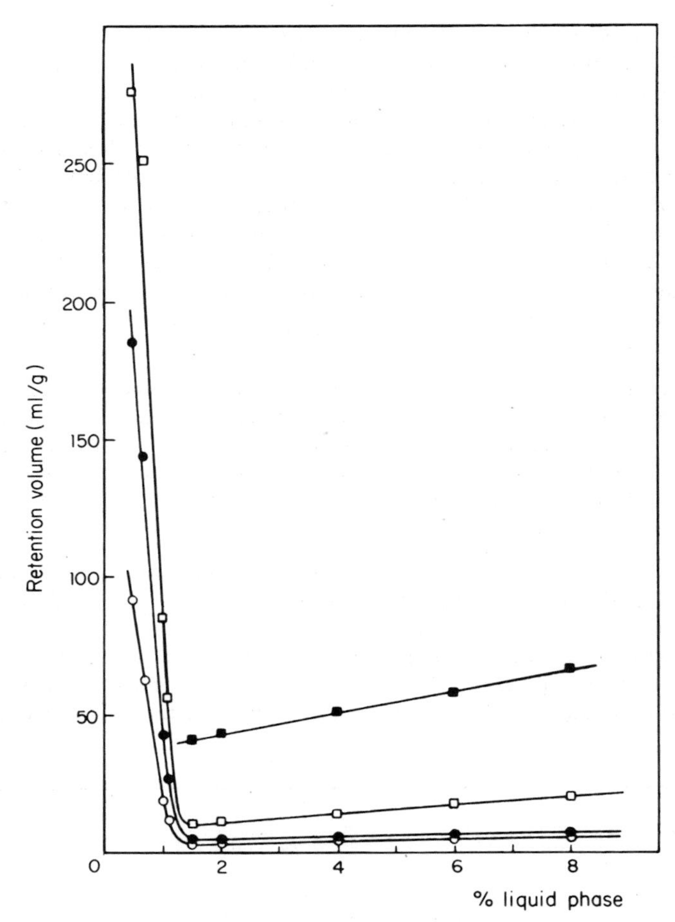

FIG. 23. Variation of the specific retention volume of various compounds as a function of percentage of Dexsil on H_2-treated Sterling FT-G. $C_6H_5N(CH_3)_2$, o; $C_6H_5N(C_2H_5)_2$, •; $(C_6H_5)_2O$, □; $(C_6H_5)_2NH$, ■ [24]. (Reproduced by permission of the Journal of Chromatography.)

close-packed monolayer of solvent, by taking 200 $\mathring{A}^2$ as the molecular cross-sectional area of squalane and 23 m^2 as the specific surface of Pelletex. As soon as a monolayer is formed, no further decrease of the retention time is observed. On the contrary, a slight but steady increase takes place as the percentage of liquid is increased. In this region of liquid molecule concentration it can be safely supposed that some kind of sorption into the layers, which are stacked roughly parallel to each other, occurs.

When considerable lateral interactions between eluate molecules and preadsorbed, nonvolatile molecules are operative, a more complex shape of retention-liquid surface coverage curves can be observed. As an example, in Fig. 24 is shown the decrease of retention volume of hexane as the amount of PEG 600 added to Graphon is increased. The dashed line shows the presumable rate of retention volume decrease if lateral interactions were inoperative [62]. In Fig. 25 are shown plots of a chromatographic parameter, i.e., the capacity ratio k, defined as the ratio between the corrected retention volume and the dead volume, versus the surface coverage of PEG 1500 molecules preadsorbed on both Sterling FT-G and Graphon [43]. These curves should be considered in connection with those reported in Figs. 6 and 8.

It should be pointed out that since the chromatographic parameter k is related to the free energy of the process of adsorption, plotting k values is the most direct, chromatographic way for obtaining meaningful information on the adsorptive modifications for gaseous adsorbates caused by the preadsorption of molecules of a liquid. As an example, let us examine the k curve for butanol adsorption on modified Graphon. Initially, filling chemical and geometric surface heterogeneities, where the adsorption heat assumes the greatest values, by preadsorption of liquid molecules, involves an increase in free energy of adsorption from the gas phase with respect to adsorption on the bare surface of the solid. This accounts for the initial decrease in retention for butanol. As the liquid/solid ratio increases, so does the retention of butanol since the adsorption heat increases by lateral

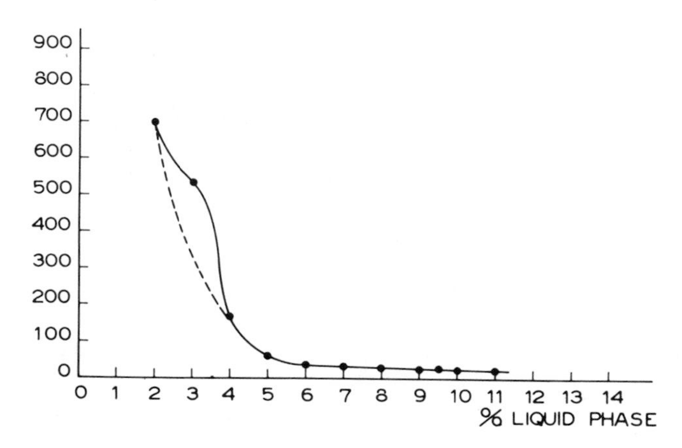

FIG. 24. Variation of specific retention volume at 60° C of hexane as a function of percentage of PEG 600 on Graphon.

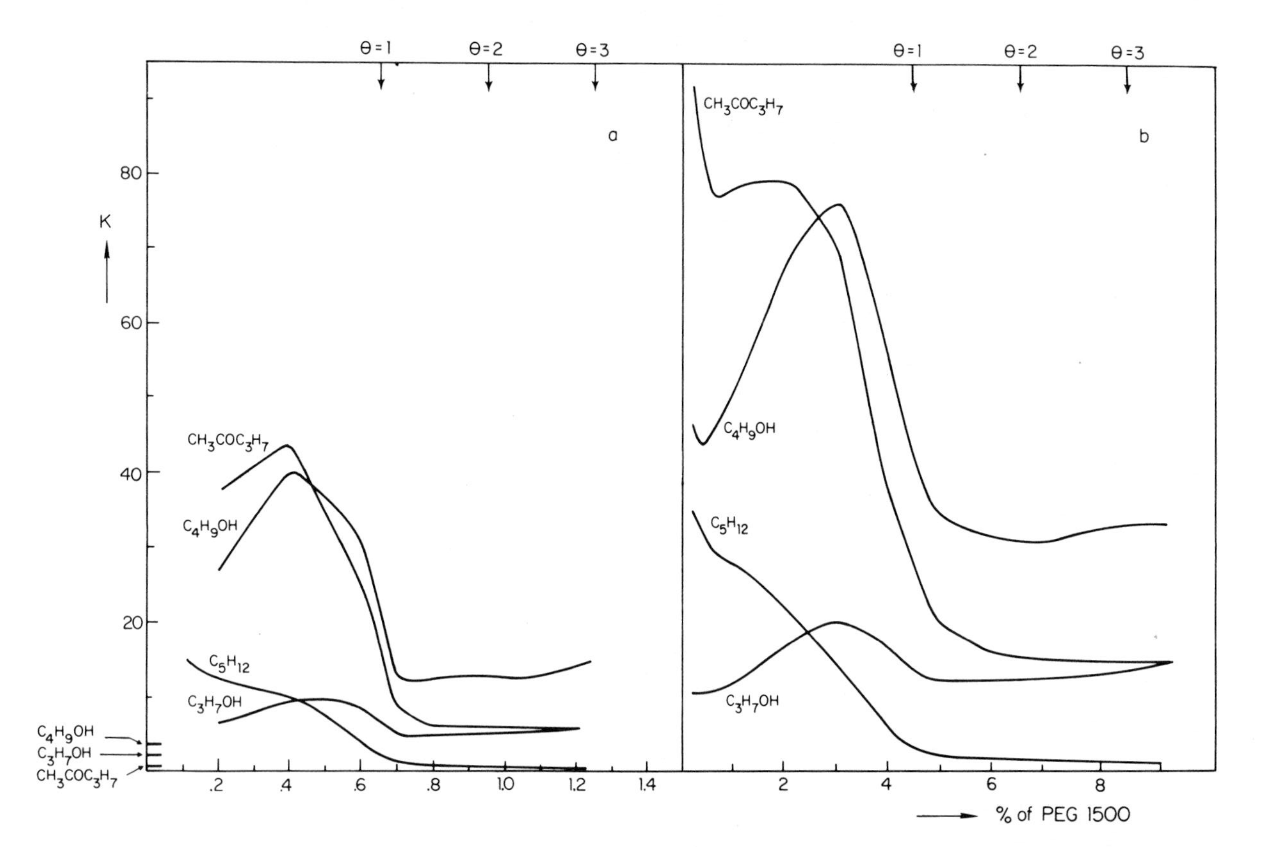

FIG. 25. Variation of capacity ratios at 89° C of some compounds as a function of percentage of PEG 1500 on (a) Sterling FT-G and (b) Graphon. Capacity ratios at the same temperature on a gas-liquid column of PEG 1500 are indicated as short lines [43]. (Reproduced by permission of Analytical Chemistry.)

interactions displayed between the adsorbed eluate molecules and the functional groups of PEG 1500. By reducing the solid surface further, a sharp fall in the retention of butanol is observed, though the adsorption heat is still increasing. This behavior is rationalized by assuming that, in this region of surface coverage, a sharp decrease of the adsorption configurational entropy takes place. After one layer of PEG 1500 is completed, the capacity ratio of butanol tends to increase slightly; this effect is probably caused by the contribution of sorption among the deposited layers of PEG 1500. Capacity ratio-coverage curves for the other compounds on Graphon as well as for adsorption on the Sterling FT-G + PEG 1500 system have a similar trend. The observed differences can be accounted for in terms of different strengths of the mutual attractions between the liquid phase and eluate molecules in the former case, and to the higher degree of surface homogeneity in the latter case.

By comparing these plots with that reported in Figs. 22 and 23, it appears that beyond a certain liquid surface coverage these curves resemble each other. However, in some cases unexpected effects of liquid load on retention may be observed. An example is given in Fig. 26, which shows capacity ratio-coverage curves for some compounds eluted on Sterling FT-G modified with both squalane and glycerol [44]. As can be seen, whereas the behavior of hydrocarbons on squalane-modified GCB is a typical one, an uninterrupted drop in retention is observed when the same compounds are adsorbed on GCB modified with increasing amounts of glycerol. In contrast, on this adsorbing system, a continuous increase in retention as the liquid is increased is observed for alcohols. The absence of the minimum in the retention for hydrocarbons can be explained assuming that, at least up to the surface coverage examined, the carbon surface is not yet completely shielded by glycerol molecules. The behavior of alcohols, on the other hand, can be explained by assuming that they are preferentially adsorbed on preadsorbed glycerol molecules.

In conclusion, some general considerations arising from the observation of the plots reported can be made on the effect of the liquid load on retention. Generally, in the submonolayer region of liquid concentration, increasing the amount of liquid does not always involve a corresponding decrease of eluate retention. In extreme cases, a GLS column can directly behave as a GL column in that the retention time continuously increases as the amount of liquid is increased. Moreover, in GLSC it is not possible to define a stereotyped shape of retention-liquid coverage curves. From this it follows that in GLSC the choice of the best amount of a given modifying liquid should also be made by a detailed evaluation of the influence of liquid load on retention time with a view to optimizing gas-chromatographic analysis.

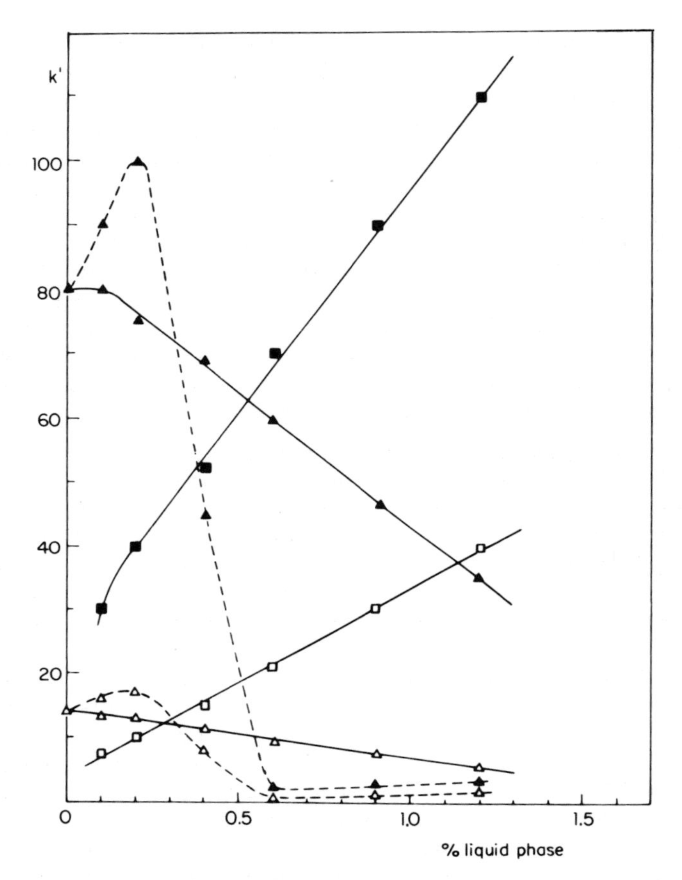

FIG. 26. Capacity ratios at 46° C of some compounds as a function of percentage of squalane (dashed lines) and glycerol (solid lines) on Sterling FT-G: butane (Δ), pentane (▲), ethanol (□), propanol (■) [44]. (Reproduced by permission of Analytical Chemistry.)

VIII. EFFECT OF LIQUID LOAD ON COLUMN SELECTIVITY

The variation of the separation factor of a given pair of substances with the variation of the liquid/solid ratio has been recognized as the most characteristic feature of GLSC since its early applications [63-65]. A comprehensive study of this effect should be made by measuring differences of the thermodynamic quantities involved in the process of adsorption for some selected pairs of substances as a function of the surface concentration of the nonvolatile molecules. An example of this procedure is summarized in Fig. 27 [43]. Here, the $(\Delta G' - \Delta G)/T$ term, which is equal to $R \ln \alpha$, where α is the separation factor, for the chosen pairs of compounds is graphically represented as resultants from the plots of the $\Delta S - \Delta S'$ and $(\Delta H' - \Delta H)/T$ terms, according to the Gibbs equation $\Delta G = \Delta H - T\,\Delta S$.

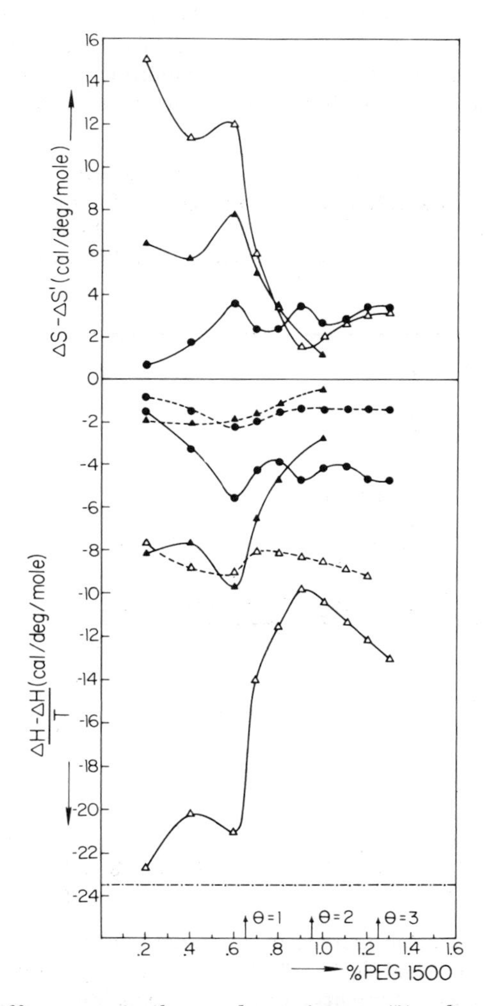

FIG. 27. Differences in thermodynamic quantity changes upon adsorption at 89° C of some isomeric pairs as a function of percentage of PEG 1500 on Sterling FT-G: 1-propanol + 2-propanol (Δ), 2-pentanone + 3-pentanone (●), 3-methylbutanal + 2-methylbutanal (▲) [43]. (Reproduced by permission of Analytical Chemistry.)

More simply, direct information on the effect of liquid load on column selectivity can be obtained by plotting the separation factor of a given pair of compounds as a function of the relative amount of the liquid added to an adsorbing medium. An example is given in Fig. 28, where separation factor-liquid surface coverage curves are reported for some selected pairs of organic acids eluted on Sterling FT-G modified with FFAP, which is a high-boiling, acid liquid [51]. For the sake of comparison, separation factors obtained by eluting the same pairs on an FFAP column are also reported. Modifications in the separation factors of acidic pairs as the FFAP amount is varied are meaningful in illustrating some features of the working mechanism of GLS columns.

Let us first examine the elution of chlorobenzoic acids. The acid liquid phase separates primarily according to differences in the acidic strength of the three isomers. Therefore, in GLS columns with relatively high liquid/solid ratios, the elution sequence is the same as in the GL

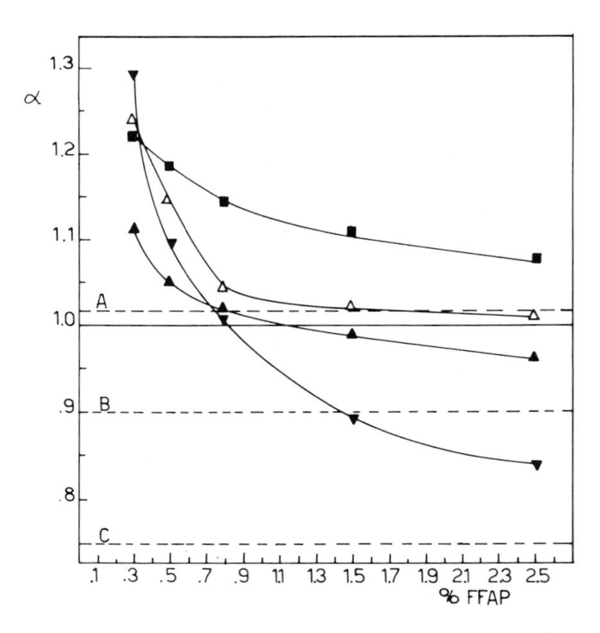

FIG. 28. Separation factors of some acid pairs as a function of percentage of FFAP on Sterling FT-G (60-80 mesh): p-toluic + m-toluic; 3-methylbutyrin + 2-methylbutyric; p-chlorobenzoic + m-chlorobenzoic; m-chlorobenzoic + o-chlorobenzoic. Dashed lines indicate the values attainable by the FFAP gas-liquid column. A, p-Toluic + m-toluic; B, p-chlorobenzoic + m-chlorobenzoic; C, m-chlorobenzoic + o-chlorobenzoic [51]. (Reproduced by permission of Analytical Chemistry.)

column, that is, para-, meta-, ortho-chlorobenzoic acid. As the relative amount of FFAP is lessened, the influence of the solid surface force centers in the process of adsorption gradually increases. At low surface coverages, the differences in local electron density distribution of acidic eluates are attenuated. Thus, differences in the geometric configuration and orientation relative to the adsorbing surface play a predominant role in the chromatographic process and reverse the order of elution.

The separation of p-toluic and m-toluic acid is a clear-cut example of the greater flexibility of GLS columns with respect to GL columns. As can be seen from Fig. 28, the elution sequence on the FFAP column for the acid pair considered is reversed with respect to differences in their acidic strength. Moreover, the separation factor attainable with such columns is poor. This may be explained by taking into account that the effect due to nonspecific interactions, which always occur between a polar liquid phase and a polar solute, acts in opposition to the effect due to specific interactions. From a theoretical point of view, then, the separation of the p-toluic, m-toluic acid pair could be improved by carrying out the elution on a nonpolar GL column. However, this procedure is impractical because of the effect of peak tailing which invariably accompanies the elution of strongly polar compounds on nonpolar liquid-coated chromatographic supports. In this case, the action of the carbon surface is to strengthen the effect due to nonspecific interactions, so that the separation is greatly improved. From a practical point of view, such an improvement permits complete separation in a short analysis time. In fact, it can be calculated that, by use of the FFAP column, an efficiency equal to about 30,000 plates is required to get a complete separation of the m-toluic and p-toluic acids at 213° C. At the same temperature and with a similar capacity ratio, a column filled with Sterling FT-G and 0.8% FFAP is able to yield the same resolution with an efficiency of only 2300 plates.

The separation of polar molecules with very slight differences in their electron density distribution is very difficult to carry out with GLC. However, if there are differences in either the geometric structure or polarizability, the separation of such pairs of substances can be made easier by a properly modified GCB, bearing in mind that the adsorption potential on this material depends mainly on the polarizability and geometric structure of the adsorbate. In the case of 2-methylbutyric and 3-methylbutyric acids, for example, any difference in retention times is really precluded when eluted on the FFAP column. On the other hand, by taking advantage of the difference in their geometric structure, the separation of this pair is made possible by the use of FFAP-modified GCB, as shown by the relative separation factor curve.

Generally, it can be stated that where the liquid and the adsorbing medium act in opposition to each other, and the latter is far more selective than the former increasing the amount of liquid causes a steep decrease in the selectivity power of GLS columns. In such cases, apart from other considerations, the best amount of a selected, modifying liquid is the minimum just necessary to block surface chemical inhomogeneities. Because of the use of Sterling FT-G modified with only 0.1% FFAP, a rapid separation of m-cresol from p-cresol can be obtained [66], as shown in Fig. 29. Also, by adding only a slight amount of PEG 1500 to Sterling FT-G, a partial resolution of 2-methylbutanal and 3-methylbutanal is made possible (Fig. 30) [43]. Further, as shown in Fig. 31, a fast gas-chromatographic

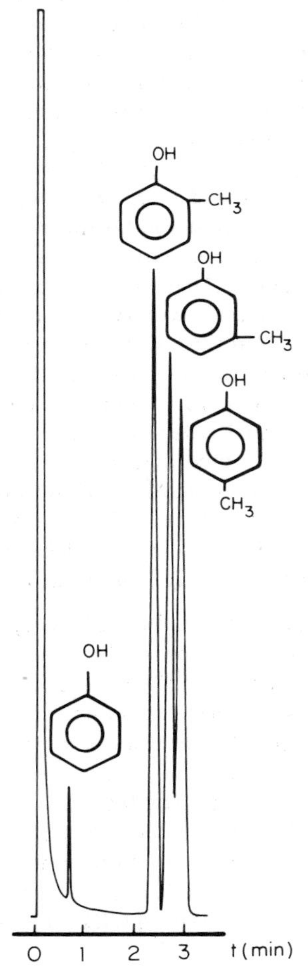

FIG. 29. Chromatogram showing the separation of cresols. Column, 1.7 m × 2 mm; Sterling FT-G (60-80 mesh) + 0.1% FFAP; temperature, 280° C; linear carrier gas velocity, 12 cm/sec [66]. (Reproduced by permission of the Journal of Chromatography.)

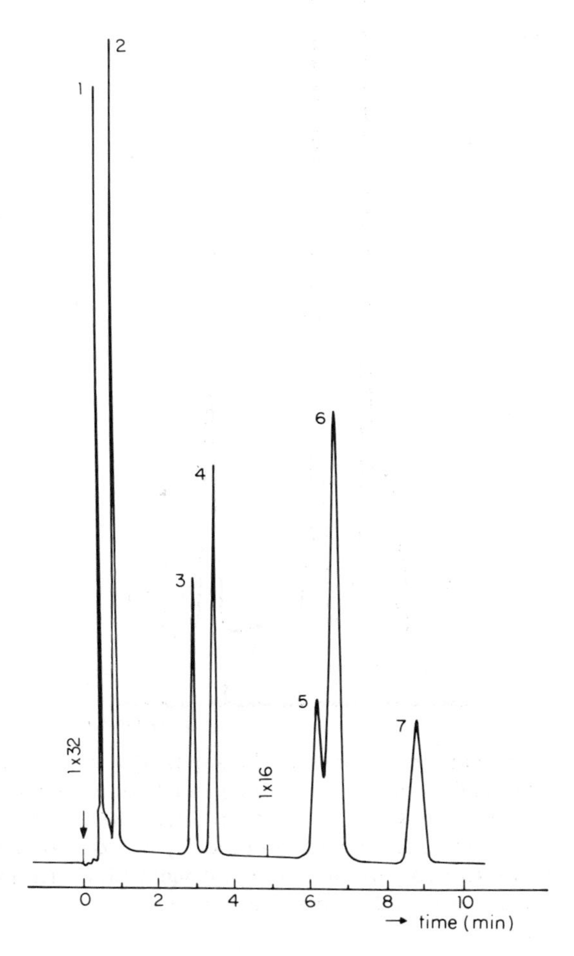

FIG. 30. Chromatogram showing the separation of a water solution of C_2-C_5 aliphatic aldehydes. Column, 1.4 m × 2 mm; Sterling FT-G (60-80 mesh) + 0.2% PEG 1500; temperature, 101° C; linear carrier gas velocity, 11 cm/sec; injected amount of each componet, ~30 ng. 1, Ethanal; 2, propanal; 3, 2-methylpropanal; 4, butanal; 5, 2-methylbutanal; 6, 3-methylbutanal; 7, pentanal [43]. (Reproduced by permission of Analytical Chemistry.)

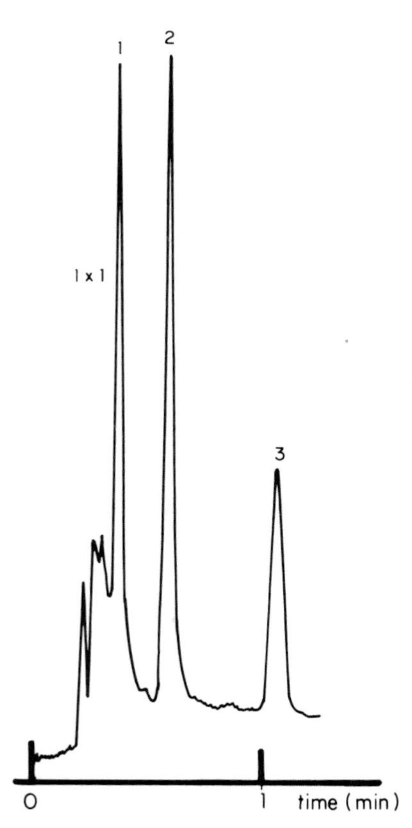

FIG. 31. Chromatogram showing the separation of methylamine (1), dimethylamine (2), and trimethylamine (3), in a water solution. Column, 1.4 m × 2 mm; Carbopack C (80-100 mesh) + 0.3% KOH + 0.5% PEI 40 M; temperature, 68° C; linear carrier gas velocity, 8.5 cm/sec; sample size, 0.8 μl containing about 0.5 ppm of each component [55]. (Reproduced by permission of the Journal of Chromatographic Science.)

analysis of an aqueous solution containing methylamine, dimethylamine, and trimethylamine, each in concentration as low as 0.5 ppm, can be performed using Carbopack C (Supelco Inc., Bellefonte, Penn.), which is an adsorbing material similar to Sterling FT-G, modified with 0.3% KOH and 0.5% PEI 40 M [55].

In other cases, the effect of the liquid can act in the same direction as that of the solid surface so that the column selectivity can be enhanced by properly choosing the liquid/solid ratio. Let us consider a pair of compounds made of a hydrocarbon and its deuterated derivative. Such an isotopic pair displays a slight difference in polarizability, which can be duly exploited by adsorption on a graphitized carbon surface with a view to attaining their separation. As shown in Fig. 32, in the submonolayer concentration as the amount of squalane is increased, so is the column selectivity [44]. This effect can be accounted for by considering that lateral interactions taking place between the liquid molecules and the two adsorbates via dispersion forces cause an increase in the difference of the adsorption free energies and give a maximum in the separation factor-liquid surface concentration curve which corresponds to the maximum intensity of lateral interactions.

The concept of mutual aid between the liquid and the adsorbing medium on column selectivity should not be misinterpreted. Let us take, for example, a pair of geometric isomers such as pentane and isopentane. Since the former is retained to a much greater degree than the latter on a squalane column as well as on a GCB column, an increase in the separation factor would be expected when squalane is deposited on the GCB surface. Conversely, as shown in Fig. 33, the separation factor for pentane and

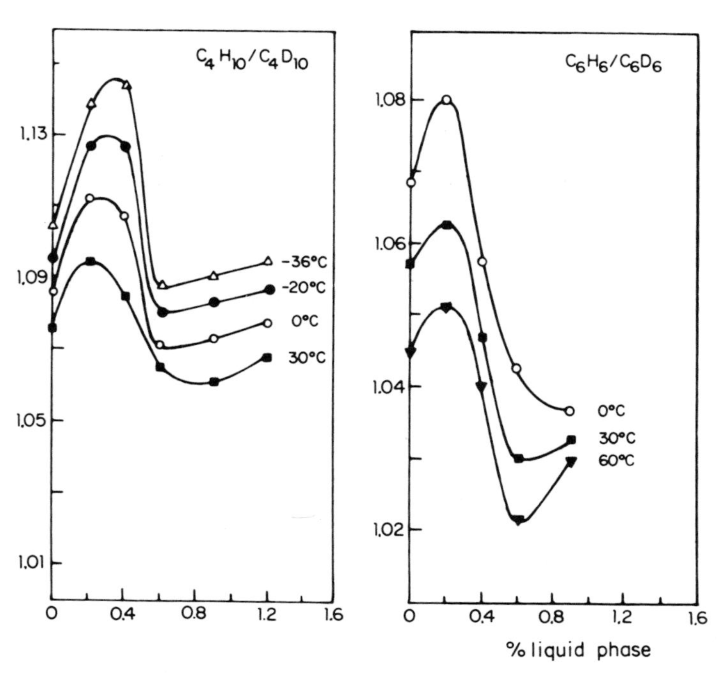

FIG. 32. Separation factors of two isotopic pairs as a function of percentage of squalane on Sterling FT-G [44]. (Reproduced by permission of Analytical Chemistry.)

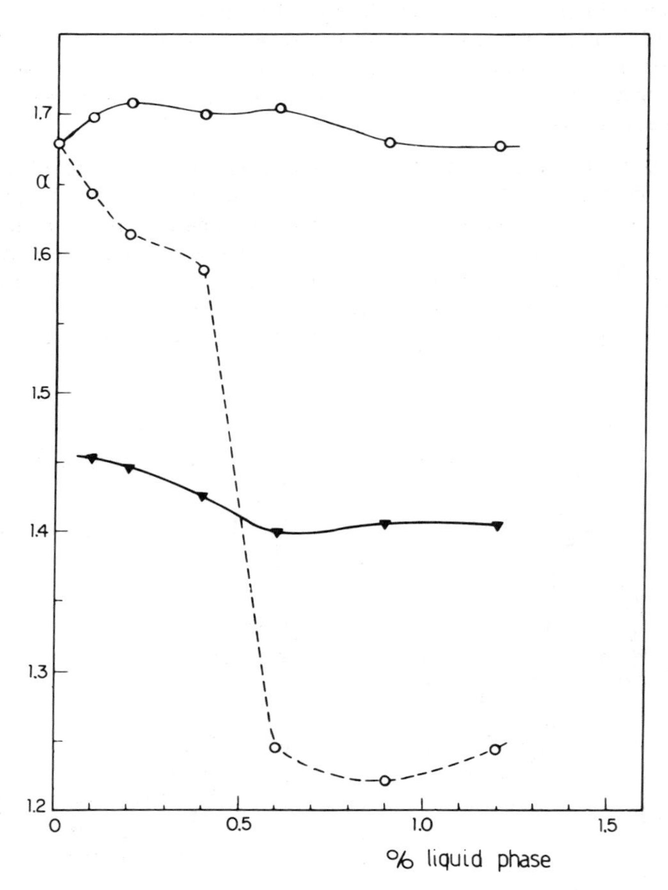

FIG. 33. Separation factors at 46° C of two isomeric pairs as a function of percentage of squalane (dotted lines) and glycerol (solid lines) on Sterling FT-G: pentane + isopentane (o), propanol + isopropanol (▼) [44]. (Reproduced by permission of Analytical Chemistry.)

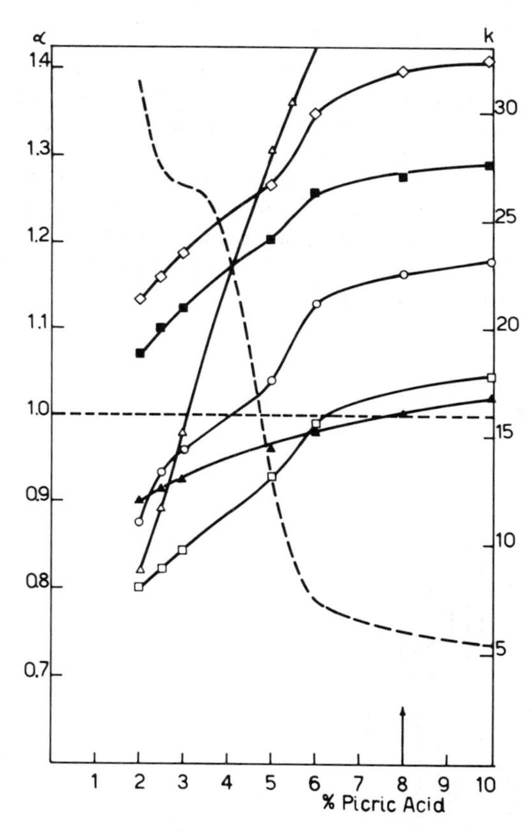

FIG. 34. Separation factors of some hydrocarbon pairs of interest (solid lines) and capacity ratio of trans-2-butene (dashed line) at 50° C versus the relative amount of picric acid added to Carbopack B. The arrow in the abscissa indicates roughly the completion of one layer of picric acid. Acetylene + ethane (Δ), ethylene + ethane (□), propene + propane (■), 1-butene + butane (○), isobutene + butane (◊), cis-2-butene + trans-2-butene (▲) [57]. (Reproduced by permission of the Journal of Chromatography).

isopentane decreases continuously as the amount of squalane is increased [44]. This effect can be explained by considering that to some extent lateral interactions cause modifications in the original orientation of the two adsorbates with respect to the basal graphite plane and definitively cause a decrease in the difference of the adsorption potential.

The example of squalane-modified GCB shows clearly that varying the liquid load involves contrasting effects on the column selectivity, depending on the particular pair of compounds to be separated. Hence, when dealing with the separation of a complex mixture by GLSC it may be difficult to make predictions as to the best liquid/solid ratio to yield the highest selectivity power. In Fig. 34, is shown a plot of separation factors for some light hydrocarbon pairs of interest obtained by modifying Carbopack B (Supelco Inc.), which is similar to Graphon, with varying amounts of picric acid, which is a strong π acid [57].

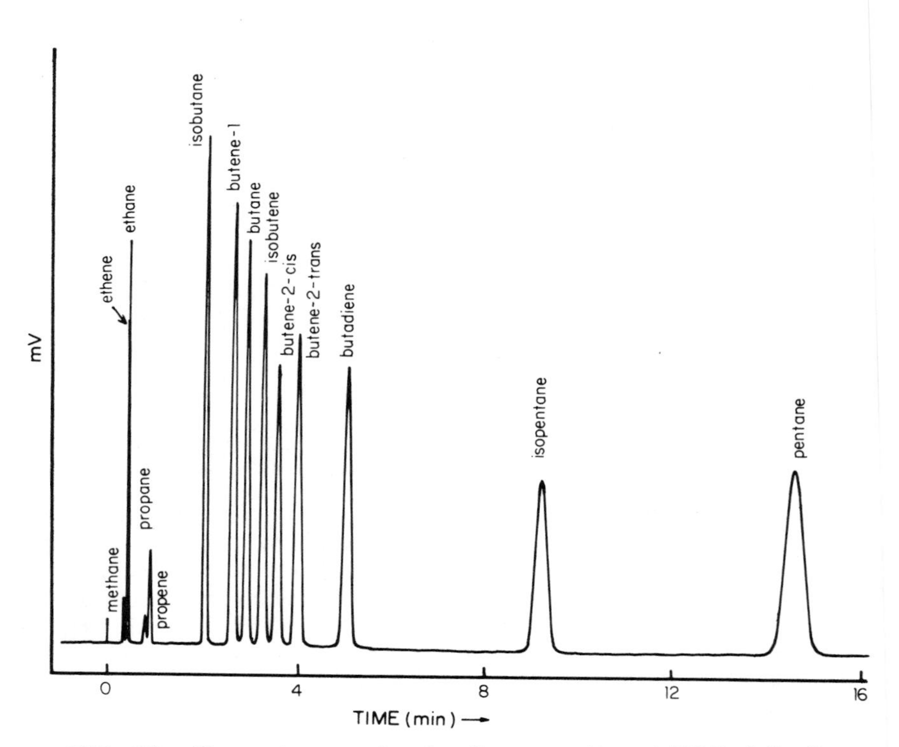

FIG. 35. Chromatogram showing the separation at 50° C of C_1-C_5 hydrocarbons. Column, 2.2 m × 2 mm stainless steel; Carbopack C (100-120 mesh) + 0.19% picric acid; pressure drop, 4.8 kg/cm^2 [67]. (Reproduced by permission of Analytical Chemistry.)

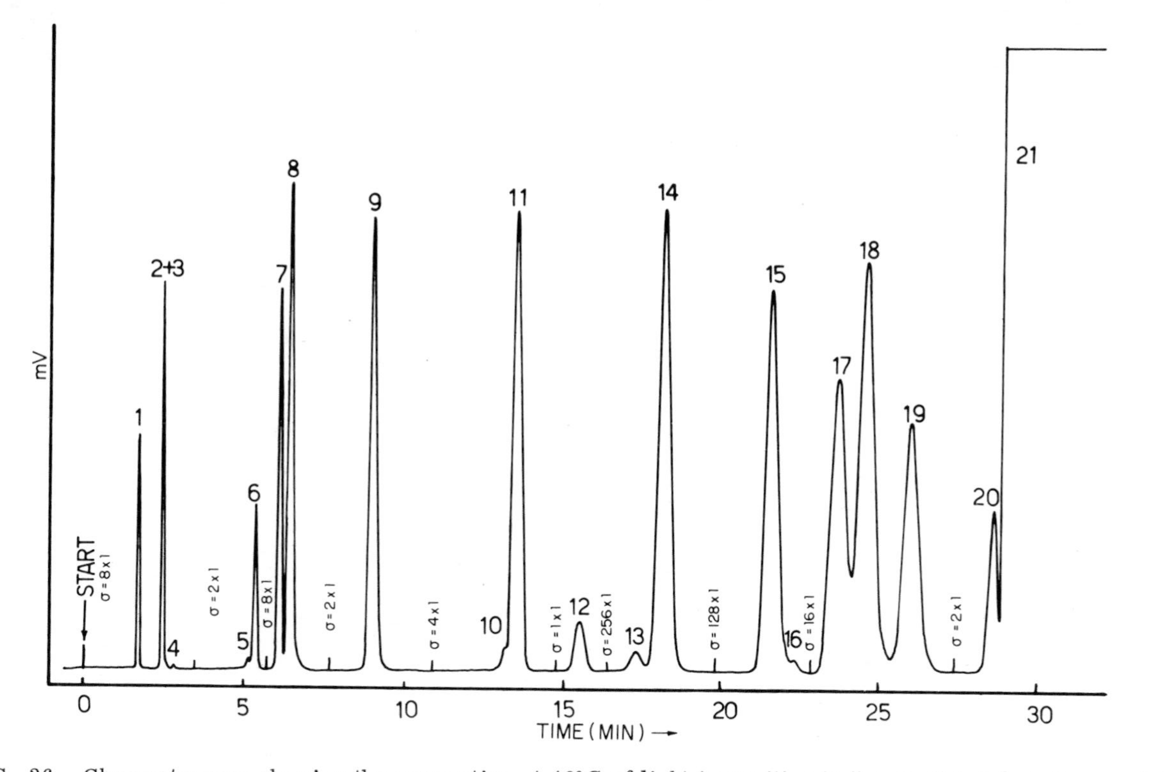

FIG. 36. Chromatogram showing the separation at 46°C of light impurities in "puram" 1,3-butadiene. Column, 4 m × 2 mm stainless steel; Carbopack B (100-200 mesh) + 4.8% picric acid; pressure drop, 4.9 kg/cm^2. 1, Methane (10 ppm); 2, ethylene (12 ppm); 3, ethane (7 ppm); 4, acetylene (0.3 ppm); 5, cyclopropane (0.4 ppm); 6, propane (5 ppm); 7, propene (46 ppm); 8, propadiene (70 ppm); 9, propine (13 ppm); 11, isobutane (61 ppm); 13, butane (210 ppm); 14, 1-butene (5500 ppm); 15, isobutene (2500 ppm); 17, cis-2-butene (270 ppm); 18, trans-2-butene (390 ppm); 21, 1,3-butadiene; 10, 12, 16, 19, 20 are unidentified peaks [57]. (Reproduced by permission of the Journal of Chromatography.)

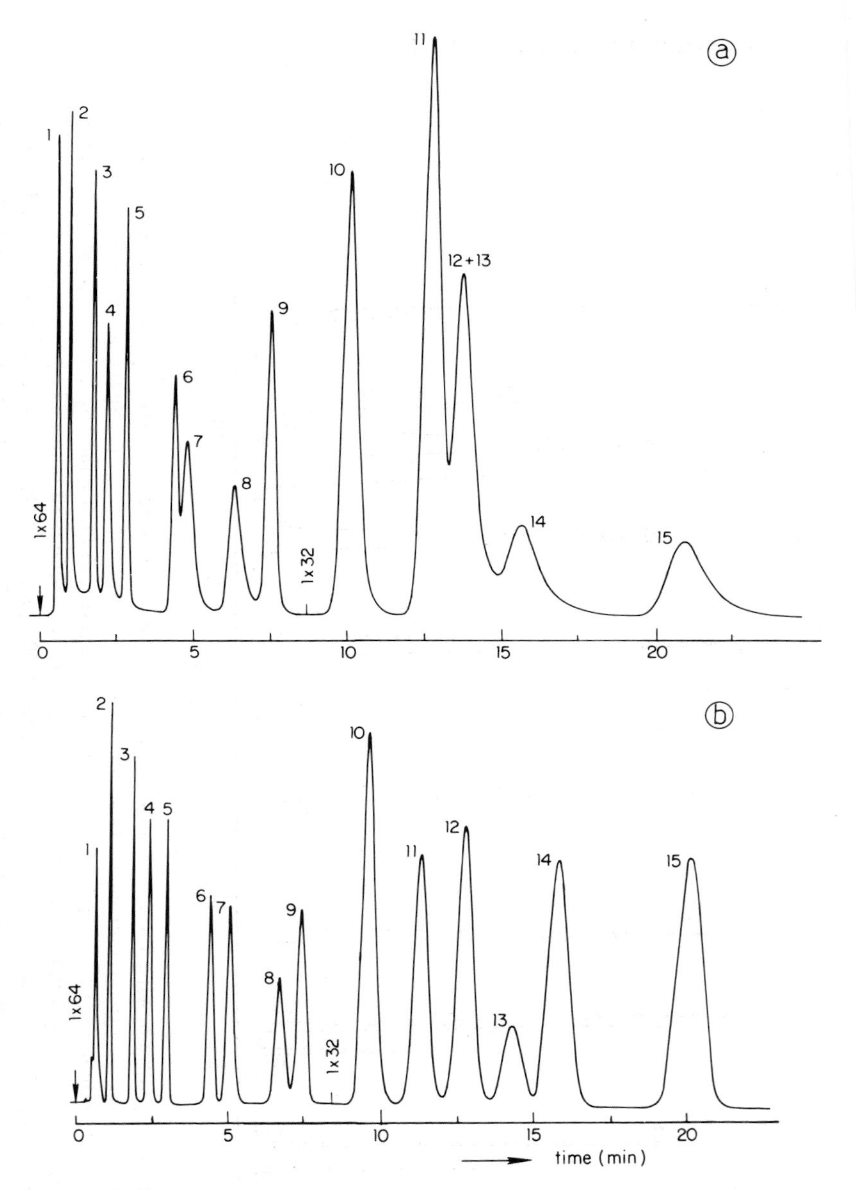

FIG. 37. Chromatogram showing the elution of C_1–C_5 alcohols in water solution. Column, 1.4 m × 2 mm; sample size of each component, 30 ng; linear carrier gas velocity, 10 cm/sec; (a) Graphon + 0.6% PEG 1500; temperature, 126° C; (b) Graphon + 2% PEG 1500; temperature, 131° C;

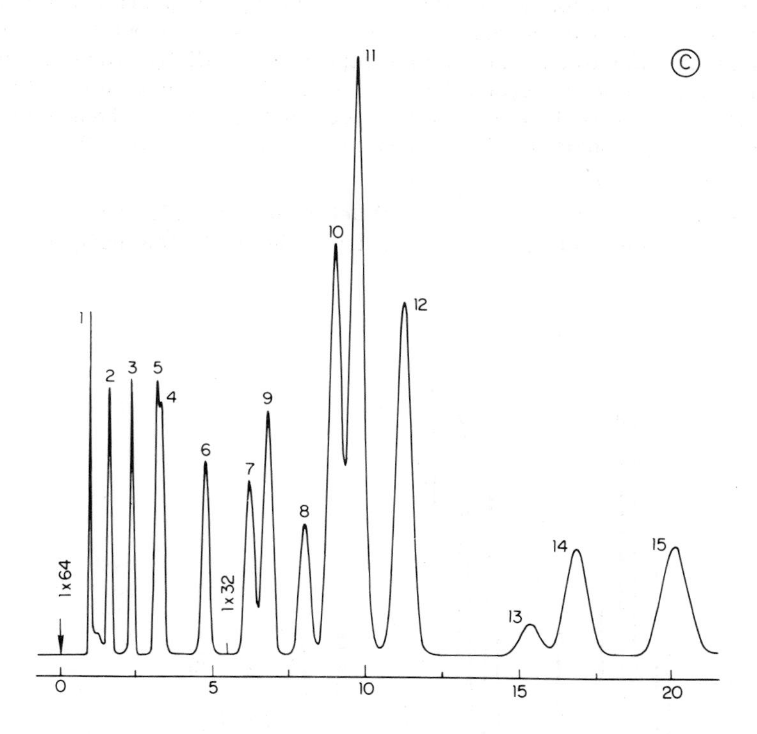

(c) Graphon + 5% PEG 1500; temperature, 98° C. 1, Methanol; 2, ethanol; 3, 2-propanol; 4, 1-propanol; 5, 1-methyl-2-propanol; 6, 2-butanol; 7, 2-methyl-1-propanol; 8, 1-butanol; 9, 2-methyl-2-butanol; 10, 3-methyl-2-butanol; 11, 3-pentanol; 12, 2-pentanol; 13, 2-methyl-1-butanol; 14, 3-methyl-1-butanol; 15, 1-pentanol [43]. (Reproduced by permission of Analytical Chemistry.)

It may be seen that preadsorbed Lewis-acid-type molecules on the graphitized carbon surface can considerably modify not only separation factors but also the elution sequence for a mixture containing both unsaturated and saturated C_1-C_4 hydrocarbons. In such a situation, the adsorbing characteristics of the carbon surface are largely modified by strong lateral interactions coming into play between the π acceptor and unsaturated hydrocarbons through partial charge transfer.

From observation of the plot just described, it appears that Carbopack B with a surface coverage of picric acid equal to about 0.25 should be the most selective column for the separation of all C_4 hydrocarbons. In fact, the use of Carbopack C modified with 0.19% picric acid, which corresponds to a surface coverage of about 0.25, reached the aim of separating a C_4 hydrocarbon mixture commonly encountered in refinery operations (Fig. 35) [67].

On the other hand, in the case of determination of light impurities in purum-grade butadiene, the most favorable picric acid percentage to be

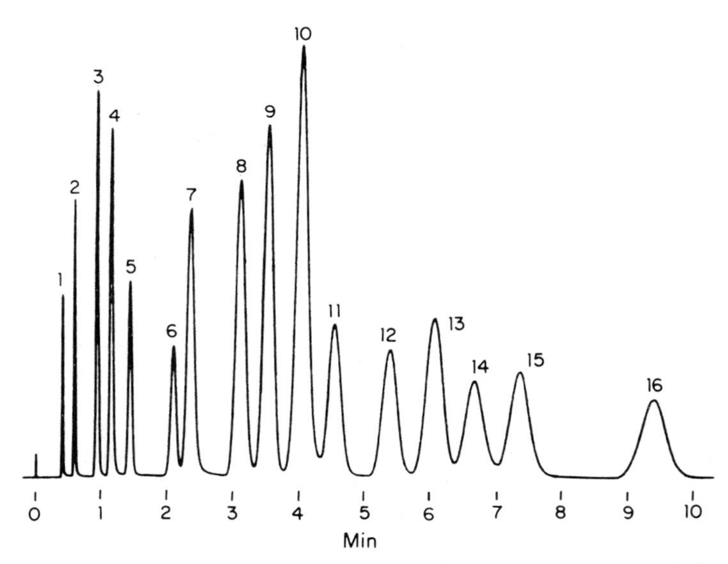

FIG. 38. Chromatogram showing the elution of C_1-C_5 alcohols. Column, 6 ft × 2 mm; Carbopack A + 0.4% PEG 1500; temperature, 135° C; flow rate, 20 ml/min N_2; sample size, 0.05 μl [68]. 1, Methanol; 2, ethanol; 3, 2-propanol; 4, 1-propanol; 5, 2-methyl-2-propanol(t-butyl); 6, 2-butanol(sec-butyl); 7, 2-methyl-2-butanol(t-amyl); 10, 2,2-dimethyl-1-propanol; 11, 3-methyl-2-butanol; 12, 3-pentanol; 13, 2-pentanol; 14, 2-methyl-1-butanol(active); 15, 3-methyl-1-butanol; 16, 1-pentanol.

added to a surface of Carbopack B was found to be 4.8% in order to obtain the best column performance (Fig. 36) [57].

By continuously adjusting the liquid/solid ratio, a "tailor-made" column able to yield the best fractionation of a given mixture can be prepared. A clear-cut example is given in Fig. 37, which shows chromatograms on the elution of a diluted water mixture of all aliphatic alcohols up to C_5 except neopentanol, eluted on Graphon modified with three different amounts of PEG 1500 [43]. As can be seen, a baseline separation for all components in achieved in 20 min only by using Graphon + 2% PEG 1500 as packing material. With the same analysis time, neither the carbon surface modified with exactly the proper amount of PEG 1500 needed to eliminate the effect of peak tailing nor a PEG 1500 monolayer supported on the carbon surface is able to separate all components of the mixture. Also, by using a graphitized carbon with a smaller surface area, such as Carbopack A (Supelco Inc.), modified with 0.4% PEG 1500 to obtain similar surface coverage as above, the same mixture can be fractionated with an elution time of only 10 min (Fig. 38) [68]. The great usefulness of GLSC as an analytical tool is once more emphasized by considering that a difficult fractionation such as the complete separation of isomers of pentanol can be achieved only by the use of capillary columns coated with blended mixtures [69, 70].

IX. EFFECT OF LIQUID LOAD ON COLUMN EFFICIENCY

Halasz and Heine [30], Giddings [71], and Kiselev and Yashin [5] stated that the mass transfer coefficient in GSC is much smaller than that in GLC. This difference arises from the fact that in the former the speed with which the adsorption equilibrium is reached is usually very small, whereas in the latter the passage through the interface and diffusion within the liquid film are the limiting factors which cause peak broadening when high flow rates of carrier gas are used. Thus, GSC offers an obvious advantage over GLC as the loss of performance is restrained at high linear carrier gas velocities so that high-speed analysis is allowed.

In this respect, it can be safely expected that a GLS column behaves similarly to a GSC column. In Fig. 39 is shown a comparison between an FFAP-coated Chromosorb W column and an FFAP-modified Sterling FT-G column in terms of efficiency as the linear carrier gas velocity is increased [51]. It may be seen that at high carrier gas velocities the efficiency of the GLS column is more preserved by far than in the case of the GL column.

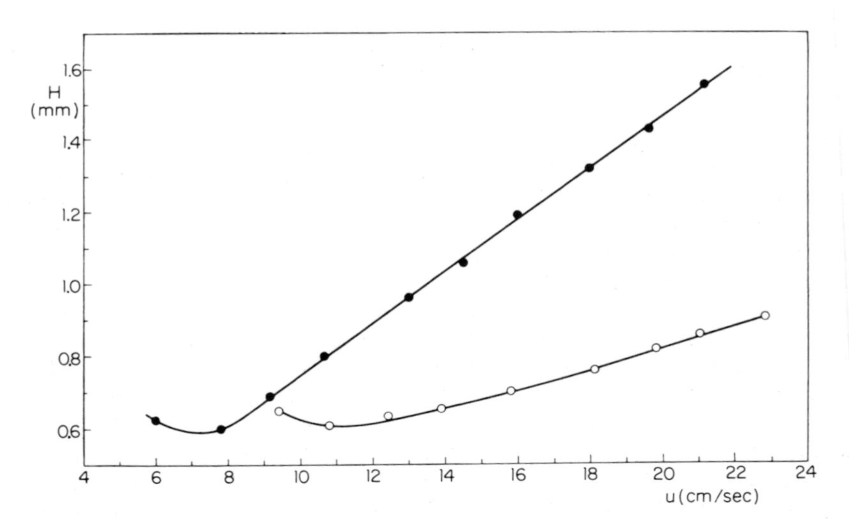

FIG. 39. Plot of the column efficiency versus the linear carrier gas velocity corresponding to the elution of caprylic acid on 10% FFAP + 1% H_3PO_4 on Chromosorb W at 190°C (●); 0.3% FFAP on Sterling FT-G at 205°C (○) [51]. (Reproduced by permission of Analytical Chemistry.)

De Boer [72] pointed out that the presence of any preadsorbed substance causes a decrease in the speed with which the adsorption equilibrium is attained during adsorption from the gas phase. In addition, it can be reasonably expected that the extent of this decrease is dependent on the surface concentration of the preadsorbed substance. Thus, in GLSC the shape of the right-hand branch of the Van Deemter curve should vary with the variation in the liquid/solid ratio. The effect of both the type and amount of liquid on column efficiency has been extensively investigated by Bruner et al. [44, 52].

Figure 40 shows some Van Deemter curves pertaining to adsorption of pentanone on Graphon modified with various amounts of PEG 1500 [43]. As expected, there is a steady increase in the slope of the right-hand branch of the Van Deemter curve as the amount of PEG 1500 is increased. From the plot it appears that GLS columns generally exhibit good efficiency. By using finer particle sizes of PEG 1500-modified Carbopack B, at the optimum carrier gas velocity an efficiency as high as 3300 plates per meter of column can be readily obtained [57].

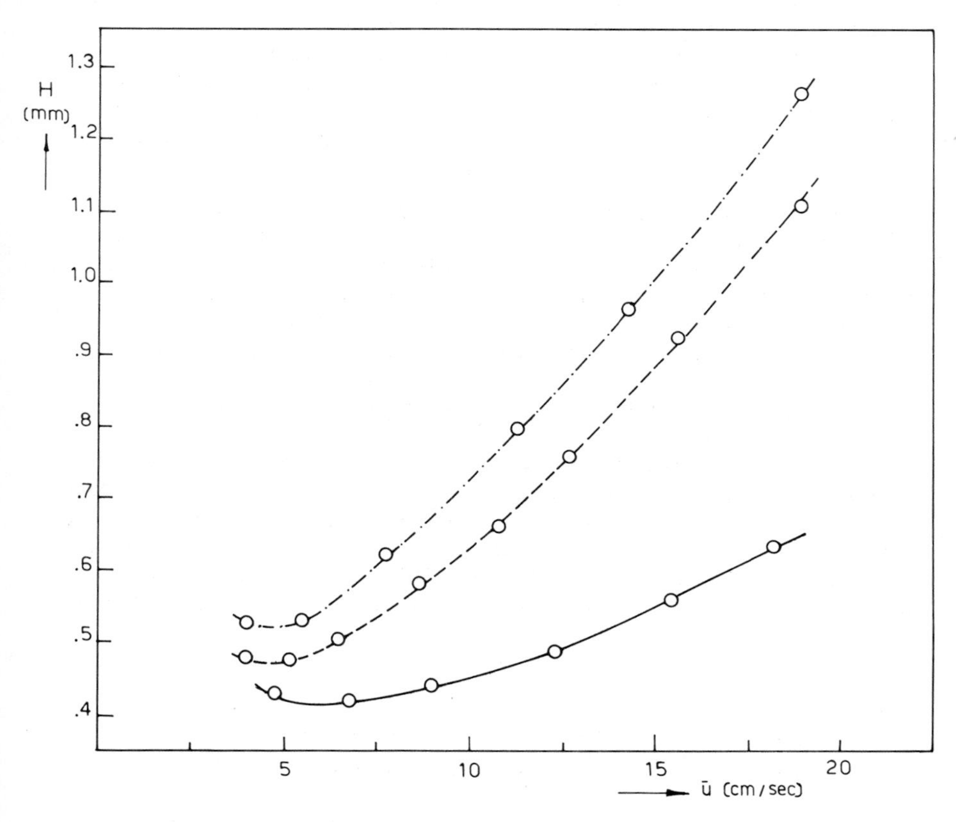

FIG. 40. Column efficiency versus the linear carrier gas velocity concerning the elution of 2-pentanone on Graphon + 1% PEG 1500 (solid line), Graphon + 4% PEG 1500 (dashed line), and Graphon + 9% PEG 1500 (dash-dot line) [43]. (Reproduced by permission of Analytical Chemistry.)

X. EFFECT OF LIQUID LOAD ON COLUMN THERMAL STABILITY

Heats of adsorption are generally higher than heats of condensation. Therefore, volatility of molecules in the adsorbed state can be much lower than that of molecules in the corresponding liquid state. As a result, an ordinary partitioning liquid adsorbed as a monolayer on a solid surface can be employed at temperatures higher than the upper limit of temperature at which the corresponding GL column can operate. Kiselev and his co-workers [73] showed that when monoethanolamine is deposited as a single monolayer on silica gel the upper limit of working temperature could be readily shifted more than 200° C (Fig. 41).

Since mutual attractions among adsorbed molecules involve an additional contribution to the adsorption heat, the volatility of the adsorbed molecules should also be dependent on their surface concentration. Experimental evidence of this assumption was obtained by measuring the background ion currents of a flame ionization detector of some chromatographic columns packed with Graphon modified with various amounts of PEG 1500 [43]. As can be seen from Fig. 42, at any liquid/solid ratio, column "bleeding" is considerably lower than that of the corresponding GL column. Moreover, thermal stabilities of GLS columns vary with the percentage of PEG 1500. The highest thermal stability is shown by the column filled with 4% PEG 1500-modified Graphon. This finding can be accounted for by considering that, as previously shown in Fig. 6, at this percentage one monolayer of PEG 1500 is almost virtually filled. In such a situation, a maximum in the adsorption heat of PEG 1500 can be assumed to occur when the contribution to the heat from adsorbate-adsorbate interactions is highest.

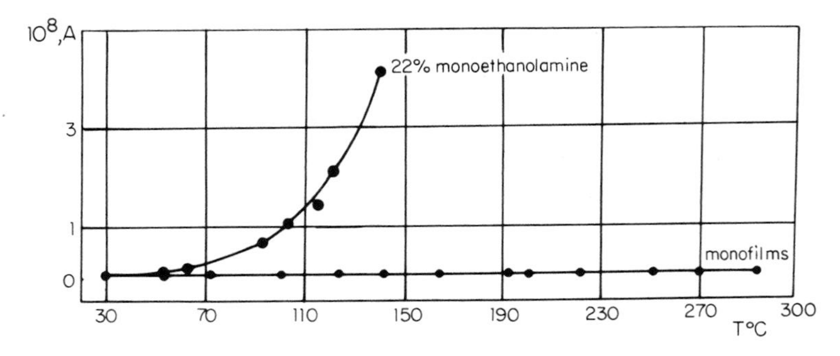

FIG. 41. Relation of background ion current to temperature for thick and monomolecular films of monoethanolamine on macroporous silica gel containing alumina [73]. (Reproduced by permission of Izd. Niite Khim.)

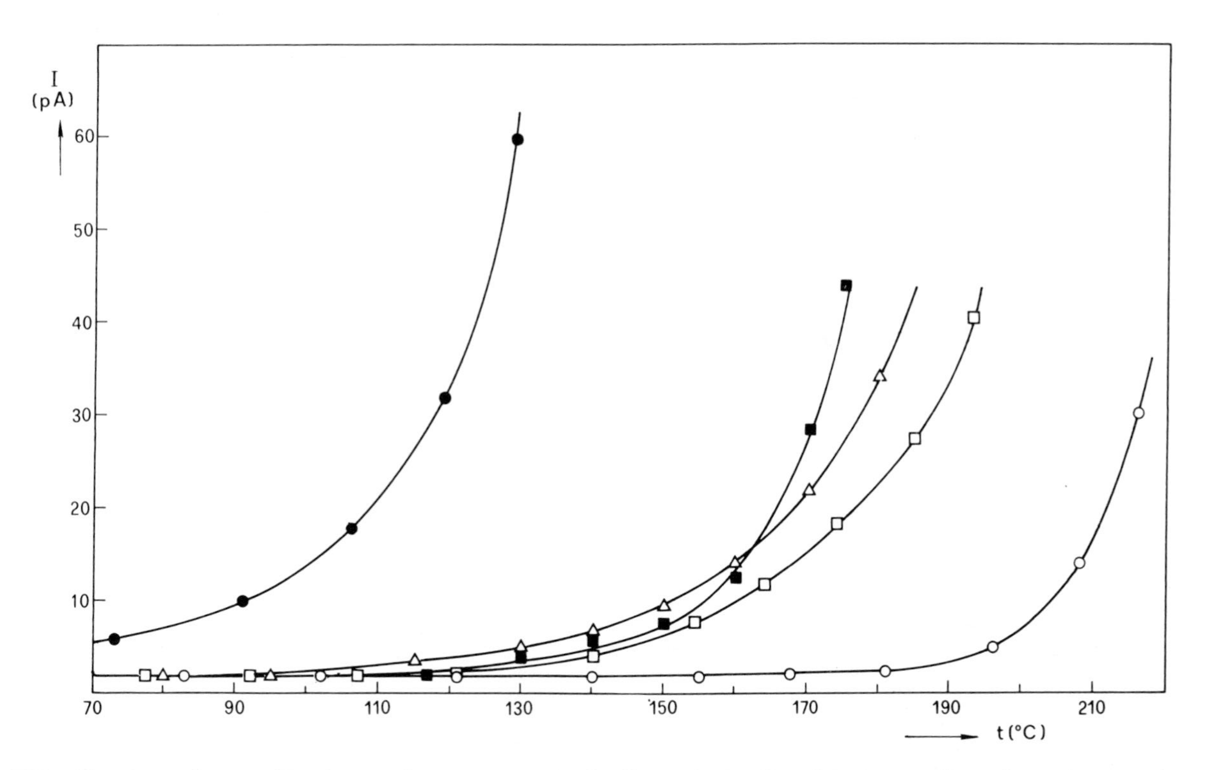

FIG. 42. Dependence of background ionic current of a flame ionization detector on the column temperature for Chromosorb W + 3% PEG 1500 (●), Graphon + 0.6% PEG 1500 (□), Graphon + 4% PEG 1500 (○), Graphon +5% PEG 1500 (■), Graphon + 9% PEG 1500 (Δ) [43]. (Reproduced by permission of Analytical Chemistry.)

XI. CONCLUSIONS

This survey of adsorption on a liquid-modified adsorbing medium has shown that GLSC can be a very powerful tool since it has a number of advantages over both GL and GS methods, especially for the analysis of low- and medium-boiling substances. The results discussed in this chapter show that some long-standing problems relative to the analysis of highly polar compounds can be easily resolved with the aid of a homogeneous adsorbing medium suitably modified by a nonvolatile liquid. It has also been shown that the best liquid/solid ratio for the optimization of all chromatographic parameters cannot be established a priori. Hence, preliminary experiments on the chromatographic behavior of any liquid-solid system are required in order to assess beforehand optimum conditions for a chromatographic analysis. However, a knowledge of the theoretical principles involved could considerably reduce the time and effort spent in trials.

As far as the applicability of gas chromatography to studies of adsorption of nonvolatile molecules is concerned, much more work remains to be done. However, the few results discussed in this paper indicate that gas chromatography could become a very sensitive and fast method of investigating the mode of deposition of very large molecules on a solid surface and the interactions taking place between adsorbed molecules of different composition.

APPENDIX: PREPARATION OF LIQUID-MODIFIED GCB FOR GLSC

It has been pointed that the preparation of GLS columns may involve some problems. Indeed, there is no difficulty in obtaining high-resolution GLS columns and in duplicating them, provided some care is observed during their preparation.

After graphitized carbon particles have been sieved to obtain the selected mesh range, column packings should be prepared by putting the adsorbing medium into a properly chosen flat dish so that the layer of material is about 0.5 cm thick. Then weighed samples of modifying liquid needed to obtain the desired liquid/solid ratio are dissolved in a suitable solvent and added to the solid so that the solid is uniformly covered by few millimeters of solution. In some cases, it may be desirable to use a Teflon sheet-lined dish to avoid loss of material which is inclined to stick on glass surfaces [55].

In order to allow a regular distribution on the solid surface of the modifying nonvolatile molecules, the slush should be slowly dried at room temperature. During drying, any stirring of the material must be avoided, as it causes crushing of the carbon particles.

This precaution is particularly important when working with GCBs of lower surface area whose mechanical strength is rather low as compared with the solid media commonly used as supporting materials for gas-liquid columns. In any case, it is necessary to resieve dried materials to maintain the proper mesh range.

With the materials prepared in this fashion, columns are packed by vibrating with the aid of a vibrator, taking care to avoid an excessive shaking of the column. When stainless steel tubing is used as the chromatographic column, it is necessary that the void tubing be coiled and ready for connection to the gas chromatograph.

Finally, it should be mentioned that the percentage of liquid required to obtain the selected surface coverage can be made to vary to some extent by varying the adsorbent particles size range. This effect is intuitively explained considering that the apparent (geometric) surface area of a nonporous or macroporous adsorbing medium is highest for the finest particles.

REFERENCES

1. O. L. Hollis, Anal. Chem., 38, 309 (1966).
2. C. G. Pope, Anal. Chem., 35, 654 (1963).
3. I. Halasz and C. Horvath, Nature, 197, 71 (1963).
4. A. V. Kiselev, in Gas Chromatography 1964 (A. Goldup, ed.), Inst. of Petroleum, London, 1965, p. 238.
5. A. V. Kiselev and Y. I. Yashin, Gas-Adsorption Chromatography, Plenum, New York, 1969, p. 1.
6. J. V. Hallum and H. V. Drushel, J. Phys. Chem., 62, 110 (1958).
7. W. D. Schaeffer, W. R. Smith, and M. H. Polley, Ind. Eng. Chem., 45, 1721 (1953).
8. M. H. Polley, W. D. Schaeffer, and W. R. Smith, J. Phys. Chem., 57, 469 (1953).
9. W. B. Spencer, C. H. Amberg, and R. A. Beebe, J. Phys. Chem., 62, 719 (1958).
10. D. Graham, J. Phys. Chem., 61, 1310 (1957).
11. A. A. Isirikyan and A. V. Kiselev, J. Phys. Chem., 66, 205 (1962).
12. A. Di Corcia and R. Samperi, J. Phys. Chem., 77, 1301 (1973).
13. A. C. Zettlemoyer, J. Colloid Interface Sci., 28, 343 (1968).
14. J. R. Lindsay Smith and D. J. Waddington, Anal. Chem., 40, 523 (1968).

15. C. Pierce, R. N. Smith, J. W. Wiley, and H. Cordes, J. Am. Chem. Soc., 73, 4551 (1951).

16. B. Millard, E. G. Caswell, E. E. Leger, and D. R. Mills, J. Phys. Chem., 59, 976 (1955).

17. A. Di Corcia and F. Bruner, Anal. Chem., 43, 1634 (1971).

18. A. Di Corcia and R. Samperi, J. Chromatogr., 77, 277 (1973).

19. C. G. Scott, J. Inst. Pet., 45, 115 (1959).

20. C. Vidal-Madjar, J. Ganansia, and G. Guiochon, in Gas Chromatography 1970 (R. Stock, ed.), Inst. of Petroleum, London, 1971, p. 20.

21. J. Janak, Collect. Czech. Chem. Commun., 18, 798 (1953).

22. A. Liberti, G. Nota, and G. Goretti, J. Chromatogr., 38, 282 (1968).

23. A. Nonaka, Anal. Chem., 45, 483 (1973).

24. A. Di Corcia and F. Bruner, J. Chromatogr., 62, 462 (1971).

25. A. Di Corcia, P. Ciccioli, and F. Bruner, J. Chromatogr., 62, 128 (1971).

26. F. T. Eggertsen, H. S. Knight, and S. Groennings, Anal. Chem., 28, 303 (1956).

27. F. T. Eggertsen and H. S. Knight, Anal. Chem., 30, 15 (1958).

28. O. Grubner and E. Smolkova, Gas Chromatography, 3rd Conf. Anal. Chem., Prague, Sept. 1959.

29. J. H. Purnell, in Gas Chromatography, Wiley, New York, 1962, p. 376.

30. I. Halasz and E. Heine, in Advances in Chromatography, Vol. 6 (J. H. Purnell, ed.), Wiley, New York, 1968, p. 170.

31. B. V. Derjaguin and V. V. Karassev, in Proceedings of the Second International Congress of Surface Activity, London, Vol. III, Academic Press, New York, 1957, p. 531.

32. A. M. Taylor and A. King, Nature, 132, 64 (1933).

33. K. G. Brummage, Proc. R. Soc. (London), 191A, 243 (1947).

34. S. J. Hawkes and J. C. Giddings, Anal. Chem., 36, 2229 (1964).

35. J. W. McBain, Colloid Science, Heath, Boston, 1950, Chapter 4.

36. M. Taramasso, Gas Cromatografia (F. Angeli, ed.), Milano, 1966.

37. L. D. Beliakova, A. V. Kiselev, N. V. Kovaleva, L. N. Rozanova, and V. V. Khopina, Zh. Fiz. Khim., 42, 77 (1968).

38. A. V. Kiselev, J. Chromatogr., 49, 84 (1970).

39. A. V. Kiselev, N. V. Kovaleva, and Y. S. Nikitin, J. Chromatogr., 58, 19 (1971).

40. R. A. Beebe and R. M. Dell, J. Phys. Chem., 59, 746 (1955).

41. W. B. Spencer, C. H. Amberg, and R. A. Beebe, J. Phys. Chem., 62, 719 (1958).

42. G. Crescentini, Thesis, University of Rome, Italy, 1971.

43. A. Di Corcia, A. Liberti, and R. Samperi, Anal. Chem., 45, 1228 (1973).

44. F. Bruner, P. Ciccioli, G. Crescentini, and M. T. Pistolesi, Anal. Chem., 45, 1851 (1973).

45. J. C. Henniker, Rev. Mod. Phys., 21, 322 (1949).

46. D. E. Martire, R. L. Pecsok, and J. H. Purnell, Trans. Faraday Soc., 61, 2496 (1965).

47. V. G. Berezkin, J. Chromatogr., 65, 227 (1972).

48. I. Halasz and H. Bruderreck, Anal. Chem., 36, 1533 (1964).

49. H. P. Boehm, in Advances in Catalysis, Vol. 16 (D. D. Eley, H. Pines, and P. B. Weisz, eds.), Academic Press, New York, 1966, p. 179.

50. A. Di Corcia, D. Fritz, and F. Bruner, Anal. Chem., 42, 1500 (1970).

51. A. Di Corcia, Anal. Chem., 45, 492 (1973).

52. F. Bruner, A. Di Corcia, A. Liberti, and R. Samperi, in Gas Chromatography 1972 (S. G. Perry, ed.), Applied Science, London, 1973, p. 275.

53. A. Di Corcia and R. Samperi, Anal. Chem., 46, 140 (1974).

54. A. Di Corcia and R. Samperi, Anal. Chem., 46, 977 (1974).

55. A. Di Corcia, A. Liberti, and R. Samperi, J. Chromatogr. Sci., 12, 710 (1974).

56. F. Bruner, P. Ciccioli, and F. Di Nardo, Anal. Chem., 47, 142 (1975).

57. A. Di Corcia and R. Samperi, J. Chromatogr., 107, 99 (1975).

58. F. Bruner, P. Ciccioli, and A. Di Corcia, Anal. Chem., 44, 894 (1972).

59. A. Di Corcia and A. Liberti, Trans. Faraday Soc., 66, 967 (1970).

60. A. Di Corcia, D. Fritz, and F. Bruner, J. Chromatogr., 53, 135 (1970).

61. S. Ross and J. P. Olivier, in On Physical Adsorption, Wiley (Interscience), New York, 1964, p. 156.

62. A. Di Corcia, unpublished results.

63. C. G. Scott and D. A. Rowell, Nature, 187, 143 (1960).

64. I. Halasz and C. Horvath, Anal. Chem., 36, 2226 (1964).

65. I. Halasz and E. E. Wegner, Nature, 189, 570 (1961).

66. A. Di Corcia, J. Chromatogr., 80, 69 (1973).

67. A. Di Corcia and R. Samperi, Anal. Chem., 47, 1853 (1975).

68. Chromatography/Lipids, Newsletter, Vol. VI, No. 7, Supelco Inc., Bellefonte, Pennsylvania, 1973.

69. R. D. Schwartz and R. G. Mathews, J. Chromatogr. Sci., 7, 593 (1969).

70. V. Palo and J. Hrivnak, J. Chromatogr., 59, 154 (1971).

71. J. C. Giddings, Dynamics of Chromatography, Vol. 1 (J. C. Giddings and R. A. Keller, eds.), New York, 1965, Part I, p. 120.

72. J. H. De Boer, The Dynamical Character of Adsorption, Oxford Univ. Press (Clarendon), London and New York, 1953, p. 87.

73. V. I. Kalmanovskii, A. V. Kiselev, G. G. Sheshenina, and Ya. I. Yashin, Gas Chromatography, Izd. Niite Khim., Moscow, 1967, p. 45.

Chapter 8

RETENTION INDICES IN GAS CHROMATOGRAPHY

J. K. Haken

Department of Polymer Science
The University of New South Wales
New South Wales, Australia

I. INTRODUCTION

The first workers [1] in gas chromatography recognized the characteristic nature of a substance's elution or retention behavior. Since their pioneering works the abscissa response on the usual differential chromatographic record has been extensively studied.

This chapter considers the abscissa relationship and its expression principally as retention indices for the specification of retention, and its further correlation with molecular structure for use in compound identification and in the characterization of stationary phases.

The retention volume of a compound, as we observe it on a recorder, as retention time or most commonly as retention distance following the injection of a substance, is a composite of two components. The major contribution to retention in gas-liquid chromatography is from the time the sample remains in contact with the stationary phase; in gas-solid chromatography it is the time the substance is on the surface of the absorbent, whereas the other contribution arises from the time required by the carrier gas to pass through the system.

Retention data relative to a single arbitrary standard are currently the most widely used method of data presentation [1] although the retention index introduced by Kovats [2] where the retention is relative to a series of homologous compounds is finding increasing acceptance; many authors are reporting their data in both ways. The specific retention volume V_g [3] was introduced in an attempt to produce retention data independent of experimental variable, but it has found little general application because of the experimental difficulties that are encountered.

Irrespective of the retention scheme used a knowledge of the column dead volume or holdup is necessary and may be obtained experimentally or by calculation.

II. CALCULATION OF SIMPLE RELATIVE RETENTION

Relative retention described by James and Martin [1] in 1952 in their pioneering work compared the retention behavior of the compounds concerned with that of a standard substance. Identical experimental conditions (column temperature, gas flow rate) are necessary and a calculation is made directly from the chromatogram, as shown in Fig. 1.* The distance is measured from the point of injection to the maximum of the chromatographic peak d_3, to the reference standard d_2, and to the "air peak" d_1. The relative retention volume of the substance V_R is

*There is no internationally accepted nomenclature for the terms shown in Fig. 1 and Eq. (1). An ASTM Specification [4] details V_R, d_1, d_2, d_3, $d_2 - d_1$, and $d_3 - d_1$ as $r_{3,2}$, t_M, t_R2, t_R3, t^1_R2 and t^1_R3, respectively, but the expressions used without a multiplicity of superscripts allow the relationships shown in Eqs. (2) to (11) to be more readily appreciated.

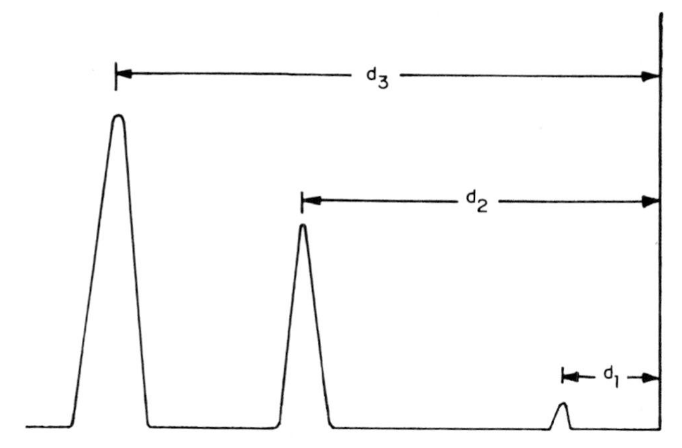

FIG. 1. Calculation of relative retention data.

$$V_R = \frac{d_3 - d_1}{d_2 - d_1} \tag{1}$$

It is necessary to subtract the retention due to the "air peak" to eliminate the effect of "dead volume" on the volume of gas in the column and connecting lines. To achieve a linear relationship by plotting V_R against carbon number for a homologous series, it is essential that the retention data be corrected for gas holdup. Some commercially available digital and reporting integrators with no provision for correction provide a result shown erroneously as relative retention.

III. DETERMINATION OF COLUMN DEAD VOLUME

The flame ionization detector does not ordinarily produce a signal with air, and methane or another unretarded substance is usually used to determine the dead volume or, alternatively, it may be calculated. The detector can produce a signal with air or other inert gas by the injection of a large volume of the gas (300-500 μl) into the column, which, when operated at high sensitivity, produces a significant negative peak. The technique described by Hilmi [5] has been used with success in this laboratory, although the explanations presented for the phenomenon are not very convincing. The use of small (i.e., 1 μl) samples with ionization detectors has recently been reported by Riedmann [6].

The dead volume may be calculated from a knowledge of the dimensions of the column and of its contents but for practical purposes this procedure is of limited value as the necessary data are not readily available.

Procedures which consider correction of experimental data may be successfully employed. Linearization of the logarithmic plot of homologous n-alkanes by graphical trial and error has been reported by Evans and Smith [7], and simple calculations of the position of the air peak from the retention of homologous compounds has been reported by Peterson and Hirsch [8] and Gold [9]. A statistical procedure for calculation of the dead volume, the b values of the n-alkane plot, and the retention indices has been reported by Grobler and Balizs [10].

The calculation of Peterson and Hirsch requires that the chromatogram of three evenly spaced homologous compounds be considered, and employs the linear relationship that exists between the logarithm of the adjusted retention time and the number of carbon atoms in a homologous series:

$$n_2 = m \log(d_2 - d_1) + b \tag{2}$$

where n_2 is the chain length of the homolog.

By the use of an arbitrarily placed line, which for simplicity coincides with the peak maximum of the second homolog, a series of three equations is obtained which on solution allows calculation of the air peak.

The procedure of Gold [9] allows calculation of the air peak using any three known homologs of chain length X, Y, and Z:

$$n_X = m \log(d_2 - d_1) + b \tag{3}$$

$$n_Y = m \log(d_3 - d_1) + b \tag{4}$$

$$n_Z = m \log(d_4 - d_1) + b \tag{5}$$

By subtraction of the equations, b is eliminated and a solution for m is obtained:

$$m = \frac{n_Y - n_X}{\log[(d_3 - d_1)/(d_2 - d_1)]} \tag{6}$$

$$m = \frac{n_Z - n_X}{\log[(d_4 - d_1)/(d_2 - d_1)]} \tag{7}$$

Equations (6) and (7) may be solved by the method of successive approximations where various assumed values of d_1 are plotted versus the corresponding calculated values of m. The common solution is the point where the two plots cross, and it gives the value of the air peak d_1.

It is obvious that the calculated value of the air peak is dependent on the linearity of the plot of the homologous compounds. The use of lower polar homologs which deviate significantly from linearity should be avoided and higher members of a series of polar compounds or, preferably, hydrocarbons should be used.

Regression analysis carried out with a computer [10] has several advantages in that a series of homologs can be considered to provide a rapid and statistical estimate of the dead volume. The b value which may be calculated simultaneously has been shown by Evans [11] to be very sensitive to variations in column temperature and flow rate and may be employed as a means of control of the experimental conditions.

IV. KOVATS RETENTION INDEX

The retention index was introduced by Kovats [2] in 1958 in the first of a series of foreign language works [2, 12-18]. Some impression was apparently made in Europe during the formative years but application in the United States dates from about 1964. At this time Kovats and Keulemans had reported on the system at the 2nd International Meeting on Advances in Gas Chromatography and Ettre [19] in a particularly fine report was able to integrate the available works incorporating a basic modification and to detail a series of elementary rules relating the indices and chemical structure.*

The retention index utilizes the linear relationship that exists between the logarithm of retention time and the number of carbon atoms within a homologous series, a condition existing with the n-alkanes on most stationary phases. The scheme is in part identical to the R_{x9} index or theoretical nonane number scheme of Evans and Smith [20, 21], who recommended that retention data be compared with n-nonane as the primary standard by first determining the data relative to the closest eluted normal paraffin.

Some refinement of the index has been effected [19] since its introduction; retention behavior is expressed on a uniform scale related to a series of closely related standards, i.e., n-paraffins. The index is the

*For another early treatment see E. Kovats, "Gas Chromatographic Characterization of Organic Substances in the Retention Index System, Advances in Chromatography, Marcel Dekker, New York, Vol. 1, 1965, pp. 229-247.

interpolated logarithm of the retention value related to the series and is calculated according to

$$I = 100 \frac{\log V_{R(\text{substance})} - \log V_{R(n-c_z)}}{\log V_{R(n-c_{z+1})} - \log V_{R(n-c_z)}} + 100Z \tag{8}$$

where V_R is the net retention data, $n - c_Z$ is n-paraffin with Z carbon atoms, $n - c_{Z+1}$ is n-paraffin with Z + 1 carbon atoms, Z is an integer by definition, and

$$V_{R(n-c_z)} \leqq V_{R(\text{substance})} \leqq V_{R(n-c_{z+1})}$$

According to the equation, the retention index of the normal paraffins will be 100 times the carbon number, i.e., 100, 200, etc., for methane, ethane, etc. Calculation of indices in the region 0 to 100 is made by considering hydrogen as "zero carbon."

Ettre and Billeb [22] rearranged Eq. (8) and Eq. (9) and applied simplifications as shown in Eqs. (10) and (11):

$$I = 100 \frac{\log V_{R(\text{substance})}/V_{R(n-c_z)}}{\log V_{R(n-c_{z+1})}/V_{R_{(n-c_z)}}} \tag{9}$$

$$I = 100 \frac{\log V_{R(\text{substance}/n-c_z)}}{\log V_{R(n-c_{Z+1}/n-c_z)}} \tag{10}$$

where

$V_{R(\text{substance}/n-c_z)}$ = retention of substance relative to the n-paraffin with Z carbon atoms

$V_{R(n-c_{z+1}/n-c_z}$ = retention of the n-paraffin with Z + 1 carbon atoms relative to the n-paraffin with two carbon atoms

$$V_{R(n-c_{z+1}/n-c_z} = V_{R(n-c_{z+1})} V_{R(n-c_z)} \tag{11}$$

will be constant for a particular phase and temperature independent of the value of Z. The constancy of relative retention of adjacent n-paraffins was shown on squalane at temperatures between 60° and 90° C. Some simplification in the calculation of indices may be achieved by using established values of the relative retention of the n-paraffins.

While the retention indices may be determined graphically from a simple n-paraffin plot this is of restricted accuracy; some improvement has been reported by the use of a nomograph [23] but the availability of simple programmable calculators makes such considerations redundant.

The retention index I by convention has the column temperature and the stationary phase as subscript and superscript. A retention index determined at 130° C on a silicone SE-30 column is shown as I_{130}^{SE30}.

The methylene units (MU) of VandenHeuvel and his co-workers [24] are essentially identical to the retention indices. A calibration line using n-docosane and n-tetracosane was used and the methylene unit value multiplied by 100 is the retention index.

In temperature-programmed gas chromatography the simple expression of elution behavior as retention relative to an arbitrary standard is not applicable, and peak position is conveniently expressed as the time required for elution or the column temperature at this time.

With temperature-programmed operation an approximately linear relationship exists between elution temperature of n-alkanes and their carbon number, provided that the initial column temperature is low and that only a relatively limited range of carbon numbers is considered [25, 26]. Whereas the theoretical background of this linearity is somewhat obscure it has been shown independently by Vanden Dool and Kratz [27] and Guiochon [28] that sufficient experimental data are available to allow a generalization of the retention index to include this method of operation.

This form of the retention index is

$$I = 100 \frac{T_{R(substance)} - T_{R(n\text{-}c_z)}}{T_{R(n\text{-}c_{z+1})} - T_{R(n\text{-}c_z)}} + 100Z \qquad (12)$$

where T_R is the retention temperature of the three components considered.

The detailed theoretical and experimental aspects of temperature-programmed gas chromatography have been reported by Harris and Habgood [29], with the limitations of the relationship shown in Eq. (12) and the reduced reproducibility of programmed data.

Associated with the use of retention indices is the reduced retention of hydrocarbons on highly polar phases and the resulting comparison of low-boiling polar solutes with n-alkanes of much higher boiling points.

This problem has been considered by several workers [30-32] and various alternative reference materials have been proposed, both on theoretical and practical grounds.

The use and suitability of homologous series other than normal paraffins as a calibration line are discussed in the following section.

V. ALTERNATIVE LOGARITHMIC INDEX SYSTEMS

Since the linearity of the plot of the logarithm of retention and carbon number is evident with the majority of homologous compounds, reference series other than the n-paraffins have been proposed and have found considerable acceptance in several restricted areas.

The methyl esters of the linear saturated acids from the basis of the carbon number (CN) system or the independently developed equivalent chain-length (ECL) system were developed by Woodford and van Gent [33] and Miwa and his co-workers [34], respectively. The esters are equivalent to the hydrocarbons used by Kovats and have found acceptance with lipid chemists although a refinement of the scheme employing the methyl esters of monounsaturated esters and providing MECL (modified equivalent chain length) values has found little use [35].

The steroid number (SN) considers a calibration line using two hydrocarbons with a steroidal structure (i.e., androstrane and cholestane), and containing 19 and 27 carbon atoms such that SN values of 19 and 27 are plotted against the appropriate retention values [36]. Calculation of the steroid number has been demonstrated by the summation of contributions due to the steroid skeleton and to the functional groups present. The relationship in common with procedures applicable to fatty compounds has some value in tentative identification, and as discussed later it is evident that the schemes are applicable only where intermolecular interactions are small or theoretically assumed to be zero.

Grobler [31] has outlined the limited practicability of the Kovats retention index scheme with polar stationary phases due to the small retention of hydrocarbons on these phases and has argued that the gas chromatography of compounds with highly different boiling points cannot be performed favorably under strictly identical conditions.

A scheme employing a series of secondary standards, i.e., the primary straight-chain alcohols, was suggested but their unsuitability has been demonstrated [30]. Hawkes [30] indicated the probability of adsorption of alkanes with polar phases and has shown that the n-alkanols provide the opposite extreme effect to the n-alkanes such that on polar phases many compounds would have much lower boiling points than the alcohols with which they are chromatographed; the n-propyl ethers were recommended as reference materials, with ketones and aldehydes as alternatives.

The effect of sorption phenomena on retention indices in gas chromatography has been examined by Lorenz and Rogers [37] who found considerable errors when the polarities of the solute and solvent were widely variant. Essentially the same effect had been observed previously by VandenHeuvel and Horning [36] who recommended that the polarities of two species should be similar.

A carbon number system using the 2-alkanones has been reported by Dymond and Kilburn [38] and the merits of using these ketones as secondary references standards have been detailed by Ackman [32].

The difficulty in selecting an acceptable reference series is readily apparent, particularly with polar compounds either as solute or solvent or where particularly strong specific interactions occur. The difficulty or impossibility of obtaining a universally acceptable reference series is apparent by a consideration of the behavior of compounds of modest polar character (e.g., simple esters) on various stationary phases [39].

If the retention indices for the simple n-alkyl acetates are considered on SE-30, it is apparent that methyl acetate (b.p. 57.1° C) has a retention index of 599, i.e., equivalent to n-pentane (b.p. 36.2° C) while hexyl acetate (b.p. 169.2° C) has an I value of 988, i.e., almost equivalent to n-decane (b.p. 174.0° C). With this nonpolar stationary phase the solutes are of higher boiling point than the comparable paraffins; however, with DC-710, a material of modest polarity, the situation is reversed, with methyl acetate (I = 609) equivalent to n-hexane (b.p. 69.0° C) and hexyl acetate (I = 1092) almost equivalent to undecane (b.p. 196° C). With the weaker acceptor phases (i.e., XE-60 and OV-225), methyl acetate (I = 737) is equivalent to a boiling point of approximately 220° C, while with the more polar acceptor phases methyl and hexyl acetates are equivalent to n-paraffins of boiling points approximating 140° and 235° C. The much greater variations in boiling points at either end of the scale reflect the differences in the slopes of the paraffins and the esters on the various phases.

By considering the simple esters as alternate standards on a calibration line where methyl acetate with a carbon number of 3 has a nominal value of 300 and the higher acetates increase by 100 units per carbon number, a more realistic situation is achieved. The retention behavior of a particular ester is immediately evident by simple comparison in the same way as is possible with ECL values, while with other homologous series the problems associated with the n-hydrocarbons are minimized. Symmetrical esters with the same number of carbon atoms on each side of the carbonyl group have also been used as a reference line; the index values are slightly higher than with an acetate calibration series.

With both ester series the polar calibration lines are more suitable for use with polar solutes than the normal paraffins, proposals that have been made previously by Lorenz and Rogers [37] and VandenHeuvel and Horning [36]. While the ester materials as secondary reference are more suitable, as suggested, than highly polar materials, it is unlikely that any single series could be satisfactory for all solutes. There is no reason, however, that retention relative to several reference series should not become more popular and be produced without difficulty with the increasing use of minicomputers as data processing units.

VI. LINEAR INDEX SCHEMES

Vigdergauz [40, 41] has described a linear index scheme (J) which considers the uncorrected retention of the n-paraffins as a basis for interpolation. The relationship is

$$J = \frac{d_{substance} - d_{n-c_z}}{d_{n-c_{z+1}} - d_{n-c_z}} + Z \tag{13}$$

where $d_{substance}$, d_{n-c_z} are gross retention values for the substance and n-paraffins of Z and Z + 1 carbon numbers, respectively.

The same relationship was subsequently introduced as the arithmetic retention index I_A by Harbourn [42] who also stressed its simplicity of application.

These schemes have found some minor acceptance [43, 44] but are considered to be of limited general use. The linear values cannot be applied in considerations of molecular structure as discussed in the next section. The index values of a compound must be determined by interpolation between the two appropriate n-paraffins; extrapolation outside these limits, as with retention indices, is not possible.

VII. RETENTION INDICES AND STRUCTURE

The early works of Kovats [2, 13-18] included observations concerning the relationship of retention indices and structure and their use in retention prediction. These elementary rules were first assembled by Ettre [19], then later in a slightly more detailed form by Kovats [45], and as discussed here are shown generally to be oversimplifications of the situations considered.

1. Incremental increases of 100 index units per homolog containing five or more carbon atoms were reported for some series of esters, aldehydes, alcohols, and ketones on a polar and nonpolar phase by Wehrli and Kovats [2, 13, 14]. The same workers [45] later stated that "in any homologous series the retention index of higher members increases by 100 index units per methylene group introduced." This implied quite erroneously that the slopes of the conventional data plots of all homologous series are the same although index contributions from a methylene group of less than 90 index units had been reported earlier by Zulaica and Guiochon [46] with dibasic esters.

   The variability of the slopes of plots of the n-paraffins and n-alkyl acetates on a variety of polysiloxane stationary phases is shown in Table 1 where it is evident that an incremental increase of 100 index units per methylene group would be a chance occurrence [47].

TABLE 1

Slopes of Plots of n-Paraffins and n-Alkyl Acetates at 150° C

Stationary phase	n-Paraffins	Acetate esters
SE-30	0.545	0.527
OV-7	0.555	0.557
DC-710	0.557	0.547
OV-25	0.542	0.488
100% phenyl	0.524	0.470
DC-230	0.552	0.555
DC-530	0.514	0.492
XE-60	0.470	0.462
XF-1150	0.473	0.409
OV-225	0.487	0.464
SILAR 5CP	0.479	0.443
F-400	0.533	0.500
F-500	0.509	0.490
QF-1	0.438	0.440

The retention behavior of homologous esters which were conveniently represented as

$$R-\overset{\overset{\displaystyle O}{\|}}{C}-O-R'$$

where the carbon numbers of the acid and alcohol chains are R and R', has been reported [48]. It was apparent that the slopes of plots representing homologous esters with the same number of carbon atoms in the acid chain (R) decrease as the value of R is increased; similarly, for the same number of carbon atoms in the alcohol chain (R') the slopes decrease as the value of R' increases with the phases considered. Tables 2 and 3 show these effects as

TABLE 2

Effect of Incremental Changes in the Alcohol Chain (R') for n-Saturated Acid Esters[a]

Stationary phase	Ester series					
	Formate	Acetate	Propionate	Butyrate	Pentanoate	Hexanoate
SE-30	105.0	99.6	96.1	94.6	92.3	90.2
OV-7	105.8	103.0	101.0	97.1	93.7	93.3
DC-710	108.6	101.7	99.7	96.0	94.0	92.0
OV-25	98.7	94.8	93.3	91.4	90.1	88.6
100% phenyl	95.2	93.0	93.5	91.0	89.7	90.3
DC-230	104.5	105.1	103.7	98.9	96.2	94.3
DC-530	99.9	99.9	99.2	96.1	93.3	92.4
XE-60	105.2	102.3	101.6	99.2	97.0	91.9
XF-1150	92.5	93.5	89.6	85.7	84.8	83.5
OV-225	102.0	98.9	101.8	97.3	92.4	91.9
SILAR 5CP	98.7	97.3	97.0	92.4	91.2	88.0
F-400	101.2	103.9	100.7	97.5	96.4	95.0
F-500	98.4	100.9	100.0	97.9	95.4	94.3
QF-1	100.4	104.6	101.2	98.0	95.8	94.1

[a]From Ashes and Haken [47].

TABLE 3

Effect of Incremental Changes in the Acid Chain (R) for n-Saturated Alcohol Esters[a]

Stationary phase	Ester series					
	Methyl	Ethyl	Propyl	Butyl	Pentyl	Hexyl
SE-30	103.5	93.8	91.4	89.0	86.8	85.9
OV-7	103.4	97.6	93.0	91.2	87.3	86.3
DC-710	101.6	94.8	90.8	88.9	86.5	85.3
OV-25	95.5	87.8	86.6	85.0	83.0	81.7
100% phenyl	88.5	85.4	83.5	81.6	81.9	81.4
DC-230	100.4	99.3	97.4	93.5	90.1	91.1
DC-530	93.8	95.2	91.7	89.7	88.9	87.5
XE-60	95.8	89.9	86.3	85.4	84.1	82.7
XF-1150	81.5	81.3	78.8	74.9	72.6	75.5
OV-225	93.6	90.4	87.7	85.7	82.7	81.7
SILAR 5CP	92.0	85.8	85.5	81.2	79.0	78.1
F-400	97.9	97.3	92.5	90.2	88.9	87.5
F-500	96.8	92.8	90.2	88.7	87.1	85.5
QF-1	94.4	96.5	94.3	91.1	87.7	86.6

[a]From Ashes and Haken [47].

index increments where it is evident that the methylene group has a greater effect on retention when in the alcohol chain (R') than in the acid chain (R), this effect having also been observed with α-alkyl acrylic esters [49].

The effect of increment changes with variation of the alcohol (R') and acid (R) chains as shown in Tables 2 and 3. From Table 2 it is apparent that the incremental changes are maximized with the lower members and decrease as R increases. With the methylene increments associated with increasing the acid chain (R) of the alcohol ester plots, the values shown in Table 3 are calculated from the lines of best fit of the homologs since a loss of linearity occurs, particularly with stationary phases containing highly polarizable substituent groups. However, with the curvilinear plots trends were clearly evident, and with Tables 2 and 3 incremental ranges of 108.6 to 88.0 and 103.5 to 78.1, respectively, were recorded, there being little evidence of any constancy with the systems considered.

In the same studies the retention increment due to the carboxyl group of n-alkyl esters on various stationary phases was calculated following an earlier report of Zulaica and Guiochon [46]. The increment was found as the intercept of the linear relationship, i.e., R = R' = 0, calculated from the line of best fit determined by regression analysis of an ester plot where R = R'. The values which increase with the polar character of the phase, i.e., as the polarizability of the solvent species is increased, are shown in Table 4. As expected, the value with SE-30 is somewhat greater than that in the report of Zulaica and Guiochon [46] because of the lower operating temperature.

TABLE 4

The Retention Increment of the Carboxyl Group of n-Alkyl Esters[a]

Stationary phase	Increment
SE-30	324
OV-7	363
DC-710	417
OV-25	507
100% phenyl	548
DC-230	320
DC-530	377
XE-60	539
XF-1150	718
OV-225	558
SILAR 5CP	663
F-400	417
F-500	473
QF-1	571

[a]From Ashes and Haken [47].

Retention indices of saturated and unsaturated hydrocarbons and of methyl esters of isomeric aliphatic monocarboxylic acids have been considered by Schomburg [50]. Relationships between retention, structure, and boiling point of normal and branched-chain methyl esters were demonstrated. The equations of simple plots were determined and empirical relationships reported which allowed calculation of either retention indices or boiling points of the esters using the particular experimental conditions. Subsequent work with methyl and ethyl esters of α- and β-alkyl, α, β, γ, ..., ω-methyl-substituted carboxylic acid esters [51] allowed the formulation of rules for the elution sequence of different isomers by considering the retention behavior on two standard nonpolar phases.

A concept termed homomorphy which considered the retention difference of two substances differing by one or two structural elements was developed as follows:

$$H^A \text{ (homomorphy factor)} = I_{\text{substance}} - I_{\text{n-paraffin}} \tag{14}$$

The value H^A with n-alkyl compounds is similar to the functional retention index (FRI) used earlier by Swoboda [52] in studies with esters:

$$\text{FRI} = H^A = I_{\text{ester}} - I_{\text{hydrocarbon skeleton}} \tag{15}$$

The superscript A follows from the European usage (i.e., apolar), and is equivalent to nonpolar such that the expression shown in Eq. (15) is substantially free of polar interactions. A similar relationship utilizing a polar phase was proposed as

$$H^P = I_{\text{substance}} - I_{\text{n-paraffin}} \tag{16}$$

but with the significant contributions of polar interactions the values of H^P are of limited use. The familiar relationship ΔI of Kovats was combined with the homomorphy expression of Eq. (14) such that the influence of structural variations on the ΔI values could be investigated

$$\delta(\Delta I) = \Delta I_{\text{substance}} - \Delta I_{\text{homomorph}} \tag{17}$$

Values of H^A, $\delta(\Delta I)$, and retention index differences (ΔI) on polar and nonpolar phases were shown for many ester structures.

A consideration of H^A and ΔI has subsequently been proposed for use in tentative identification [53, 54].

2. Kovats noted application of the work of Evans and Smith [55] to retention prediction where the retention index of an asymmetrically substituted compound A-X-B could be determined from a knowledge of the index values of the compounds A-X-A and B-X-B on the same phase according to the simple relationship

$$I_{A-X-B} = \frac{1}{2}(I_{A-X-A} + I_{B-X-B}) \tag{18}$$

With simpler compounds where X is a single molecule (i.e., ethers) or where it is a group of atoms (i.e., ketones) where the total symmetry of the molecule is maintained the relationship is generally applicable. However, when the central group destroys the overall symmetry, common groups on each extremity are not equivalent as has been demonstrated with alkyl esters. Deviations also often occur with the first members in the nature of accentuated retention due to the well-known methyl effect and reduced retention with isopropyl compounds [56].

3. With the functional retention index (FRI) of Swoboda [52] similar substitution in similarly constituted compounds produced similar increases in the retention indices on a particular phase. With some compounds, particularly on nonpolar phases, remarkably similar increments were obtained but the observations are not generally applicable.

4. A similarity of retention index values of a compound on nonpolar and polar phases [2, 13, 14] has been observed, and while the behavior of paraffins might be expected to follow that of n-paraffins, a generalization of the concept is opposed to the practical operation of stationary phases that exhibit specific interactions.

5. Where a molecule contains more than one functional or structural group the ΔI value of the compound may be obtained by a simple summation of individual index increments for the various functional or structural groups. This type of approach has been developed with some success with larger molecules containing essentially noninteracting substituents using the restricted index schemes applicable to steroids and fatty esters [2, 13, 14].

A variation of this approach has been reported by Schomburg and Dielmann [54] who reported an increment equivalent to the interaction of a functional group in a compound as related to its equivalent hydrocarbon on a nonpolar stationary phase. These authors determined the retention indices of hexylcyclopropane and 1-octane on squalane at 80° C as 913.0

and 782.7. There are functional group increments of +13.0 and -17.3 by subtraction of 900 and 800 units appropriate to an n-paraffin of the same carbon number

$$H^{NP}_{\text{cyclopropane function}} = I^{NP}_{\text{hexylcyclopropane}} - I_{\text{n-nonane}} \quad (19)$$

$$H^{NP}_{\text{unsaturation}} = I^{NP}_{\text{1-octene}} - I^{NP}_{\text{n-octane}} \quad (20)$$

The retention index of 5-hexenylcyclopropane containing both of these functions was calculated to be

$$H^{NP}_{\text{unsaturation}} + H^{NP}_{\substack{\text{cyclopropane}\\ \text{function}}} = I^{NP}_{\text{5-hexenylcyclopropane}} - I^{NP}_{\text{n-nonane}} \quad (21)$$

$$-17.3 \quad + \quad 13.0 = I^{NP}_{\text{5-hexenylcyclopropane}} - 900$$

The value of 895.7 calculated from Eq. (21) is in substantial agreement with the experimentally determined value of 895.3. This approach is essentially the same as that proposed by Evans and Smith [55] and the calculations are again restricted by interactions that occur between functional groups. Schomburg and Dielmann [54] suggested that the interactions be described by second-order increments.

Takacs and his co-workers [57-59] have reported a procedure for the prediction of the retention index of a hydrocarbon solute on squalane based on a summation of contributions as follows:

$$I = I_a + I_b + I_c \quad (22)$$

where

I_a = an atomic term representing a contribution toward the total retention due to the bonds within the solute molecule,

I_b = a bond term representing the interatomic bonds of the solute,

I_c = an interaction term dependent on the solute, solvent, and temperature

The atomic term I_a may be represented by a sum of individual atomic contributions. Since all carbon atoms on the molecule are assumed to exhibit identical individual contributions, as are all of the hydrogen atoms, I_a may be shown as

$$I_a = n \cdot i_{a,H} + m \cdot i_{a,C} \quad (23)$$

where n and m are the total number of each type of atom present and the i_a terms are the appropriate atomic contributions which conveniently are shown as approximately one-tenth of the respective atomic weight.

The bond term I_b is also a summation of values of all the individual carbon-hydrogen and carbon-carbon bonds. The former bonding is assigned a constant value irrespective of its location with saturated, unsaturated, or aromatic carbon atoms. However, the carbon-carbon bond is assigned values varying by a factor of 30 depending on its type.

The combined interaction term I_c, being dependent on the solute, solvent, and temperature, is a very significant term and accounts for approximately 75.0% of the total retention index. Substituting this term in Eq. (22) provides a relationship suitable for the evaluation of I_b in branched-chain hydrocarbons:

$$I = \frac{I_a + I_b}{1.00 - 0.7455} \qquad (24a)$$

$$= 3.93(I_a + I_b) \qquad (24b)$$

Souter [60] has attempted to evaluate this approach of Takacs et al. [57-59] and has presented a very detailed work showing the serious limitations of the reports despite the absence of the experimental data on which the hypothesis was developed.

An important conceptual weakness of the scheme is considered to involve the question of whether both atomic and bond contributions should be invoked in Eq. (22). This situation does not appear to have been considered but is obviously of prime importance since the difficulty in distinguishing these concepts is readily apparent.

Since each hydrogen atom is uniquely related to a C—H bond it is not possible to isolate the effect contributed by a hydrogen atom. The calculations would be influenced if the I_b value for the C—H bond was increased by the I_a value assigned to the hydrogen atom; i.e., I_a for hydrogen is zero.

From Eq. (24) it seems that the interaction term I_i may be unnecessary with squalane and that the I_a and I_b terms might be simply increased by a factor of 3.93. A subsequent work by Souter [61] summarizes the criticisms of the procedure.

The work of Takacs et al. [62] has been applied to compounds containing hydroxy groups, halogens, amines, ketones, esters, heterocyclic compounds, adamantanes, silanes, and steroids on both polar and nonpolar stationary phases. Bond index contributions I_b are shown for a number of chemical structures but no calculations are shown or any worthwhile discussion included. The index contributions reported for some similar compounds seem to be somewhat unexpected and considering the criticisms of the earlier work [60, 61], the need for consideration of many experimental values as suggested by Ettre [63] seems prudent. Following the criticisms of Souter [60], Vanheertum [115] applied the procedure of Takacs [57-59] to a series of saturated and unsaturated paraffins. A significant number of compounds showed a calculated retention error exceeding five index units, and as an accuracy of a few index units is required for identification it was concluded that the method lacked reliability. An improvement of reliability was suggested by the use of an increased number of bond contributions rather than by simplification as suggested by Souter.

The inclusion of interaction between functional groups has been considered in a procedure employing the additivity of incremental values [64] for a series of branched-chain paraffins on nonpolar phases. Linear relationships between I and n, the carbon number, were observed with 10 series of paraffins and were represented by the relationship

$$I = E_0 + E_1 \cdot n \tag{25}$$

where E_0 and E_1 are constants calculated by a least-squares method. Calculation of the retention index for a member of a series is thus possible, if the constants are known, by a characteristic procedure of restricted application. Of somewhat more general application is the use of a relationship based on the linearity of the isoparaffin plot of carbon number and molecular volume.

VIII. STEROID NUMBER AND STRUCTURE

Calculation of the steroid number (SN) as a summation of contributions from the steroid skeleton and from the noninteracting functional groups that are present has been previously mentioned; the development of the concept as shown is equivalent to the concepts discussed in the preceding section.

Clayton [65] reported a series of retention ratios from data of unsaturated sterol methyl ethers:

$$\frac{\text{Relative retention time of compound with } \Delta S}{\text{Relative retention time of analog lacking } \Delta S} \tag{26}$$

The results show that the introduction of an isolated double bond ΔS into a stanol s will change its relative retention time r_s to a new value r_{s+x}, according to the following relationship:

$$r_{s+x} = r_s \times k_x \tag{27}$$

where k_x is a constant (retention constant) for the double bond in position x, which is virtually independent of molecular weight. Further, for two noninteracting double bonds at positions x and y (e.g., Δ^7 and Δ^{22} in the ergostane series), the relative retention time of the new $\Delta^{x,y}$-dienol is given by

$$r_{s+x+y} = r_s \times k_x \times k_y \tag{28}$$

The equation in its most general form may be applied to steroids, so that the retention time of a polysubstituted steroid in which intramolecular group interactions are negligible can be expressed as

$$r_{n+a+b+c+\cdots} = r_n \times k_a \times k_c \times \cdots \tag{29}$$

where $r_{n+a+b+c+\cdots}$ is the retention time of the total structure, r_n is the retention time of the unsubstituted nucleus, and $k_{a,b,c,\ldots}$ are group retention factors for a series of noninteracting substituents at positions a, b, c, ···.

The relationship was used in a different form by VandenHeuvel and Horning [66] to obtain the steroid number (SN) as

$$SN = S + F_1, \cdots, F_n \tag{30}$$

where S is the number of carbon atoms in the steroid skeleton and $F_1, \cdots, F_n$ are values characteristic of the functional groups present.

In practice, a different and much more useful way of expressing this relationship is to use the carbon number concept. A calibration line using andostane (SN 19.0) and cholestane (SN 27.0) was drawn, and F_n values for representative functional groups were determined. While the early work [66] was carried out using SE-30, it was soon shown [67] that the concept was applicable with selective stationary phases (NGS and QF-1), and that values obtained with these phases may be used to indicate the number, nature, position, and stereochemical arrangement of the functional groups that are present.

It has been shown [68] that the steroid number can be expressed as

$$SN_i = 19 + 8 \log\left(\frac{r_{ti}}{r_{tc}}\right) \tag{31}$$

where r_{ti} and r_{tc} are retention times of steroid i relative to both androstane and cholestane, respectively. The use of this relationship has been shown to provide a more accurate estimate of the SN than the graphical procedure.

The selective retention behavior of different functional groups on QF-1 has also been studied by Knights and Thomas [69], who also showed the retention of a compound as a summation of contributions from the nucleus and the individual groups present.

The concept of increments or additive contributions has been studied with a large number of steroids by VandenHeuvel and Court [70]. Remarkable agreement between calculated and observed values was found using a nonpolar dimethylpolysiloxane stationary phase. With polar phases, interactions between the phase and the steroid were reported, and the accuracy of the incremental additions was reduced. These authors suggest that the claimed selectivity of polar phases is not justified.

Increments allowing prediction of retention from structural parameters are indicated by the linear relationship

$$13^3 \times \log r_i = A_i + 10^3 + B_i T^{-1} \tag{32}$$

If $\Delta 10^3 \times \log r_f$ represents an addition increment corresponding to a structural feature of the molecule, then

$$10^3 \times \log r_i = \Sigma 10^3 \times \log r_f \tag{33}$$

expresses an additive situation. By considering Eqs. (32) and (33) it becomes apparent that

$$\Delta 10^3 \times \log r_f = \Delta A_f + 10^3 \times B_i T^{-1} \tag{34}$$

and $A_i = \Sigma A_f$ and $B_i = \Sigma B_f$.

It is possible to predict ΔA_f and ΔB_f, which when evaluated in Eq. (33) allows estimation of additive incremental factors for a particular structural parameter at any temperature.

It has been suggested that steroid numbers have greater validity in interlaboratory comparisons than relative retention times, and that a particular advantage lies in the relative constancy of these values with respect to changes in temperatures of the determination [71].

$$T = \frac{r_{x(s)} - r_{x(n)}}{r_{x(n)}} \tag{35}$$

where $r_{x(s)}$ is the relative retention time of compound x observed with a selective phase, and $r_{x(n)}$ is the relative retention time observed with a nonselective phase. Both retention time values should be determined with equivalent amounts of liquid phase and at the same temperature. These values are temperature dependent: A linear relationship was found in several instances for log T and temperature over a short temperature range.

VandenHeuvel and Court [70] have considered temperature-retention time relationships with many steroids and shown application of

$$10^3 \log r_i = A_i + 10^3 + B_i T^{-1} \tag{32}$$

where with steroid i, A_i and B_i are constants independent of T, the absolute temperature. Similarly, with a standard steroid s

$$10^3 \times \log r_s = A_s + 10^3 \times B_s T^{-1} \tag{36}$$

From Eqs. (32) and (35)

$$10^3 \times \log\left(\frac{r_i}{r_s}\right) = A_i - A_s + 10^3 (B_i - B_s) T^{-1} \tag{37}$$

The constants are independent of temperature, and were expressed as $\bar{A}_{i,s} = A_i - A_s$ and $\bar{B}_{i,s} = B_i - B_s$.

With cholestane as a primary standard, the constants were defined as retention constants (RC).

To obtain retention constants for a steroid, its relative retention times with cholestane as standard are obtained at two temperatures, and the solution of a pair of equations as shown in Eq. (36) is carried out. VandenHeuvel and Court [70] have discussed the value of retention constants [71] as a means of identifying steroid materials.

IX. FATTY ESTERS AND RETENTION

The first significant advance in the development of a correlation between the structure of unsaturated esters and retention was reported by Ackman [72] although a grid of parallel lines of varying unsaturation established with known esters [73] had been reported as a means of determining the number of double bonds and the carbon number of an unknown ester. Various other earlier reports [74] based on limited data indicated that linear plots could be obtained.

The work of Ackman showed the following:

1. Esters of monounsaturated fatty acids with the same end-carbon chain lengths z exhibit a linear relationship between the logarithm of their corrected relative retention times log V_R and the total chain length x. Here, end-carbon chain length refers to the number of carbon atoms in the section of the chain after the center of the double bond furthest removed from the carboxylic group.
2. Normal methylene-interrupted polyunsaturated acid esters, with both the number of double bonds and the end-carbon chain-length constant, also have log V_R linear with total chain length. Further, the slope of these lines was reported to be the same as that for the monounsaturated acids. Later data [75] using a capillary column allowed a reexamination of the parallel line concept, and with the more refined data, it was observed with the shortest end chain examined, i.e., $\omega = 3$,* that esters with four and five double bonds dipped slightly with increasing chain length relative to the monounsaturated reference line.
3. For esters of a common chain length and with the same number of double bonds, log V_R increases as the end-carbon chain becomes shorter.

Linear plots of the available data were used to demonstrate the conclusions together with a method based on separation factors, where it was apparent that prediction of retention on the basis of structure was possible, and was of value as a means of tentative identification.

Three types of separation factors described as types I, II, and III were proposed by Ackman:

Type I: Relationship between relative retention times or volumes (V_R) of pairs of esters of fatty acids of the same chain length

---

*The abbreviated ω notation has its primary use in showing metabolic relationships with the double bonds counted from the terminal methyl group. Since the polyunsaturated fatty esters considered are methylene interrupted the formula may be simply shown by the position of the first double bond. Thus, 9,12,15-octadecatrienoic acid becomes 18: ω3, 6, 9 or more conveniently 18: 3ω3.

The general case may be shown as x: yωz where x, y, and z are, respectively, the total chain length, the number of methylene-interrupted double bonds, and the end-carbon chain, which is defined as the number of carbon atoms in the chain after the center of the double bond furthest from the carboxylic group.

and the same end-carbon chain, but with varying number of methylene-separated double bonds.

Type II: Relationship between V_R of esters of fatty acids with the same chain length, varying number of double bonds, and different end-carbon chains.

Type III: Relationship between V_R of esters of fatty acids of the same chain length, the same number of double bonds, and different end-carbon chains.

A further type of separation factor which may be described as type IV has been reported by Haken [76].

Type IV: Relationship between V_R of pairs of esters of different total chain length but with the same number of double bonds and the same end-carbon chain.

The four types may be shown by the relationships:

Type I $$\frac{V_{R(x,y+1,z)}}{V_{R(x,y,z)}}$$

Type II $$\frac{V_{R(x,y+1,z)}}{V_{R(x,y,z+3)}}$$

Type III $$\frac{V_{R(x,y,z)}}{V_{R(x,y,z+3)}}$$

Type IV $$\frac{V_{R(x+2,y,z)}}{V_{R(x,y,z)}}$$

where x, y, and z indicate the total chain length, the number of double bonds, and the end-carbon chain, respectively.

The type IV separation factors are with the saturated esters identical to the methylene separation factors reported by James and Martin [1], but a constancy is apparent with a variety of series of esters independent of unsaturation, providing that the end-carbon chain remains constant in any series. This relationship was later reported by Jamieson and Reid [77] without comparison to the saturated esters as "retention time ratios of similarly unsaturated derivatives."

A relationship similar to that developed by Clayton has been reported by Haken [76], where V_R of the methyl esters of fatty acids of different chain length and degree of unsaturation may be calculated, the end-carbon chain in a particular case being constant.

The retention data of esters of the type (x + 2, y + 1) may be calculated from the esters (x, y) by

$$V_{R(x+2,y+1)} = V_{R(x,y)} \frac{V_{R(x+2,y)}}{V_{R(x,y)}} \frac{V_{R(x,y+1)}}{V_{R(x,y)}} \tag{38}$$

where x and y are the total chain length and the number of methylene-interrupted double bonds, respectively, and $V_{R(x+2,y)}/V_{R(x,y)}$ and $V_{R(x,y+1)}/V_{R(x,y)}$ are separation factors of types IV and I, respectively.

The values of the separation factors used in the development of the relationship were 1.82 and 1.12, respectively, so that the retention value of the desired ester may be calculated by

$$V_{R(x+2,y+1)} = V_{R(x,y)} \times 1.82 \times 1.12$$

Data for esters of the type (x + 4, y + 1) may be calculated by

$$\begin{aligned} V_{R(x+4,y+1)} &= V_{R(x,y)} \times V_{R(x+2,y)} \times V_{R(x+2,y)} \times V_{R(x,y+1)} \\ &= V_{R(x,y)} \times 1.81 \times 1.81 \times 1.12 \end{aligned} \tag{39}$$

The deviations from the experimentally determined values are ±1.20 and ±1.35%, respectively. Clayton has shown that with bifunctional steroids agreement within +4.0% is obtained, and with certain sapogenins calculated from the work of VandenHeuvel and Horning, agreement of ±2.0% is possible.

Relative retention data of fatty esters are frequently shown in terms of equivalent chain length (ECL) or modified equivalent chain length (MECL) values. Jamieson and Reid [77] have determined their values using the logarithmic form of Eq. (15) as follows:

$$ECL_{(x+2)(y+1)} = ECL_{(x,y)} + 2 + k_1 \tag{40}$$

$$MECL_{(x+2)(y+1)} = MECL_{(x,y)} + 2 + k_1 \tag{41}$$

where k_1 is the difference in ECL or MECL values of the pairs of esters used to calculate type I separation factors.

The same workers have shown calculated ECL and MECL values for esters of the type (x + 4, y + 2), while equations for esters of the type (x + 4, y + 1) and (x + 2, y + 2) have since been reported [78].

The most detailed study of ECL values is that of Hofstetter et al. [79] who suggested that the values are rather independent of the operating conditions (i.e., gas flow rate, column dimensions, and proportion of stationary phase). The values are only slightly altered by temperature, and column age caused variations of less than 0.1 ECL units. However, Ackman reported that both the column temperature and the age of the column are significant in determining ECL values [75].

When ECL values of esters were plotted against carbon atoms of the acids, parallel straight lines were obtained for the homologous series x: 1ω7, x: 1ω8, x: 1ω9, x: 2ω5, x: 2ω6, x: 3ω5, x: 3ω6, and x: 5ω3. The lines for x: 3ω9, x: 4ω5, and x: 4ω6 were not parallel to the others nor to themselves, but they were straight lines.

The two major components of the ECL value are contributed by London forces associated with the hydrocarbon chain and by polar attraction of the double bonds and ester group with the stationary phase. These effects are additive with the polar polyester stationary phases. The nonpolar stationary phases produce ECL values where the contribution from chain length is again positive but that from unsaturation is negative. The difference between ECL values determined on a polar and nonpolar stationary phase minimizes the chain-length component and accentuates the unsaturation component. The subtraction of ECL values has been studied by Hofstetter and co-workers [79] using ethylene glycol succinate and Apiezon L stationary phases. The ECL differences of long-chain fatty esters are approximately proportional to the number of nonconjugated double bonds with an incremental rise of approximately 0.84 ECL units per double bond independent of their position in the chain. The increment is slightly increased with the shorter chain esters because of the greater effect of unsaturation.

With monoynes, methylene-interrupted eneynes, and tetramethylene-interrupted diynes and triynes the triple bonds contribute an increment approximately equal to three double bonds. Methyl n-decadiene-2,8,-diyne-4,6-oate, where the carbonyl, diene, and diyne are all conjugated, has an even higher increment.

The esters mainly considered in the studies described have been those characteristic of marine products, where in common with other naturally occurring systems, the unsaturation tends to be located toward the center of the fatty acid chain rather than at one end. With the availability of data for complete series of isomeric fatty esters, it is apparent that the interactions which occur with esters that have very short or very long end chains are significant and systematic procedures based on the end-carbon chain relationships are not generally applicable to complete homologous series [78].

The significant effect of interactions experienced with some fatty esters has been considered in the calculation of ECL values for multiple-branched-chain esters by the determination of fractional chain length (FCL) contributions from unsaturation or a substituent group rather than the use of the concept of a summation of contributions from the skeleton and of noninteracting substituents [80]. In common with the unsaturated esters, the methyl group in the 2-5 and the 12-17 positions of a chain of length 18 showed FCL values significantly different from the values of the 6-11 positions. The respective groups were designated as carboxyl end C_x, terminal end W_y, and the central positions m_m.

X. STATIONARY PHASE CHARACTERIZATION AND RETENTION INDICES

Rohrschneider proposed a scheme applicable to retention prediction and to stationary phase classification based on the additivity of intermolecular forces. The scheme is essentially the representation of the retention behavior as retention indices related to retention on a nonpolar standard phase.

Following studies by Pullin and Werner [81, 82] it was assumed that for each type of polar intermolecular interaction, the interacting energy is proportional to a value a, ..., e characteristic of each solute, and to a value x, ..., s characteristic of the stationary phase. The index difference ΔI is thus compiled of products as follows:

$$\Delta I = ax + by + cz + du + es \tag{42}$$

where a, b, c, d, and e are characteristics representative of benzene, ethanol, methyl, ethyl ketone, nitromethane, and pyridine, and x, y, z, u, and s are characteristics of the stationary phase [83].

Rohrschneider first attempted to describe index differences in terms of the first three product terms [84] but found it necessary to consider five terms, which explains the lack of sequence of the solvent terms.

Rohrschneider obtained data for 23 phases and 30 substances. The phases were characterized by the five index differences of the standard substance, the solute-specific constants a, ..., e being determined by a least-squares procedure. The five substance polarities were defined as measures of orientation forces (factor c), charge transfer forces, i.e., donor (factor a) and acceptor (factor d), and hydrogen bonding, i.e., hydrogen donor (factor b) and hydrogen acceptor (factor e). Although it has become apparent that the standard substances are far from a simple measure of an isolated interaction, the utility of the scheme and the suitability of the reference substances remain. The substances characteristic of x, y, and z together with cyclohexane were previously recommended by Averill [85] for the qualitative classification of stationary phases.

In theory Eq. (42) is not limited by temperature but in practice there is a self-imposed temperature limit of between 100° and 120° C because of the volatility of squalane. It is thus difficult or impossible to evaluate either substance or stationary phase polarity factors for data obtained at higher temperatures.

McReynolds [96] has recommended variation of some of the standard substances suggested by Rohrschneider, namely, n-butanol, 2-pentanone, and nitropropane replacing ethanol, 2-butanone, and nitroethane, respectively. These particular changes are of considerable practical value since on polar phases the retention times of the lower homologs are very small. The addition of 2-methyl-2-pentanol is designed to improve the prediction of retention of branched-chain compounds, particularly of alcohols. For each stationary phase, $X' = \Delta I$ (benzene), $Y' = \Delta I$ (n-butanol), $Z' = \Delta I$ (2-pentanone), $u' = \Delta I$ (nitropropanone), $S' = \Delta I$ (pyridine), and $H = \Delta I$ (2-methyl-2-pentanol).

A number of reports have appeared concerning the number of reference substances necessary to evaluate the Rohrschneider type of summation. Weiner et al. [86-89] have proposed a scheme based on factor analysis which utilizes eigenvectors to select physically and chemically significant parameters which best fit the problem. In this manner,

$$I_{i,\alpha} = \sum_{i=1}^{n} c_{j,\alpha} P_{j,i} \tag{43}$$

is established where $I_{i,\alpha}$ represents the retention index of solute i on stationary phase α; j corresponds to the n test factors; $c_{j,\alpha}$ is the multiplicative constants associated with the solvent α; and $P_{j,i}$ is the n test parameters evaluated for solute i.

The equation is essentially the same as that developed by Rohrschneider (Eq. (42)], the major difference being that the parameters, many of which are interrelated, are statistically determined. Weiner has produced entirely different parameters which fit closely related sets of retention data. Rohrschneider adopted the more practical approach and assigned the retention behavior of five classes of compounds as the parameters. To date, this seems to be of the greatest accuracy; i.e., Rohrschneider's error is 4.1 index units [83] whereas for a set of Weiner's results [86] the average error is greater for the same number of parameters but a different data set.

A principal component analysis [90] of McReynolds data [96] showed that to reproduce the data within about 30 index units three components have to be introduced. The work described the attributes of the components used, and while advantages of the scheme are claimed they are not readily apparent.

The Rohrschneider approach for characterizing stationary phases and the extension of the solubility parameter theory of Hildebrand [91, 92] has recently been reviewed by Hartkorf [93] with the conclusion that three predominant types of solute-solvent interactions exist: dispersion, dipole interaction, and hydrogen bonding. The same author [94] then demonstrated the use of four functional probes to characterize liquid phases and demonstrated the conclusion with a four-term form of the Rohrschneider equation which allowed prediction of ΔI values with the accuracy of the five-term equation. Further consideration of the significance of various groups of probes is obviously necessary [90, 93], since 3 [94], 5 [83], 6 [96], 7, 8, and 10 [87] materials have been variously suggested with differing claims of precision.

Rohrschneider and McReynolds constants currently provide the only really practical means of classifying stationary phases. The constants are readily determined, and data for many phases have been reported by Supina and Rose [95], McReynolds [96], and in the literature of the Supelco organization. With the availability of data concerning established and newer phases, it is often possible to replace a reported phase with one that is readily available and which may possess other features, i.e., thermal stability; alternatively, a replacement phase of greater selectivity may allow a superior separation to be achieved.

Various reports have specified values of Rohrschneider constants where a particular separation is achieved. This practice, while qualitatively of value, must be viewed with caution since the constants are temperature dependent, and in reality, in many cases the values reported at 100° C are considerably different from those experienced at operating temperatures near 200° C.

The elimination of squalane as a basis for comparison has been reported [97] and should allow further development of Rohrschneider's method. Based on Eq. (42) and considering two distinct stationary phases (1 and 2), it is possible to eliminate squalane entirely, as shown by

$$I_{1,2}^{A} = I_{1}^{A} - I_{2}^{A} = \frac{a}{100}(I_{1}^{x} - I_{2}^{c}) + \frac{b}{100}(I_{1}^{y} - I_{2}^{y}) + \frac{c}{100}(I_{1}^{z} - I_{2}^{z}) + \frac{d}{100}(I_{1}^{u} - I_{2}^{u}) + \frac{e}{100}(I_{1}^{s} - I_{2}^{s}) \tag{44}$$

where I_1^A and I_2^A are the retention indices for compound A determined on the two stationary phases 1 and 2, and $I_1^x, \ldots, I_1^s$ and $I_2^x, \ldots, I_2^s$ are the retention indices for the five standard substances on columns 1 and 2, respectively. Equation (44) shows that substance-specific factors can be determined with any two stationary phases and that the original choice of a nonpolar reference is unnecessarily restrictive.

The substance-specific factors were calculated using a least-squares criterion which minimizes the sum of the squared errors for all of the stationary phases for each substance separately. The values of a, ..., e obtained in this way, using squalane as the basis for comparison, differ slightly from those obtained by Rohrschneider, whose method of calculation used a different criterion. The values determined were in agreement with those of Leary and his co-workers who used a similar approach [98]. The relative merits of the two approaches have been discussed by Souter [99].

The difference between the calculated and observed values was used to calculate the root mean square error and the average absolute error for each substance, for each column, and for each complete data set. These values are shown in Table 5 where it is evident that the errors are not increased with any particular type of stationary phase.

The use of an alternative base stationary phase has been reported by Takacs and his co-workers [100, 101]. These authors considered Rohrschneider's original schemes to have been theoretically sound but "unsuitable for following some of the more complicated processes because of the possibility of error compensation." They therefore selected a method that involved the ratios of retention indices, rather than their differences, as shown below, which applies for constant temperature:

$$\frac{I_P^A}{I_{sq}^A} = \sum_{i=1}^{n} f_i \cdot s_i \tag{45}$$

where f_i is the i-th polarity factor for the stationary phase, given by $f_i = I_P^i/I_{Sq}^i$, i is the serial number of the standard solute, and s_i is the i-th solute-specific factor characterizing the molecular structure of solute A.

The equation may then be applied to any two stationary phases as

$$\frac{I_{P1}^A}{I_{P2}^A} = \sum_{i=1}^{5} F \cdot S_i \tag{46}$$

where F_i is the i-th polarity factor for any two stationary phases, 1 and 2, i.e., $F_1 = I_{P1}^i/I_{P2}^i$ for a system not based on squalane, and S_i is the i-th solute-specific constant for the same system.

If Eq. (45) is to be compatible with Eq. (42), the solute-specific factors must be related by

$$\sum_{i=1}^{5} s_i = 1$$

TABLE 5

Errors in the Change of Base Stationary Phase Calculations[a]

Stationary phase	Error (calculated-observed) Average absolute	RMS[b]
Squalane	3.25	4.97
DC-200	3.73	5.72
Apiezon L	3.39	5.33
Diethyl hexyl sebacate	3.52	5.60
Celanese ester No. 9	3.77	5.95
Diisodecyl phthalate	3.75	5.66
DC-710	3.55	5.32
QF-1	3.22	4.97
Ucon LB-550X	3.35	5.12
Acetyl tributyl citrate	3.94	6.21
Tricresyl phosphate	4.09	6.11
Polyphenyl ether	3.67	5.56
Marlophen 87	3.47	5.46
Polypropylene sebacate	3.88	5.96
Marlophen 814	3.40	5.10
Neopentyl glycol succinate	3.27	5.08
XF-1150	3.77	6.17
Carbowax 20M	3.26	4.96
Carbowax 4000	3.23	4.90
Reoplex 400	3.42	5.17
Diethylene glycol sebacate	3.28	5.08
Ethylene glycol bis(cyanoethyl) ether	3.35	5.16
Tris(cyanoethyoxy)propane	3.20	4.90
Average	3.51	5.37

[a]From Haken et al. [97].
[b]Root mean square.

This relationship appears to hold within 1 or 2%, indicating that the representation shown in Eq. (45) is reasonable, although for accurate work a variation of this magnitude could be quite significant.

At present the approach using retention index differences rather than retention index ratios is preferred since it incorporates the well-established x, ..., s values and leads to a, ..., e values similar to those of Rohrschneider, rather than introducing two completely new and different sets of values for essentially similar quantities. A decision on the merits of the two representations may eventually be possible when further details of Takacs' results are available.

The use of Rohrschneider constants as a means of investigation of solid support activity has been reported by Golovistikov [114]. Retention indices for the standard substances were determined with squalane and polyethylene glycol 400 on acid-washed Chromosorb W, acid-washed Chromosorb W with 0.1% Alkaterge, Teflon D, tin, and aluminum and it was found that the highest constants for each substance corresponded to maximum inertness of the support.

XI. TEMPERATURE DEPENDENCE OF RETENTION INDICES

The relationship between retention behavior and operating temperature has been the subject of several investigations. Several workers [102-104] have found that a linear relationship exists between relative retention and temperature for the several classes of compounds they examined, while others [105, 106] have found that the relative retention is inversely proportional to absolute temperature T. Chovin and Lebbe [107] derived the following relationship between relative retention and temperature:

$$\log V_R = a + \frac{b}{T} \tag{47}$$

where a and b are constants. This was later expounded by Ettre and Billeb [108], who found that a linear relationship existed between I and T for hydrocarbons on a nonpolar phase. Takacs et al. [109, 110] have shown that the retention index I is related to the absolute temperature by

$$I_{\text{substance}}^{\text{stationary phase}}(T) = A + \frac{B}{T + C} \tag{48}$$

where A, B, and C are constants defined by the substance involved and the properties of the stationary phase. This relationship is suggested to allow an accurate estimate of I to be made over a large temperature range and has been extensively developed by these workers. However, the constants A, B, and C are not easily obtained since they require a knowledge of the molar heats of evaporation of the substances to be characterized as well as fairly extensive experimentation and computation [110].

Saha and Mitra [104] have made use of temperature versus retention index plots of hydrocarbons on a variety of stationary phases, while Ettre and Billeb [108] have used similar plots with various compounds on squalane and polyethylene glycol 400. The results of these plots are of great use for data prediction and also for selecting conditions for the optimum analysis of multicomponent mixtures, since the retention indices may change positively or negatively, depending on the relevant conditions.

Temperature coefficients of Kovat's indices have been proposed by Saha and Mitra [104] as a parameter in identifying peaks, especially those of the hydrocarbons.

The effect of temperature on the polarity of stationary phases has been examined to a lesser extent than its effect on retention indices, but since the common polarity classification schemes are based on this retention index system, many of the results would appear to be applicable. Of all the polarity schemes, those developed by Rohrschneider [83, 84, 111] appear to be the most widely accepted. Lapkin and Makina [112] and Petsev [116] examined the influence of temperature on the polarity of stationary phases based on an early Rohrschneider classification system [111], where a polarity factor P was determined as the difference between the logarithm of the ratio of retention of butadiene and n-butane on a polar and nonpolar phase. From these results, it was apparent that the polarity varied linearly with temperature, but as expected the rate of change varied with the nature of the stationary phase examined.

The effect of temperature on polarity of the stationary phases classified according to the latest scheme of Rohrschneider [83] and McReynolds [96] has recently been reported on several polysiloxane phases [113] where retention data were recorded at temperatures between 90° and 150° C for the principal reference substances of McReynolds. At the higher temperatures, the column life of squalane was short owing to high column bleed, but the experimental time at these higher temperatures was short and with frequent calibration the results were extremely reproducible, namely, ±1 unit, which enabled reliable results to be obtained.

Linear relationships were observed for all of the standard substances on the various phases when log V_R was plotted against both T and 1/T. The correlation coefficients for the benzene plots are shown in Table 6.

Plots of retention indices versus temperature are shown in Fig. 2 for the five phases. It is apparent that a linear relationship exists between I and T similar to that found by Ettre and Billeb [108] for various hydrocarbons on nonpolar phases. As they suggest, the I versus T relationship

TABLE 6

Correlation Coefficients for Benzene Plots

Stationary phase	V_R versus T	Log V_R versus 1/T
Squalane	0.997	0.999
SE-30	1.000	1.000
DC-710	0.997	0.997
QF-1	0.998	0.997
XF-1150	0.932	0.935

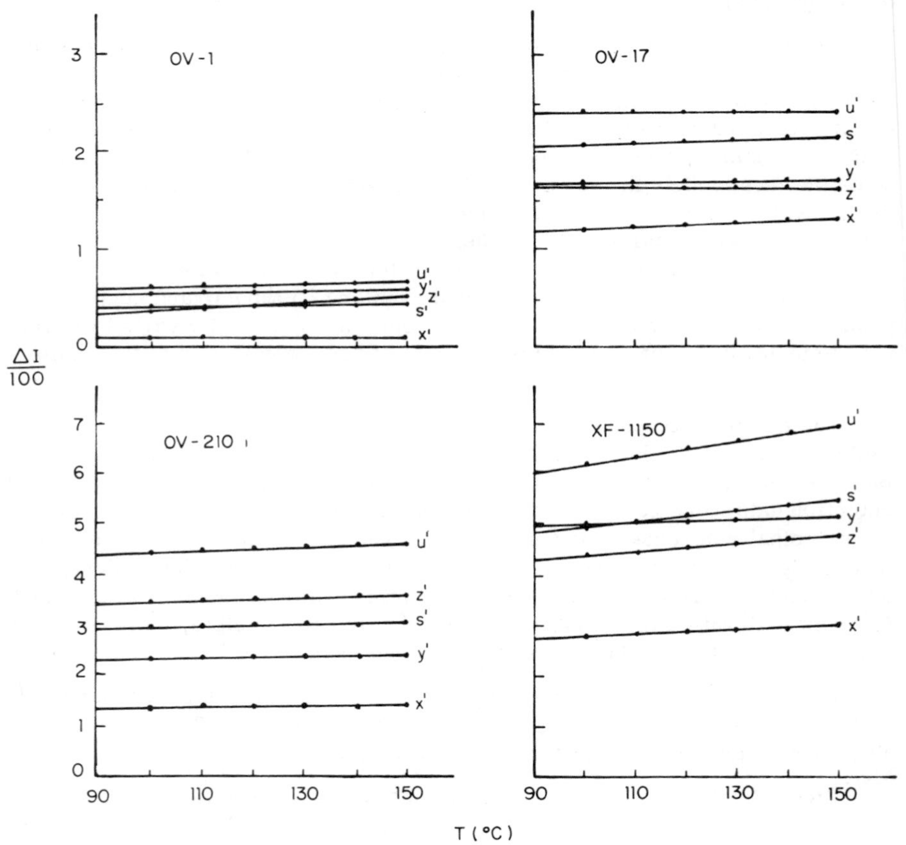

FIG. 2. Plots of McReynolds' constants versus temperature (° C) for polysiloxane phases.

appears to be valid for other classes of compounds on polar phases, although this behavior has only been evaluated on four polysiloxanes.

The effect of column polarity on the retention indices and elution order was essentially similar to that observed when the effect of polarity on the relative retention volumes was considered. This retention index system is an empirical method of representing relative retention volumes in terms of normal paraffins. Therefore, although the slopes of the corresponding V_R and I versus temperature plots will change, owing to a difference in temperature coefficients between the hydrocarbons and standard substances on phases on varying polarity, the magnitude of the retention indices and the elution order will be very similar to those observed with the V_R plots.

The change in stationary phase polarity with temperature can be found from the differences, divided by 100, between the values for the standard substances on squalane and the stationary phase to be characterized. These values are shown in Table 7 and Fig. 2.

TABLE 7

Variation of Modified Rohrschneider Constants with Temperature

Stationary phase	Column temperature (°C)	Rohrschneider-type constants				
		X'	Y'	Z'	U'	S'
SE-30	90	0.09	0.54	0.34	0.57	0.39
	100	0.09	0.54	0.37	0.58	0.39
	110	0.09	0.55	0.39	0.58	0.40
	120	0.09	0.55	0.41	0.59	0.40
	130	0.09	0.55	0.43	0.59	0.41
	140	0.09	0.56	0.45	0.60	0.41
	150	0.08	0.56	0.47	0.61	0.42
DC-710	90	1.17	1.64	1.63	2.38	2.01
	100	1.19	1.64	1.63	2.38	2.03
	110	1.20	1.64	1.63	2.37	2.05
	120	1.22	1.65	1.63	2.37	2.06
	130	1.24	1.65	1.63	2.37	2.08
	140	1.26	1.66	1.63	2.37	2.10
	150	1.28	1.66	1.63	2.37	2.12

TABLE 7 (Continued)

Stationary phase	Column temperature (° C)	Rohrschneider-type constants				
		X'	Y'	Z'	U'	S'
QF-1	90	1.28	2.24	3.38	4.35	2.87
	100	1.30	2.27	3.42	4.41	2.91
	110	1.32	2.30	3.46	4.47	2.95
	120	1.34	2.33	3.51	4.52	2.99
	130	1.36	2.35	3.55	4.58	2.03
	140	1.39	2.38	3.60	4.64	2.07
	150	1.41	2.41	3.64	4.70	2.11
XF-1150	90	2.77	4.99	4.34	6.12	4.88
	100	2.82	5.03	4.42	6.27	5.00
	110	2.88	5.08	4.51	6.43	5.11
	120	2.91	5.12	4.59	6.59	5.22
	130	2.96	5.17	4.67	6.74	5.33
	140	3.01	5.21	4.75	6.90	5.44
	150	3.06	5.26	4.84	7.06	5.55

The effect of temperature on column polarity is variable, SE-30 and DC-710 showing the least variation of the four phases evaluated. As expected, the greatest variation was obtained on the more polar phase, XF-1150, where the U' factor showed a positive rate of change of ±0.157 units per 10° C. In comparison, the smallest rate of change was found with Z' on DC-710, where a negative change of 0.004 units per 10° C was obtained. With almost all phases the polarity increased with temperature, except for several cases with SE-30 and DC-710 phases, where the negative change was small enough to be considered constant over the temperature range considered.

It is apparent that if the polarity of phases is to be considered at higher (or lower) temperatures than that of published data, especially for polar or nonpolar phases that differ markedly in structure from squalane, consideration should be given to temperature effects when standard substances are used to characterize the stationary phase to be examined.

REFERENCES

1. A. T. James and A. J. P. Martin, Biochem. J., 50, 679 (1952).
2. E. Kovats, Helv. Chim. Acta, 41, 1915 (1958).
3. A. B. Littlewood, C. S. G. Phillips, and D. T. Price, J. Chem. Soc., 1955, 1480.
4. ASTM Specification E355-68, Standard Recommended Practice for Gas Chromatography Terms and Relationships, A.S.T.M., Philadelphia, 1968.
5. A. K. Hilmi, J. Chromatogr., 17, 407 (1965).
6. M. Reidmann, Chromatographia, 7, 59 (1974).
7. M. B. Evans and J. F. Smith, J. Chromatogr., 9, 147 (1962).
8. M. L. Peterson and J. Hirsch, J. Lipid Res., 1, 132 (1959).
9. H. J. Gold, Anal. Chem., 34, 174 (1962).
10. A. Grobler and G. Balizs, J. Chromatogr. Sci., 12, 57 (1974).
11. M. B. Evans, Chromatographia, 6, 301 (1973).
12. P. Toth, E. Kugler, and E. Kovats, Helv. Chim. Acta, 42, 2519 (1959).
13. A. Wehrli and E. Kovats, Helv. Chim. Acta, 42, 2709 (1959).
14. K. Kovats, Z. Anal. Chem., 181, 351 (1961).
15. D. Felix, G. Ohloff, and E. Kovats, Liebigs Ann. Chem., 652, 126 (1962).
16. E. Kugler and E. Kovats, Z. Anal. Chem., 46, 1480 (1963).
17. D. Felix, G. Melera, J. Seibl, and E. Kovats, Helv. Chim. Acta, 46, 1513 (1963).
18. E. Kovats, Helv. Chim. Acta, 46, 2705 (1963).
19. L. S. Ettre, Anal. Chem., 36 (8), 31A (1964).
20. J. F. Smith, Chem. Ind. (London), 1960, 1024.
21. M. B. Evans and J. F. Smith, J. Chromatogr., 6, 293 (1961).
22. L. S. Ettre and K. Billeb, J. Chromatogr., 30, 1 (1967).
23. K. P. Hupe, J. Gas Chromatogr., 3, 12 (1965).
24. W. J. A. VandenHeuvel, W. L. Gardiner, and E. C. Horning, Anal. Chem., 36, 1550 (1964).
25. J. C. Giddings, J. Chromatogr., 4, 11 (1960).

26. H. W. Habgood and W. E. Harris, Anal. Chem., 32, 450 (1960).

27. M. Vanden Dool and P. D. C. Kratz, J. Chromatogr., 11, 463 (1963).

28. G. Guiochon, Anal. Chem., 36, 661 (1964).

29. W. E. Harris and H. W. Habgood, Programmed Temperature Gas Chromatography, Wiley, New York, 1966.

30. S. J. Hawkes, J. Chromatogr. Sci., 10, 535 (1972).

31. A. Grobler, J. Chromatogr. Sci., 10, 128 (1972).

32. R. G. Ackman, J. Chromatogr. Sci., 10, 536 (1972).

33. F. P. Woodford and C. M. van Gent, J. Lipid Res., 1, 188 (1960).

34. T. K. Miwa, K. L. Mikoljczak, F. R. Earle, and I. A. Wolft, Anal. Chem., 32, 1739 (1960).

35. R. G. Ackman, J. Am. Oil Chem. Soc., 40, 558 (1963).

36. W. J. A. VandenHeuvel and E. C. Horning, Biochim. Biophys. Acta, 64, 41 (1962).

37. L. J. Lorenz and L. B. Rogers, Anal. Chem., 43, 1593 (1971).

38. H. F. Dymond and K. D. Kilburn, in Gas Chromatography 1966 (A. B. Littlewood, ed.), Inst. of Petroleum, London, 1967, p. 353.

39. J. K. Haken, J. Chromatogr., 99, 329 (1974).

40. M. S. Vigdergauz, Gas Chromatographie 1968, 6th Symposium on Gas Chromatographie, Akademie-Verlag, Berlin (H. G. Struppe, ed.), 1968, p. 625.

41. M. S. Vigdergauz and A. A. Martynov, Chromatographia, 4, 463 (1971).

42. C. L. A. Harbourn, Cited by A. G. Douglas, J. Chromatogr. Sci., 7, 581 (1969).

43. G. W. K. Cavill and E. Houghton, Aust. J. Chem., 26, 1131 (1973).

44. G. W. K. Cavill and E. Houghton, Aust. J. Chem., 27, 879 (1974).

45. E. Kovats, Adv. Chromatogr., 1, 229 (1965).

46. J. Zulaica and G. Guiochon, Bull. Soc. Chim. Fr., 1963, 1242.

47. J. R. Ashes and J. K. Haken, J. Chromatogr., 101, 103 (1974).

48. J. R. Ashes and J. K. Haken, J. Chromatogr., 60, 33 (1971).

49. J. K. Haken, T. R. McKay, and P. Souter, J. Gas Chromatogr., 3, 61 (1965).

50. G. Schomburg, J. Chromatogr., 14, 157 (1964).

51. G. Schomburg, Sep. Sci., 1, 339 (1966).

52. P. A. T. Swoboda, in Gas Chromatography 1962 (M. van Swaay, ed.), Butterworth, London, 1962, p. 273.

53. G. Schomburg, Advances in Chromatography, Vol. 6, Marcel Dekker, New York, 1968, p. 211

54. G. Schomburg and G. Dielmann, J. Chromatogr. Sci., 11, 151 (1973).

55. M. B. Evans and J. F. Smith, J. Chromatogr., 5, 300 (1962).

56. I. D. Allen and J. K. Haken, J. Chromatogr., 49, 409 (1970).

57. J. Takacs, C. Szita, and G. Tarjan, J. Chromatogr., 56, 1 (1972).

58. J. Takacs, I. Talas, I. Bernath, Gy. Czako, and A. Fischer, J. Chromatogr., 67, 203 (1972).

59. J. M. Takacs, J. Chromatogr. Sci., 11, 210 (1973).

60. P. Souter, J. Chromatogr. Sci., 12, 418 (1974).

61. P. Souter, J. Chromatogr. Sci., 12, 424 (1974).

62. J. M. Takacs, E. Kocsi, E. Garamvolgyi, E. Eckart, T. Lombosi, Sz. Nyiredy, Jr., I. Borbely, and Gy. Krasznai, J. Chromatogr., 81, 1 (1973).

63. L. S. Ettre, Chromatographia, 7, 39 (1974).

64. G. Gastello, M. Lunardelli, and M. Berg, J. Chromatogr., 76, 31 (1973).

65. R. B. Clayton, Biochemistry, 1, 357 (1962), and references cited therein.

66. W. J. A. VandenHeuvel and E. C. Horning, Biochim. Biophys. Acta, 64, 41 (1962).

67. R. J. Hamilton, W. J. A. VandenHeuvel, and E. C. Horning, Biochim. Biophys. Acta, 70, 679 (1963).

68. F. A. VandenHeuvel, G. J. Hinderks, and J. C. Nixon, J. Am. Oil Chem. Soc., 42, 283 (1965).

69. B. A. Knights and G. M. Thomas, J. Chem. Soc., 1963, 3477.

70. F. A. VandenHeuvel and A. S. Court, J. Chromatogr., 39, 1 (1969), and references cited therein.

71. F. A. VandenHeuvel and A. S. Court, J. Chromatogr., 38, 439 (1968).

72. R. G. Ackman, Nature, 194, 970 (1962), and later references.

73. A. T. James, J. Chromatogr., 2, 552 (1959).

74. J. K. Haken, Reve. Pure Appl. Chem., 17, 133 (1967), and references cited therein.

75. R. G. Ackman, Lipids, 2, 151 (1967).

76. J. K. Haken, J. Chromatogr., 23, 375 (1966); 26, 17 (1967).

77. G. R. Jamieson and E. H. Reid, J. Chromatogr., 39, 71 (1969).

78. J. K. Haken, J. Chromatogr., 43, 487 (1969); 39, 245 (1969).

79. H. H. Hofstetter, N. Sen, and R. T. Holman, J. Am. Oil Chem. Soc., 42, 537 (1965).

80. R. G. Ackman, J. Chromatogr., 28, 278 (1967).

81. I. A. Pullin and R. L. Werner, Nature, 206, 393 (1965).

82. I. A. Pullin and R. L. Werner, Spectrochem. Acta, 21, 1257 (1965).

83. L. Rohrschneider, J. Chromatogr., 22, 6 (1966).

84. L. Rohrschneider, J. Chromatogr., 17, 1 (1965).

85. W. Averill, in Gas Chromatography (N. Brenner, J. E. Callen, and M. D. Weiss, eds.), Academic Press, New York, 1962, p. 1.

86. P. H. Weiner, C. J. Dack, and D. G. Howery, J. Chromatogr., 69, 249 (1972).

87. P. H. Weiner and D. G. Howery, Can. J. Chem., 50, 448 (1972).

88. P. H. Weiner and D. G. Howery, Anal. Chem., 44, 1189 (1972).

89. P. H. Weiner and J. F. Parcher, J. Chromatogr. Sci., 10, 612 (1972).

90. S. Wold and K. Andersson, J. Chromatogr., 80, 43 (1973).

91. J. H. Hildebrand and R. L. Scott, The Solubility of Non-Electrolytes, 3rd ed., Dover, New York, 1964.

92. J. H. Hildebrand, J. M. Prausnitz, and R. L. Scott, Regular and Related Solutions, Van Nostrand-Reinhold, Princeton, New Jersey, 1970.

93. A. Hartkorf, J. Chromatogr. Sci., 12, 113 (1974).

94. A. Hartkoff, S. Grunfeld, and R. Delumyea, J. Chromatogr. Sci., 12, 119 (1974).

95. W. R. Supina and L. P. Rose, J. Chromatogr. Sci., 8, 214 (1970).

96. W. O. McReynolds, J. Chromatogr. Sci., 8, 685 (1970).

97. J. K. Haken, J. R. Ashes, and P. Souter, J. Chromatogr., 92, 237 (1974).

98. J. J. Leary, S. Tsage, and T. L. Isenhour, J. Chromatogr., 82, 366 (1973).

99. P. Souter, J. Chromatogr., 92, 231 (1974).

100. J. Takacs, Zs. Szentirmai, E. B. Molnar, and D. Kralik, J. Chromatogr., 65 121 (1972).

101. Zs. Szentirmai, G. Tarjan, and J. Takacs, J. Chromatogr., 73, 11 (1972).

102. R. G. Ackman, J. Gas Chromatogr., 1, 16 (1963).

103. P. N. Breckler and T. J. Betts, J. Chromatogr., 53, 163 (1970).

104. N. C. Saha and G. D. Mitra, J. Chromatogr. Sci., 8, 84 (1970).

105. J. Bricteux and G. Duykaerts, J. Chromatogr., 22, 221 (1966).

106. B. D. Blaustein, C. Zahn, and G. Patages, J. Chromatogr., 12, 221 (1963).

107. P. Chovin and J. Lebbe, in Separation Immediat et Chromatographie 1961 (J. Tranchant, ed.), G.A.M.S., Paris, 1962, pp. 90-103.

108. L. S. Ettre and K. Billeb, J. Chromatogr., 30, 1 (1967).

109. J. Takacs, M. Rochenbauer, and I. Olacsi, J. Chromatogr., 42, 19 (1959).

110. E. B. Molnar, P. Moritz, and J. Takacs, J. Chromatogr., 66, 205 (1972).

111. L. Rohrschneider, Z. Anal. Chem., 170, 256 (1959).

112. L. M. Lapkin and G. N. Makina, Zh. Anal. Khim., 24, 1753 (1969).

113. J. R. Ashes and J. K. Haken, J. Chromatogr., 84, 231 (1973).

114. Yu. N. Golovistikov, Zavod Lab., 39, 1071 (1973).

115. R. Vankeertum, J. Chromatogr. Sci., 13, 150 (1975).

116. N. Petsev, J. Chromatogr., 59, 21 (1971).

ADDENDUM

Chapter 4: Note Added in Proof

This literature review is more or less complete through mid-1974, when Chapter 4 was submitted. A number of papers have appeared since, adding mainly new data plus a few refinements in theory and technique to several of the areas discussed in this chapter. Unfortunately these cannot be cited here.

However, significant developments have occurred in the area of complexation studies. Purnell and Srivistava [130] compared the GLC method with UV and NMR spectroscopic methods; for the systems considered, only GLC gave consistent, reliable results. Martire [332] resolved the obvious inconsistency, and simultaneously cast suspicion on many GLC-derived equilibrium constants. GLC, Martire showed, actually yields a constant which is the sum of an association complex constant for 1:1 complexes plus an equilibrium constant reflecting solute (A) -reactive solvent component (B) interactions of a noncomplexing nature. The latter is interpretable in terms either of solution nonideality of the uncomplexed solute (A in B), or of so-called contact-pairing between A and B. Depending on the relative magnitudes of the two contributions, GLC may not give an unequivocal value of K_1. On the other hand, spectroscopic K_1 values reflect only chemical interactions, that is, specific complex formation; in all cases where GLC and spectroscopic K_1 values disagree, the GLC values are the higher ones, as Martire's model would predict. Martire further showed that mixed solvency effects can produce serious errors in the spectroscopic measurements. Martire's contribution has presumably finally put the GLC technique for determining equilibrium constants on a firm theoretical foundation.

I say presumably because of recent efforts by Purnell, who originally devised the GLC methods for study of complexation [255], to show that it is possible to account for nearly all of the data obtained on so-called complexing systems without resort to complexation at all. Specifically, Purnell, with Vargas de Andrade and Laub [333-336], has used a model based on local, microscopic immiscibility of the additive plus inert solvent mixture. This leads to an equation

$$K_R = \phi_B K^0_{R,B} + \phi_I K^0_{R,I}$$

where ϕ_B and ϕ_I are the volume fractions of ligand, B, and inert solvent I in the stationary phase, and K_R refers as before to the distribution coefficient of the solute in pure B or pure I. This model thus predicts a linear plot of K_R versus c_B, as observed. The surprising aspect of this equation is that it is interpreted to mean that "complexation" is merely a manifestation of solute dissolving in two microscopically immiscible liquids, B and I, a result not obviously consistent with current theories of solution.

Martire [347] has taken understandable umbrage at this development, since it is inconsistent with his model. Martire has shown theoretically that the local immiscibility model is itself not self-consistent and cannot, as his model can, reconcile the GLC- and NMR-derived association constants of organic complexes. The apparent efficacy of the immiscibility model is in fact shown to be merely an artifact of the insensitive method of data testing used by Purnell et al.

This controversy is far from settled, and will not be until additional, independent measurements are made not only of both the chemical and physical interaction effects postulated by Martire, but also of the presence of local molecular aggregates, as required by the Purnell model.

REFERENCES

332. D. E. Martire, Anal. Chem., 46, 1712 (1974).

333. J. H. Purnell and J. M. Vargas de Andrade, J. Am. Chem. Soc., 97, 3585 (1975).

334. J. H. Purnell and J. M. Vargas de Andrade, J. Am. Chem. Soc., 97, 3590 (1975).

335. R. J. Laub and J. H. Purnell, J. Am. Chem. Soc., 98, 30 (1976).

336. R. J. Laub and J. H. Purnell, J. Am. Chem. Soc., 98, 35 (1976).

337. D. E. Martire, Anal. Chem., 48, 398 (1976).

AUTHOR INDEX

Numbers in parentheses are reference numbers and indicate that an author's work is referred to although his name is not cited in the text. Underlined numbers give the page on which the complete reference is listed.

A

C

D

E

F

L

N

Q

R

S

T

Y

Z

SUBJECT INDEX

A

B

C

D

M

Q

R

T